PERGAMON INTERNATIONAL LIBR
of Science, Technology, Engineering and So
The 1000-volume original paperback library in aic
industrial training and the enjoyment of
Publisher: Robert Maxwell, M.C.

THE STRUCTURE OF THE BRITISH ISLES

Second Edition

OTHER RELATED PERGAMON TITLES OF INTEREST

Books

ANDERSON
The Structure of Western Europe

BOWEN
Quaternary Geology

CONDIE
Plate Tectonics and Crustal Evolution

GRIFFITHS & KING
Applied Geophysics for Geologists and Engineers - The Elements of Geophysical Prospecting (*a new edition of 'Applied Geophysics for Engineers and Geologists'*)

OWEN
The Geological Evolution of the British Isles

ROBERTS
Geotechnology - an Introductory Text for Students and Engineers

ROBERTS
Applied Geotechnology - a Text for Students and Engineers on Rock Excavation and Related Topics

Journals

Computers & Geosciences

Geochimica et Cosmochimica Acta

Journal of Structural Geology

Organic Geochemistry

THE STRUCTURE OF THE BRITISH ISLES

by

J G C ANDERSON
Emeritus Professor of Geology, University College, Cardiff

and

T R OWEN, MSc, FGS
Professor of Geology, University College of Swansea

Second Edition

PERGAMON PRESS

OXFORD · NEW YORK · TORONTO · SYDNEY · PARIS · FRANKFURT

U.K.	Pergamon Press Ltd., Headington Hill Hall, Oxford OX3 0BW, England
U.S.A.	Pergamon Press Inc., Maxwell House, Fairview Park, Elmsford, New York 10523, U.S.A.
CANADA	Pergamon of Canada, Suite 104, 150 Consumers Road, Willowdale, Ontario M2J 1P9, Canada
AUSTRALIA	Pergamon Press (Aust.) Pty. Ltd., P.O. Box 544, Potts Point, N.S.W. 2011, Australia
FRANCE	Pergamon Press SARL, 24 rue des Ecoles, 75240 Paris, Cedex 05, France
FEDERAL REPUBLIC OF GERMANY	Pergamon Press GmbH, 6242 Kronberg-Taunus, Hammerweg 6, Federal Republic of Germany

First edition 1968
Reprinted 1969
Second edition 1980

British Library Cataloguing in Publication Data
Anderson, John Graham Comrie
The structure of the British Isles. - 2nd ed.
(Pergamon international library).
1. Geology - Great Britain
I. Title II. Owen, Thomas Richard III. Series
554.1 QE261 80-41075

ISBN 0-08-023998-6 (Hardcover)
ISBN 0-08-023997-8 (Flexicover)

In order to make this volume available as economically and as rapidly as possible the author's typescript has been reproduced in its original form. This method has its typographical limitations but it is hoped that they in no way distract the reader.

Printed in Great Britain by A. Wheaton & Co. Ltd., Exeter

Preface

This book is a second edition of that first published in 1968. The mode of treatment remains unchanged but portions have been updated and some parts have been extensively rewritten (notably those on the Scottish Highlands and South-West England). A new chapter has been added on the geology of the seas around Britain.

The authors wish to express their gratitude to Mrs. D. J. Nuttall, who typed the manuscript. They also wish to thank Mrs. M. Millen, Mr. G. B. Lewis, Mr. J. U. Edwards and Mr. A. H. Smith for drawing the illustrations.

Finally, the authors wish to record their thanks to Plenum Press, Allen and Unwin, Graham Trotman Dudley Publishers Ltd. and the Geological Society for their kind permission to reproduce or adapt certain illustrations; also to the Director, the Institute of Geological Sciences for permission to use certain photographs.

ANDERSON & OWEN: STRUCTURE OF THE BRITISH ISLES, 2nd EDITION

0 08 023998 6 Hard
0 08 023997 8 Flexi

Errata

Page 29 Caption and figure: *For* Invernian *read* Inverian

Page 82 Caption B: *For* Silisian *read* Silurian

Page 92 Caption: *For* Southern Islands *read* Southern Uplands

Page 126 Caption: *For* Section *read* Sections

Page 132 Caption 7.9: *For* Carinia *read* Caninia

Page 140 Caption 7.13: *For* Troter *read* Trotter

Page 192 Line 36: *For* fig. 8.5d *read* fig. 8.6
Line 39: *For* fig. 8.6d *read* fig. 8.6

Page 196 Line 32: *For* fig. 8.7b *read* fig. 8.7
Line 39: *For* fig. 8.7a *read* fig. 8.7

Contents

Introduction

This book aims to provide a background to the geological structure of the British Isles, but, it is hoped, in sufficient detail to suit the needs of students studying for a university degree in Geology and Geography. The broad tectonic evolution of the whole area is first described, followed by regional descriptions of the major structural units. An attempt has been made to describe structures in order of decreasing age (for example, the Scottish Highlands are described before the Lake District or Wales) but it will be understood that there are many cases where a complex geological and structural history has resulted in a considerable variability in the time of origin of the tectonic elements.

The book should be read with the aid of the appropriate published geological maps, particularly the 10-inch (two-sheet) map and the 1-inch and 1:50,000 (1¼-inch) maps, published by the Institute of Geological Sciences, together with the Tectonic Map of Great Britain and Northern Ireland (25 miles to one inch), published in 1966.

List of Illustrations

Plates

CHAPTER 1

The Structural Framework of the British Isles

The British Isles, although forming only a very small part of the earth's land surface, reveal most divisions of the stratigraphical succession. Moreover since early Precambrian times the region has been one of the tectonically more active sections of the crust. In fact the rocks of the British Isles have been affected by nearly all the major structural events of geological history.

It is not surprising therefore, that several of the major "breakthroughs" in stratigraphy (William Smith) and structural geology were made in these islands. In the structural field two historical examples may be quoted. That strata had been distorted by immensely powerful crustal forces had been realised before Hutton but it was this Edinburgh geologist (1795), by interpreting unconformities, who made it possible to date these events. About 80 years later the recognition that major reversals of the original succession had taken place in North-West Scotland (Ch 5.1), along with the developing nappe-theory in the European Alps, laid the foundation of modern tectonics.

The British Isles are of course, geologically part of Europe and lie entirely within the continental shelf. The relationship of British Geology to European geology as a whole, and in fact to that of northernmost Africa, should therefore be considered.

Two extensive kratogenic blocks (parts of separate plates) of older Precambrian, covered by relatively undisturbed younger strata, occupy much of Russia and the Baltic region on the one side and North Africa - Arabia on the other. The Precambrian floor of the Russian Platform is exposed on the surface over the so-called Baltic Shield, whose western limit runs along the middle of Scandinavia (fig. 1.1).

Remnants of a third kratogenic block occur in the extreme north-west of Scotland, the Precambrian being covered here by undisturbed Lower Palaeozoic strata. Structures due to several Precambrian orogenies are evident in this Scottish block where the oldest tectonothermal event dates back at least 2700 million years.

The Baltic and Scottish blocks are part of the "Eo-Europa" of some authors (e.g. Ager, 1975). Framed by the two European blocks and that of North Africa there occurs a former mobile belt, an area of geosynclinal deposition and subsequent orogenic deformation. Remnants of three major geosynclines (deformed into three great mountain chains) can be traced within this belt: (i) the Lower Palaeozoic Geosyncline (the Caledonides or Palaeo-Europa"); (ii) the Upper Palaeozoic or mid-European Geosyncline (the Hercynides or Variscides, also termed "Meso-Europa"); (iii) the Mesozoic-Tertiary or Tethys Geosyncline (the Alpides or "Neo-Europa"). A broad southward migration with time is represented by the distribution of the above belts.

Scotland (Caledonia) gave its name to the Caledonides which underlie about three-quarters of the British Isles and come to the surface in several regions. The Caledonides are magnificently exposed for over 1000 miles in Norway. The connection between the European and British parts of the chain raises interesting problems. It was at one time assumed that their south-eastern front, seen in Shropshire, continued on a north-easterly course under the English Upper Palaeozic and Mesozoic and the North Sea to southern Norway. It now seems more probable that this front turns south-eastwards under the English Midlands round a buried Precambrian block (Turner, 1949). It may then pass under the North Sea and parts of the Low Countries to join up with the Caledonian front in Norway. Alternatively, it may not even connect with the Norwegian front but may continue eastwards to the Sudetian Caledonides. In either case a considerable part of the fold-belt must underlie Belgium, Holland, and probably some of Denmark (Von Gaertner, 1960).

The Hercynides or Variscides are today seen as eroded remnants or horsts breaking through a cover of Mesozoic and Cainozoic strata in Western Europe. From these blocks it is possible to make out two great arcs springing from the Massif Central of France. The Variscan Arc strikes N.E. through the Vosges and the Schwarswald to Bohemia. The Armorican Arc trends N.W. then W.N.W. through Brittany (Armorica). The exposed Ardennes and a deeply buried chain proved by bores under the Paris Basin fill in to some extent the space between the arcs. Spain and Portugal, too, are largely made up of a part of this fold-mountain system which may have been more directly connected with Armorica before the opening of the Bay of Biscay by anticlockwise rotation of Spain.

From the Variscan Arc the name Variscides has been applied to the whole system. However, the terms Hercynides and Hercynian, used in the first edition, are retained in the present book.

The Armorican Arc crosses South-West England, the S.W. tip of Wales and Southern Ireland, and structures of Hercynian age occur almost throughout the British Isles.

Between the major Hercynian (or Variscan) and Alpine orogenies, during the Mesozoic, the Saxonian and Cimmerian structures were developed in N.W. Europe. These are comparatively weak in the British Isles although they may be detected in several Mesozoic areas. They are also of particular significance in the North Sea oil and gas fields (Ch. 10).

The Alpides are the youngest folded chains of the European mobile belt and therefore form the highest ranges today. They swing in great sinuous curves from the westernmost Mediterranean through Italy and the Alps across to Asia Minor and beyond. They cut across the Hercynides or Variscides to override, along the Carpathian Front, the southern edge of the Russian platform. Parts of the Hercynides are occasionally exposed within the Alpine chains where they have suffered Tertiary metamorphism and migmatization. During this last great deformation of the European mobile belt, Britain and northern France experienced weaker ripples of the Alpine "Storm".

About three-quarters of the total area of the British Isles lies within the Caledonian Fold-belt although this is partly hidden under younger formations and modified by younger structures. The Caledonian Orogen is bounded to the north-west by the well-known Moine Thrust (Ch. 5.1) and to the south-east and south by a less well-defined "front" in Wales, in Shropshire and under the English Midlands. Structures formed by Precambrian orogenies ranging from about 2700 to 1500 million years in age have been identified in the north-west kratogen beyond the Moine Thrust (Ch. 4). Precambrian structures can also be studied in outcrops within the fold-belt, along the south-east front and in the small inliers which show through the Upper Palaeozoic and Mesozoic formations covering the whole of the south-east kratogen (Ch. 7.3).

Caledonian structures ranging in age from early Ordovician (or perhaps late Cambrian) to mid-Devonian are dominant or important in almost the whole of Scotland and Ireland, in the Lake District and in central and north Wales (Chs. 5 & 6).

The Hercynian "front", marking the northern limit of strata intensely compressed, overfolded and, in places, metamorphosed during the Hercynian Orogeny, extends across Ireland from Dingle Bay to near Waterford (Ch. 7.7). It then crosses the Caledonian "front" in south-west Wales and continues eastwards to the Bristol district where it disappears under the Mesozoic.

Although this line marks the northern limit of intense Hercynian folding, almost the whole of the British Isles was affected by Hercynian diastrophism. Practically all the Carboniferous strata are disturbed by Hercynian folds, although north of the "front" thrusting and overfolding are uncommon. Much of the Caledonian fold-belt was modified by Hercynian folding and faulting and itself influenced the development of the younger structures. Hercynian folding had little effect in the metamorphic parts of the Caledonian Orogen, but the whole or part of the sinistral displacement along numerous, mainly north-easterly, faults (some of considerable structural significance) may be the response in relatively rigid blocks to Hercynian compression.

Alpine movements, probably most powerful in the Miocene, determined the structure of the Mesozoic and Tertiary strata of eastern and southern England (Ch. 8). The folding is strongly developed, however, only in the south of England, e.g. in the Weald, in Dorset, and in the Isle of Wight; in the last two districts vertical or even slightly over-turned beds are present. Where the Mesozoic and Tertiary rocks are present in more northerly districts, notably in north-east Ireland and on the western seaboard of Scotland, it is possible to prove the occurrence of powerful Tertiary faults and of fairly gentle folds. There is in fact evidence that Tertiary faulting, at any rate, affected practically the whole of the British Isles. The formation of Tertiary igneous complexes in western Scotland and north-east Ireland also had considerable, if local, structural results.

Their long tectonic history makes it possible to divide the British Isles into a number of natural structural regions, separated for the most part by major faults or by important unconformities. The major structural units and their boundaries will now be considered (Fig. 1.2).

The most north-westerly of the major units includes the north-west Highlands and the Outer Hebrides (Ch. 4), parts of the foreland or kratogen bordering the Caledonian mountain chain (Fig. 1.1). This is separated from the fold-belt by the Moine Thrust and underlying, individually less persistent, planes making up a Border Thrust-zone (Ch. 5.1). There is evidence (Dearnley, 1962) that the kratogen is split by a large submarine fracture-the Minch-Fault - separating the north-west Highland or mainland portion from the Outer Hebrides. Within these two parts of the kratogen, Precambrian structures predominate; Caledonian and later movements played a very minor role.

The part of the Caledonian Fold-belt which forms the Scottish Highlands extends from the Moine Thrust to the Highland Boundary Fracture-zone, traceable in a south-westerly direction across Scotland from Stonehaven to the Firth of Clyde. This part of the Caledonian Orogen is characterized by the regional metamorphism

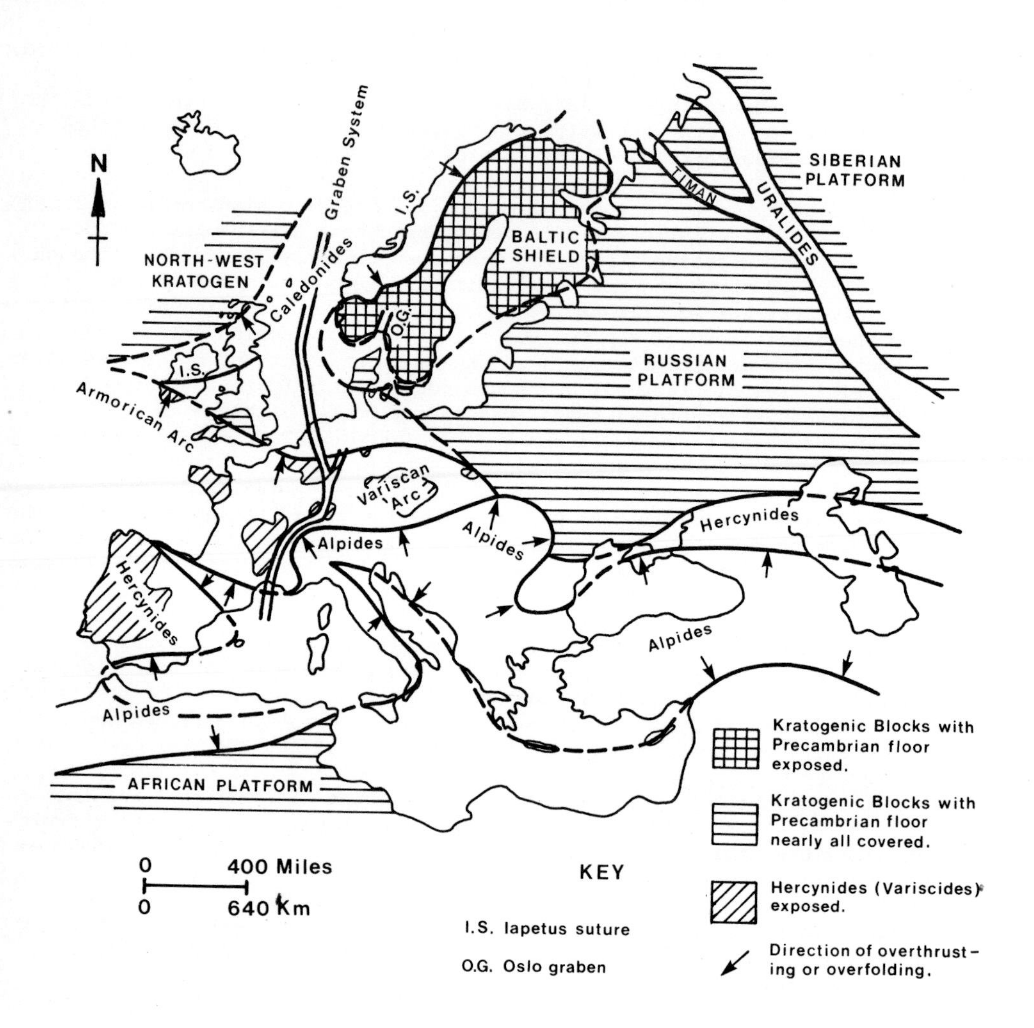

Fig. 1.1 The Major Structural Elements of Europe.

of all pre-Lower Devonian strata. The metamorphism and the folding is mainly Lower Palaeozoic and there was a late Caledonian folding phase in the mid-Devonian. Important fault-displacements are, at least in part, proto-Hercynian or Hercynian and in some districts significant Tertiary faulting can be demonstrated. Structurally, and topographically, the Scottish Highlands are divided by the Great Glen Fault into the Northern Highlands (5.2) and the Grampian Highlands (5.4). (The term Northern Highlands has also been used in a wider sense to include the north-west Highlands).

The Orkney Islands are an offshore portion of the Northern Highlands, but the structural position of the Shetland Islands (5.3) requires discussion. If the Great Glen Fault continues on its north-easterly course under the North Sea these islands lie to the north-west and should therefore be the structural continuation of the Northern Highlands. On the other hand, Flinn (1961) has suggested that the fault swings into a northerly direction and passes through the "Mainland" of Shetland, in which case the eastern part of the islands may be the structural continuation of the Grampian Highlands. They are therefore described as a separate part of the Caledonian Fold-belt. Their minimum distance from the Bergen coast of Norway (210 miles, 336 km) may be compared with that from the mainland of Scotland (103 miles, 165 km).

The Grampian Highlands continue across the north and north-west of Ireland (5.4) where they are bounded to the south-east by the continuation of the Highland Boundary Fracture Zone which can be traced to Clew Bay, although a considerable part of its course is obscured by Tertiary lavas and Lower Carboniferous sediments. These younger strata overlap far on to the Dalradian and Moinian of the Irish Grampian Highlands. In addition to the dominant Caledonian structures, it is possible to detect, to a greater extent than in the Scottish Highlands, the presence of important Hercynian and, in the north-east, Tertiary elements.

Although the suggestion has been made that the Leannan Fault (5.4) is the continuation of the Great Glen Fault, it is much more likely that the latter passes under the sea, north-west of Ireland. The Northern Highlands are, therefore, not represented in Ireland.

South-east of the Highland boundary Fracture-zone the Caledonian Fold-belt is for the most part at a much higher structural and metamorphic level. Rocks which have undergone true regional metamorphism of Lower Palaeozoic age occur only in Connemara and South Mayo in the extreme west of Ireland (5.5).

Elsewhere the Palaeozoic strata of the Caledonian Orogen, although strongly folded, have not undergone metamorphic change beyond the inducement of cleavage, noteworthy in some areas for the production of commercially important slates. Moreover, south of the Highland Boundary, the regions in which Lower Palaeozoic rocks and structures come to the surface are isolated from each other by large tracts of younger strata, usually brought down by Hercynian structures.

The Southern Uplands of Scotland (6.3) are bounded to the north-west by the Southern Uplands Fault extending in a south-westerly direction from near Dunbar to Loch Ryan. Tectonically they are characterized by dominantly north-easterly Caledonian structures affecting Ordovician and Silurian strata. Both within the lower Palaeozoic area, however, and along its southern margin there are considerable outcrops of Upper Palaeozoic rocks with Hercynian structures. The Southern Uplands extend into north-east Ireland (6.3) where they are characterized by the same structures.

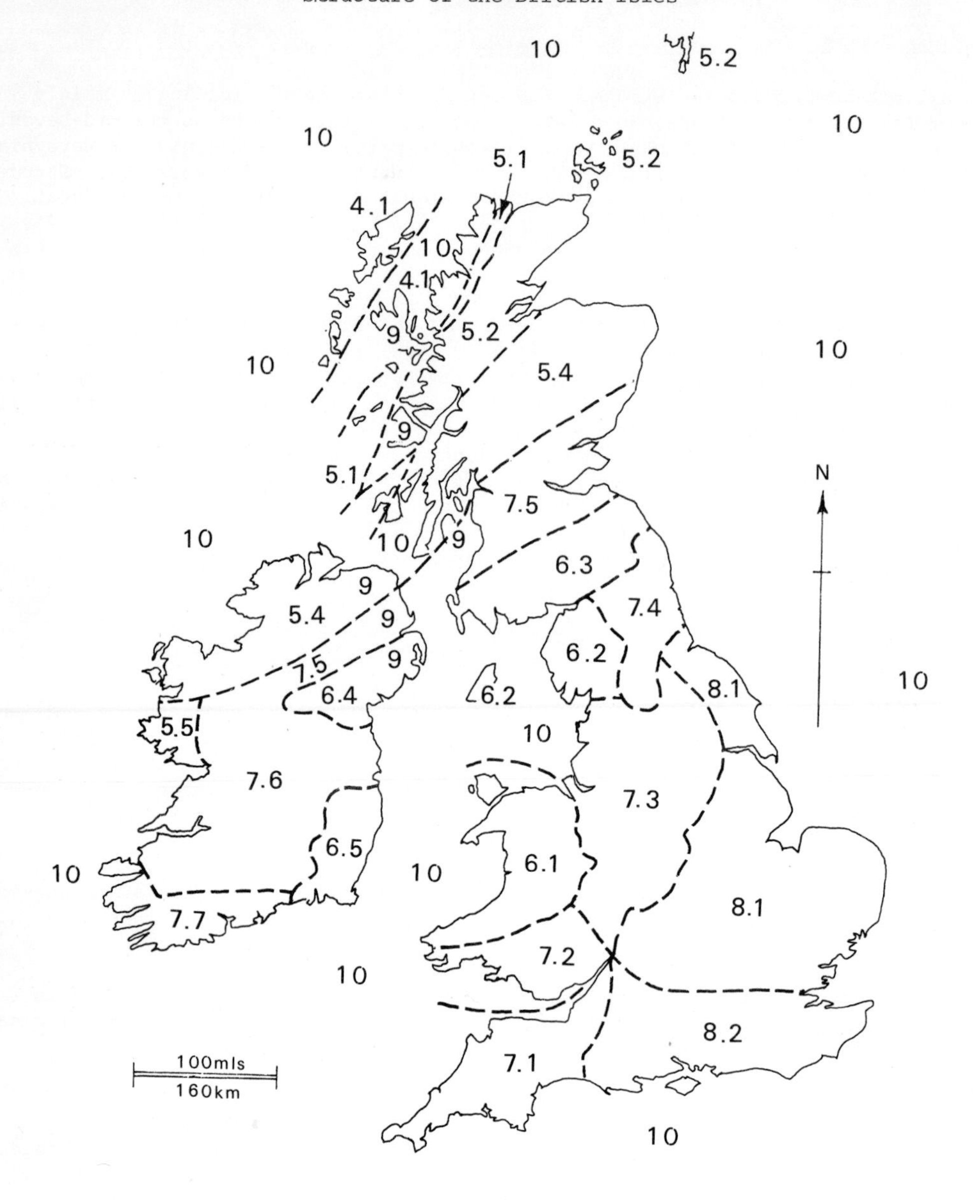

Fig. 1.2 Map Showing Distribution of Areas Described in Each Section. Key: 4.1. N.W. Highlands and part of Hebrides; 5.1. N.W. Caledonian Front; 5.2. N. Highlands and Orkney; 5.3. Shetland; 5.4. Grampian Highlands (Scotland and Ireland); 5.5. Connemara; 6.1. Welsh Block; 6.2. Lake District and Isle of Man; 6.3. S. Uplands; 6.4. S. Uplands in Ireland; 6.5. S.E. Ireland; 7.1. S.W. England; 7.2. South Wales; 7.3 Central England; 7.4. N.E. England; 7.5. Midland Valley of Scotland; 7.6. Central Ireland; 7.7. Southern Ireland; 8.1. E. England; 8.2. S. England; 9. Tertiary Igneous Terrains; 10. Seas around Britain.

To the west and south, the Irish Southern Uplands disappear under an unconformable cover of Carboniferous. The structural continuation reappears in South Mayo and Connemara but there is considerable doubt whether the Southern Uplands Fault continues thus far west (cf McKerrow and Campbell, 1960; Charlesworth, 1963). Moreover, as mentioned above, this block is characterized by Lower Palaeozoic regional metamorphism and is accordingly treated separately (5.5).

Caledonian structures with north-easterly to east-north-easterly trends are strongly developed in the Lower Palaeozoic of the Lake District and the Isle of Man (6.2). Hercynian structures affect both the surrounding Carboniferous and Lower Palaeozoic; powerful Tertiary doming as well as some faulting finally determined the structure of the Lower Palaeozoic block and the surrounding younger strata ranging up to Lias. These younger strata are, therefore, included in the Lake District structural unit which can be regarded as bounded to the north by the Solway Firth, to the east by the Pennine/Dent Faults (on the west flank of the Pennine range) and to the south by Morecambe Bay.

A wide expanse of younger strata separates the Lake District from the Lower Palaeozoic block of Wales (6.1). The structures are again dominantly Caledonian but there are also Hercynian and Tertiary elements. Comparatively small outcrops of Precambrian rocks showing Precambrian structures occur in Shropshire along the south-east "front", and also within the fold-belt, notably in Caernarvonshire and in Anglesey.

The Anglesey Precambrian forms part of the Irish Sea geanticline on which Precambrian rocks also appear in south-east Ireland. This geanticline, until Ordovician times at least, is believed to have separated the Welsh Lower Palaeozoic Basin from another in Ireland. A large block of Lower Palaeozoic strata, deposited in that basin and affected by powerful Caledonian structures, occurs in south-east Ireland (6.4). The Lower Palaeozoic strata of Scotland and of the northern part of Ireland had a different history up to the Silurian from those further south as they were formed on the opposite side of the Iapetus Ocean.

South-west England (7.1) forms a distinctive tectonic unit with strongly developed Hercynian structures, cleavage and some regional metamorphism, involving mainly Devonian and Carboniferous strata. It is bounded to the north by the Bristol Channel, under which (it is claimed by some) lies the Hercynian "front" and to the east by unconformable Permian and Mesozoic strata. Its westerly continuation forms the southern part of Ireland (7.7), bounded to the north by a well-marked "front" extending from Dungarvan in the east to Dingle Bay in the west (fig. 6.11).

To the north of the Bristol Channel the South Wales region (7.2) is dominated by Hercynian structures including the Coalfield basin with its E.W. trending. The folding is strong in the south, where there is also some thrusting, but becomes gentler in the north. The Hercynian "front" which includes the overriding of Coal Measures by Precambrian, passes through the extreme south-west of the region in Pembrokeshire, and here crosses the Caledonian "front". Precambrian and Caledonian structural elements are therefore present in this part of the region. In the south, along and near the Bristol Channel, Tertiary (or perhaps intra-Cretaceous) folds and faults can be distinguished, affecting strata up to Lower Liassic.

Central England (7.3) is another region dominated by Hercynian tectonics including not only E.-W. structures but others of varying orientation (controlled by older elements) including the N-S. Malvern and Pennine axes. Much of the boundary of the region is formed by the unconformable overlap of Permian or younger strata. The northern boundary is defined by the North Craven Fault, an important east-west fracture downthrowing to the south and also separating different sedimentary environments. Caledonian and Precambrian structures are seen in a few pre-Upper Palaeozoic inliers. Folds and faults of Tertiary age have also affected the region.

The structure of north-east England (7.4) was also essentially determined by Hercynian events. It forms a distinctive tectonic unit, characterized by ancient blocks (with granite intrusions) under the northern part of the Pennines which influenced both sedimentation and structural development over a considerable period. The Pennine/Dent Faults form the western margin and the North Craven Fault, and a fault bringing down the Magnesian Limestone, the southern and south-eastern boundaries. Caledonian structures are revealed or can be inferred along parts of the Pennines.

The Midland Valley of Scotland (7.5) is a graben clearly defined by the Highland Boundary Fracture-zone to the north-west and the Southern Uplands Fault to the south-east. The rift was formed by late Caledonian folds and faults. Nevertheless, the main structural development within the graben was Hercynian although the Caledonian structures in many cases exercised a control on the younger structures. Tertiary movements can also be demonstrated, particularly in the west.

The Midland Valley continues south westwards into Ireland (George, 1960a) but it is much less well defined for here it had largely lost its character of a graben by Lower Carboniferous times. The Highland Boundary Fracture-zone, it is true, can be traced across the country but Lower Carboniferous strata and Tertiary lavas overstep it. Moreover, the Southern Uplands Fault cannot definitely be followed beyond the centre of Ireland; further west the Lower Carboniferous of the Midland Valley merges with that of the Central Plain.

For these reasons the Central Plain of Ireland and the continuation of the Midland Valley are treated as one structural unit (7.6). As the region is largely floored by Carboniferous strata, Hercynian structures predominate with strong folding in the south (near the so-called Hercynian "front") becoming weaker further north. Caledonian structures are, however, revealed in a number of inliers of Silurian and Devonian, and Tertiary movements can be inferred where post-Carboniferous rocks are present.

South-east of a line roughly from Tynemouth to Torquay, the surface structures are for the main part of Tertiary age. In the east of England (8.1) dips are generally low but in the south of England (8.2) the folding becomes more marked with vertical beds in the extreme south. Pre-Tertiary structures are seen in a few inliers and have been proved by deep bores and by mining, e.g. in the London Basin and in the concealed east Yorkshire and Kent Coalfields. Intra-Cretaceous structures occur in Dorset and in the English Channel.

Tertiary events, primarily igneous rather than tectonic, have made parts of the west of Scotland and north-east Ireland a distinctive geological region (Ch. 9). This interpenetrates parts of five structural regions, namely the north-west kratogen, the Northern Highlands, the Grampian Highlands (in Scotland and Ireland), the Midland Valley, (mainly in Ireland) and the Southern Uplands (in Ireland). Tectonically the distinctive character of this Tertiary igneous province lies in the occurrence of important structures imposed by, or formed during, igneous intrusion and in the presence of widespread Tertiary lavas and Mesozoic sediments. These enable Mesozoic and Tertiary structures to be distinguished to a much

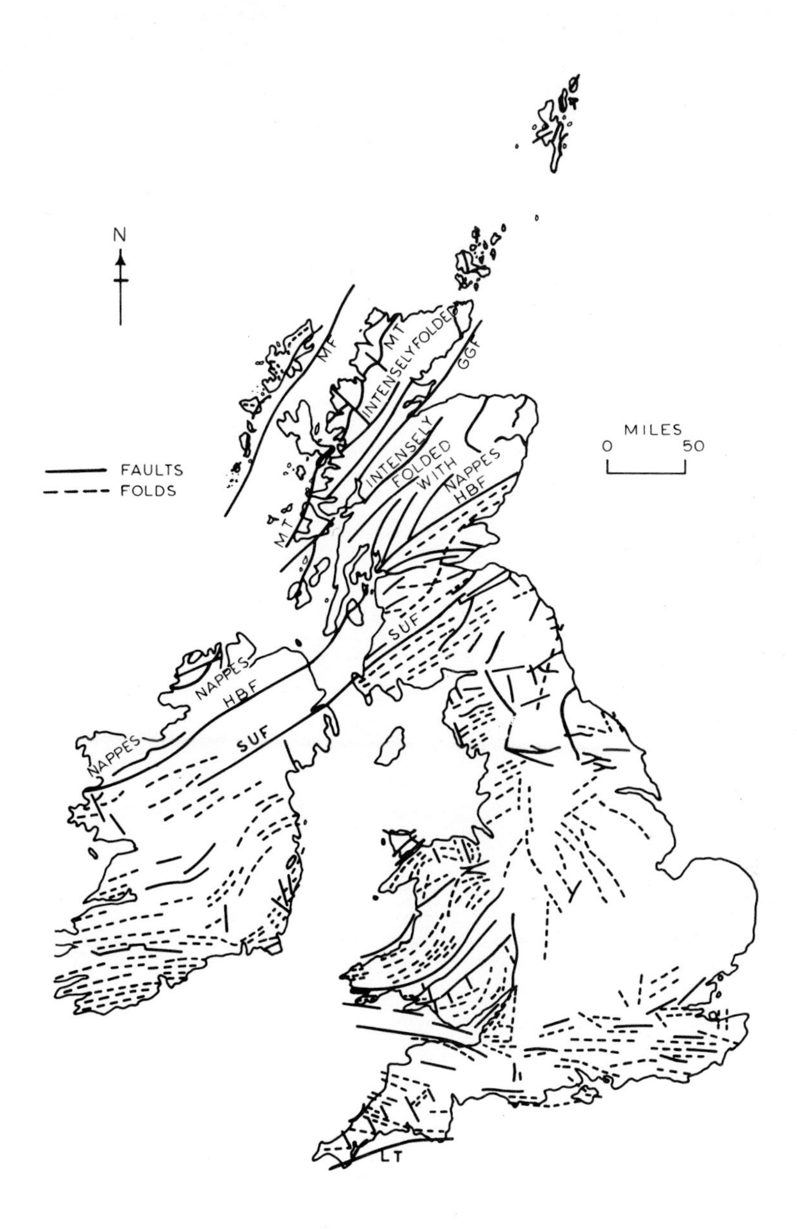

Fig. 1.3 General Structural Map of the British Isles. Key: G.G.F. Great Glen Fault; H.B.F. Highland Boundary Fracture-zone; L.T. Lizard Thrust zone; M.F. Minch Fault; M.T. Moine Thrust; S.U.F. Southern Uplands Fault.

greater extent than in other parts of the five regions mentioned above. Exploration for oil and gas under the North Sea has revealed much of the stratigraphy and structures (Ch. 10). The latter include the Viking and Central grabens, the continuation of a rift system traceable through Europe from the Mediterranean to Holland (Anderson, 1978, Fig. 10.1).

Broadly, therefore, the British Isles (Fig. 1.3) structurally consist of: a Precambrian kratogen in the north-west of Scotland; an extensively exposed part of the Caledonian Fold-belt north-west of the Highland Boundary Fault in Scotland and Ireland, characterized by regional metamorphism mainly of Lower Palaeozoic age; a broad belt north-west of the south-east Caledonian "front" (the Church Stretton-Pembroke line) in Britain and north of the Hercynian front in Ireland in which the Caledonian blocks and inliers (without regional metamorphism) are interspersed with districts dominated by Hercynian structures; the South Wales region with strong Hercynian folding; a part of the main Hercynian Orogen south of the "front" in Ireland and south-west England; and eastern and southern England dominated by mainly Tertiary structures. Hercynian structures can be distinguished in parts of the older blocks and Tertiary tectonic events, at least on a minor scale, can be proved or inferred throughout almost the whole of the British Isles.

CHAPTER 2

Tectonic Evolution of the British Isles

PRECAMBRIAN - SILURIAN

Precambrian-Archean and Lower Proterozoic

The building of the structural framework of the British Isles began in the early Precambrian or Archean; this can be deduced both from the stratigraphical evidence and from radiometric dates of metamorphic and igneous events accompanying orogenesis which go back to about 2800 million years ago. Within the British Isles it is mainly in the north-west foreland of the Caledonian chain that these ancient structures are seen. Within this block it is possible that some gneisses may be older than supra-crustal rocks affected by the event about 2800 m.y. ago and may therefore be Katarchean (Ch. 4).

When a geological history recorded in their rocks began, the British Isles formed a small part of a vast primeval continent (Laurasia) containing much of what are now North America, Europe and Asia. In another part of this old continental crust, the Baltic Shield, some very ancient Precambrian-Katarchean-radiometric dates have been recorded, notably in the Kola Peninsula of the USSR.

In the North-West Highlands of Scotland (Ch. 4) there are records of several mountain-building episodes more than 1000 million years old.

The formation of the early Scourian (Badcallian) gneisses and schists, which began about 2800 million years ago, is the earliest known metamorphic event and one which probably accompanied major orogenesis. This was the first of several such episodes in the Scourian cycle which may have ended about 2500 m.y. ago and which brought about the formation of the Scourian Complex (Ch. 4). In places the rocks affected by Scourian metamorphism are strongly migmatized and their deformation in general suggests plastic distortion. This probably took place under a thick cover, although suggestions have been made regarding some older Precambrian terrains that the Uniformitarian Principle does not completely apply and that at this early time in the history of the crust the general geothermal gradient was much steeper.

It is only remnants of the orogenic belt that can now be seen, but from the lineaments of the Scourian Complex it may be deduced that the Scourian mountains had broadly a north-easterly trend.

These Scourian structures have largely been obliterated by later Precambrian events (see below); Scourian relic structures are least obscured in the centre of the mainland portion of the foreland between Gruinard Bay and Scourie (Fig. 4.1). They have also been recognized in parts of the Outer Hebrides where, however, as in other areas, there are folds trending across the general "caledonoid" Scourian grain. Some of the mapped folds are interpreted as drag-structures and thus imply over-folds and recumbent folds of a considerably larger order; only a few of these have, however, been clearly identified.

Some 100-200 million years later began another orogeny, proved in the Lochinver area (Ch. 4) and termed the Inverian. This added a further element to the shield with major folds trending north-west. This cycle closed about 2190 m.y. ago with the intrusion of basic and ultrabasic dykes, parallel to the north-westerly trend. It has been concluded that these provided evidence of a period of uplift in a non-orogenic belt. It has also been postulated, however, that they were intruded in depth, and this view finds support in the discovery that they were formed in association with an orogeny. Their intrusion in fact may have been spread over a long period of cooling, as the earliest dykes are the most altered.

Commencing about 1850 m.y. ago, the formation of the Laxfordian Complex completed the building of the Lewisian Assemblage. It marked another major orogenic cycle taking place in three phases (Ch. 4) and was spread over 200-300 million years. It is likely that the mountains produced in this orogenic cycle had mainly a north-westerly trend as this is the principal, although by no means the exclusive, strike of the Laxfordian structures; as these are the later, this north-westerly trend is now the dominant trend of the north-west Highland foreland. The small-scale Laxfordian folds are very complex and variable in direction; tight isoclines are common, and major folds can be distinguished in places. Generally the Laxfordian folding is less of a plastic nature than that of the Scourian. At a late stage, faulting and shearing, accompanied by cataclasis, took place, mainly on north-westerly lines.

The youngest date recorded for a Laxfordian event is 1150 m.y. but this may be a diluted date and therefore too young.

Evidence for Archean and Lower Proterozoic structures in the British Isles is scarce S.E. of the Great Glen Fault. The presence of Lewisian rocks under the Midland Valley of Scotland is suggested by fragments in Carboniferous volcanic necks; the presence of similar rocks under the Southern Uplands is doubtful. Gneisses with pre-Caledonian structures within the Caledonian fold-belt in the west of Ireland may be Lewisian.

In the Channel Islands gneisses in Guernsey have been dated at 2620 m.y. (cf. Scourian) and the name Icartian has been suggested for this event. Rocks of almost the same age occur in the Cherbourg Peninsula where the Lihouan event (2000-1900 m.y., cf. Inverian) has also been detected. Penteverian (Sarnian) structures, with a northerly grain and formed some 1000-1200 m.y. ago about the end of the Lower Proterozoic, occur in Britanny and granite gneisses in the Channel Islands probably belong to this complex.

The Rosslare Complex in South-east Ireland contains metamorphic rocks formed in two adjacent cycles with radiometric dates corresponding to the Scourian and Laxfordian respectively. The possibility that Archean and Lower Proterozoic rocks and structures underlie the badly exposed kratogen beneath the English Midlands is supported by the occurrence at the present surface of the schists of Rushton and the granite-gneisses and schists of the Malverns. End-Proterozoic dates from the Malvernian may merely record an event at that time.

During the Archean and much of the Lower Proterozoic the British region broadly shared a common metamorphic and structural history with the Baltic Shield where the Pregothian (2500 m.y.) and the Gothian (1300-1750 m.y.) correspond to the Scourian and Laxfordian.

Precambrian - Upper Proterozoic

About the beginning of the Upper Proterozoic, that is about 1000 m.y. ago, two global events took place which had a profound effect on the tectonic evolution of the British Isles. Firstly, it would seem that broadly at this time a system of relatively stable shields and mobile belts became established of similar character to, although not necessarily of the same distribution as, those of the present day. As a corollary to this, thick continental sedimentary piles accumulated of which the Scottish Torridonian is an early and typical example. Secondly the great Laurasian continental block began to split with what may be termed the American Plate moving away from a Eurasian (and African) Plate. This led to the opening of the Proto-Atlantic or Iapetus Ocean. As a consequence there are important differences in the sedimentary history, organic evolution and structural development up to the Silurian between Scotland and the northern part of Ireland on the one hand and England-Wales and the southern part of Ireland on the other.

Sedimentation took place on the landward regions of these moving blocks and in shelf-seas and deeper waters on their margins. It is apparent too that the long period of sedimentation preceding the Lower Palaeozoic or Caledonian Orogeny began well before deposition of beds which can be defined as Cambrian; in fact, in some places in Britain the top of the Precambrian is not even marked by a stratigraphical break.

About 1000 million years ago the north-west Caledonian foreland became established as a kratogenic block. The fact that Cambrian or late Precambrian formations rest on deeply formed Lewisian rocks of the kratogen shows that profound denudation had occurred before, and in some cases several hundred million years before, Upper Proterozoic times. Indeed, we can perhaps envisage the Scottish area of about 1000 million years ago as having something like the aspect of the Baltic or Canadian Shields of the present day with a surface in which structures running in several directions and inherited from several orogenies had determined the lineaments of fairly subdued topographical features. Dominant trends were probably north-easterly (Scourian) and north-westerly (Laxfordian).

An early structural event in the Scottish kratogen was the uprise of a horst on the side of the Outer Hebrides probably bounded by a fault to the E. (Ch. 4). This block was probably part of the source of the molasse-type Torridonian sediments; another part, as shown by clasts, is now in Greenland.

Within the kratogen Upper Proterozoic structural events were the folding of the Lower Torridonian along N-S axes sometime between about 1000 and 800 m.y. ago, and the fairly gentle folding of the Upper Torridonian prior to the deposition of the Lower Cambrian.

The folding and metamorphism of the metasediments which make up the Caledonian fold-belt in the Scottish Highlands was mostly Lower Palaeozoic in age (see below). This applies not only to the Dalradian rocks (5.3, 5.4, and 5.5) but to part of the Moinian Assemblage (5.2). This fits in with the theory that the metasediments of part of this assemblage are the Iapetus shelf-sea equivalents of the Upper Torridonian. However, radiometric dating of another part of the Moinian shows that folding and metamorphism may have taken place about the end of Lower Proterozoic times, coeval with the formation of the Grenville metamorphic province to which of course the Scottish rocks would be close before the Iapetus split took place. This implies the existence of an "Older Moinian" with late Lower

Proterozoic structures and a "younger Moinian" with Caledonian structures and possibly some Upper Proterozoic although the break is not evident (5.6). This remains an outstanding problem of British geology.

In the south-east kratogen and in the neighbouring parts of the Caledonian fold-belt evidence of Upper Proterozoic structures is scattered, and some radiometric dates are equivocal or inconsistent with stratigraphical findings.

The Charnian provides evidence of the formation of a north-westerly fold-belt which was formed in the late Pre-cambrian. The Uriconian, and the flysch-like Eastern Longmyndian are probably the volcanics and sediments of a late Precambrian geosyncline and which were folded and eroded before the deposition of the molasse-like Western Longmyndian. A later, although still Precambrian, phase then resulted in the tight folding of the Western Longymndian into a deep NNE.-SSW. overfolded syncline (6.1).

The folding along NW. axes of the Ingletonian rocks in the Pennines is probably late Precambrian and certainly contrasts with undoubted Caledonian folding of nearby Lower Palaeozoics.

The Mona Complex of Anglesey and Lleyn (6.1) shows folding and highly variable metamorphism of late Upper Proterozoic date (a Cambrian date has also been suggested). The metamorphics were intruded by the Coedana Granite dated at 609-614 m.y. The presence of ultrabasics and glaucophane schists is evidence that the orogeny was accompanied by the subduction of a late Precambrian ocean plate but the direction of subduction has also been disputed.

The Anglesey (Monian) tectonothermal events suggest that the widespread late-Precambrian (Cadomian) orogeny may have bulged as far N. as Anglesey.

Longmyndian sedimentation may be contemporaneous with that of the Monian meta-sediments, and their folding, referred to above, may also be Cadomian. The difficulty in interpreting radiometric evidence is illustrated by a date (maximum) of 600 m.y. for the lower part of the Longmyndian which is belied by the stratigraphical and structural facts, although the date could represent an event.

Lower Palaeozoic

Lower Palaeozoic structural evolution was accompanied by the further opening of the Iapetus Ocean (perhaps as much as 2000 km wide in the N, in the Cambrian, (Fig. 2.1) and then its closure during the height of the Caledonian earth-storm.

Some time during the Silurian the ocean closed; the suture of join probably runs under the Solway Firth and south-westwards across Ireland.

Although Precambrian structural elements are present in the Scottish Highlands, the main folding and metamorphism of the Dalradian and Moinian of Scotland and Ireland took place during what may be termed the Grampian Phase of the Caledonian orogeny. This was at its peak about the mid-Ordovician, some 475 million years ago. However, both the folding and metamorphism were complex and took place in several pulses. In Connemara (5.5) there is evidence that these events started in the late Cambrian. Diachronism of the onset of tectonothermal processes in the Caledonides of the British Isles would be in line with evidence from other chains.

A. Unconformity of Torridonian Gritty Arkose on Lewisian North Shore of Loch Assynt, Sutherland, Scotland. (J.G.C.A.)

B. Folded Metadolerite Boudin in Metapsammite of Shetland Metamorphic Series, West Side of Lunna Ness, Shetland. (J.G.C.A.)

Datings around 420 m.y. suggest that further metamorphism and folding took place in the Silurian, and this may be the date of faults and higher folds detected in parts of the Highlands. As the Lower Old Red Sandstone of the region is folded it may be that folding of the metamorphic rocks extended into the Devonian. Folding and epizonal metamorphism of undoubted Wenlock strata occurred in Mayo (5.5). On the other hand it has been suggested that the 420 m.y. dates for Moinian rocks of the Northern Highlands record late cooling of metasediments altered in depth.

The Moine Thrust along the NW margin of the fold-belt was initiated early in the Caledonian orogeny, perhaps even before some of the folding, but the thrust-belt was Lower Devonian. Folding and tilting of post-Lower Ordovician date may have been the comparatively weak response in the kratogen to the Grampian Phase.

The main movement along the Highland Boundary Fracture-zone was Middle Devonian but the zone was initiated in Arenig or even earlier times. Associated serpentinites (often carbonated) and spilites were probably obducted from a subduction zone already initiated under the Highlands.

By the late-Ordovician the Highlands and their continuation into Ireland formed a major mountain chain with ocean and possibly island-arcs to the SE. Connemara also stood up as a fold-mountain region, in this case with the South Mayo trough to the N. The region SE. of the Highlands did not however, escape the mid-Ordovician disturbances for there was an important phase of folding of Ordovician age preceding the Caradoc in the Southern Uplands. The same movements occurred in the Girvan area in the Midland Valley. Here serpentinites, basic intrusions and spilitic pillow lavas may have been obducted from a subduction zone already developing in the vicinity of the Southern Uplands Fault although the main movement along this fracture was Devonian.

Movements occurred at the end of the Ordovician in the Southern Uplands, Girvan and Mayo and at the end of the Llandovery in Mayo. By the Silurian a north-easterly orientated narrow land-mass made up of Ordovician rocks - sometimes termed Cockburnland - crossed the Southern Uplands.

A major phase of Caledonian movement occurred in post-Wenlock (possibly post-Ludlow) times. This was the main fold-phase in the Southern Uplands and their continuation into Ireland and in south Mayo resulted in recumbent folding accompanied by regional metamorphism. The Southern Uplands folds and thrusts probably developed above a decollement possibly underlain by ocean-floor crust.

What now of the south-eastern margin of Iapetus? In the Lake District the first movements which can be clearly recognised were mid-Ordovician as shown by the folding of the Borrowdale volcanics. These were probably part of a Lower Ordovician volcanic arc. The main Caledonian structures, trending NE. to ENE. were developed in the late Silurian.

In Wales the development of Caledonian structures was complex and intermittent, building up to a major climax in late Silurian (or, according to O. T. Jones, in mid-Devonian) times throughout much of the Welsh area but somewhat earlier (pre-Upper Llandovery) in the Shelve-Longmynd area of Shropshire.

Intermittent movements (occasionally even involving folding, as at Comley) occurred during the Cambrian period in areas such as Llanberis, Shropshire and the Malverns, but were more intense at the close of the period when broad folding (and even faulting) along caledonoid lines occurred in north-west Wales and the Longmynd before the deposition of the Arenig (see 6.1). Mid-Ordovician unrest was

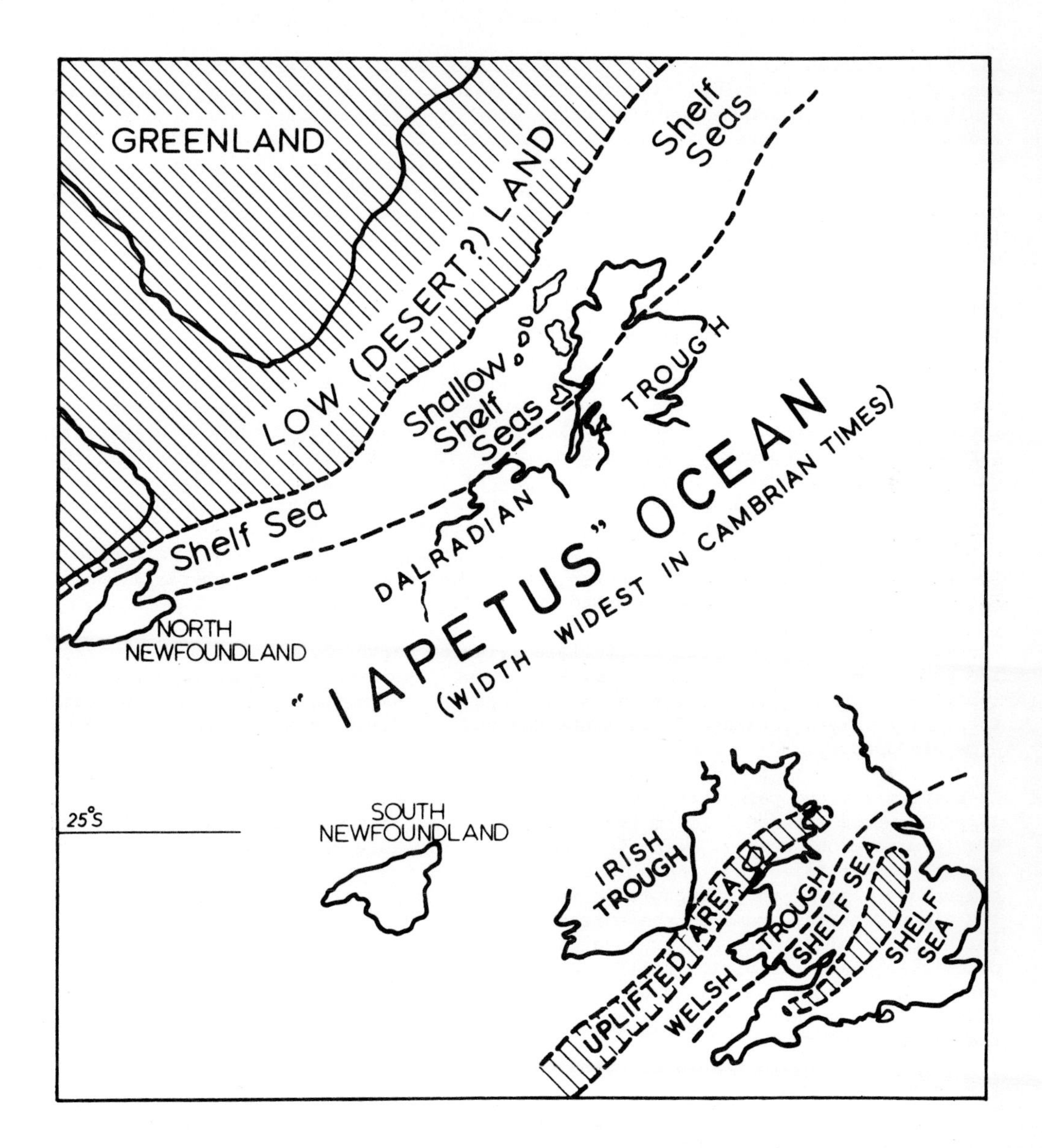

Fig. 2.1 "Iapetus" at End of Lower Cambrian Times.

widespread in Wales, and the Sub-Caradoc overstep across the Longmynd is a strong one. Pre-Ashgill movements were responsible for the restriction of Ashgillian sediments in south-central Wales and caused faulting in the Bala district.

Early in Silurian times important Caledonian movements occurred along and north-west of the Towy Anticline (already being anticipated in form), along the Benton Fault in Pembrokeshire, somewhere to the west of Cardiganshire, along the south side of the Berwyn Dome and along the Church Stretton Disturbance to Kington and beyond. In the Shelve area, intense compression folded the Ordovician sediments and volcanics on a fairly tight scale and the Corndon phaccolith was intruded. Horst-like uplift along part, at least, of the Malvern Axis removed the Cambrian (and any Ordovician) before the deposition of high Llandovery strata, as shown by unconformity at Gullet Quarry. Further, often strong, movements occurred in the vicinity of the Towy Anticline in early Salopian times, and the continued rise of a land area to the west of Wales is suggested by the inpouring of greywackes along the early lines of the Snowdon and central Wales synclinoria.

The precise timing of the Caledonian climax presents a problem, though the majority view favours a late Silurian-early Devonian date. There could, however, have been further important effects during Old Red Sandstone times, as evidenced by the fracture cleavage in those red beds of Anglesey. Other problematical features of the Caledonian climax are the appreciable swing of the folds and cleavage in north-east Wales and the complete dying away of any break at the base of the Devonian in Shropshire. The main effects, however, are obvious in north and central Wales, with the production of anticlinoria and synclinoria, marked cleavage and severe fracturing, especially in the Denbigh Moors. Basic magma welled upwards along faults in the Harlech Dome and spread out as sills elsewhere.

The major effect, of course, of the late Silurian earth movements was to transfer the area of marine deposition to the southernmost portion of the British area, the remainder becoming part of a vast, mountainous,northern continent but with, nevertheless, low-lying basins of sedimentation and, at first, even relic (isolated) arms of Iapetus.

In Eastern Ireland palaeontological contrasts support the view that the Caledonian Iapetus suture extends SE from the Solway Firth but its continuation to the Atlantic is more problematical owing to the scattered nature of the (often poorly exposed) Lower Palaeozoic rocks in the inliers of Central Ireland. This also makes the unravelling of the structural history very difficult. Not even Arenig rocks, let alone Cambrian, come to the surface in these inliers. Llanvirn strata are seen in the Chair of Kildare and are overlain by Caradoc, so that the important mid-Ordovician phase, involving uplift and erosion of areas like the Southern Uplands, was operative here too. In south-east Ireland, Lower Palaeozoic rocks are far more widespread, and in the Bray Anticlinorium an important unconformity occurs beneath the Cambrian Clara Group whilst in south-east Ireland generally, Llandeilian and occasionally Llanvirn strata are missing. The Clara Group appears to pass up conformably into the Arenig, so that the important eo-orogenic phase of pre-Arenig date in North Wales is unrepresented here in south-east Ireland.

In Central Ireland, Ashgillian rocks are seen in a number of inliers and these pass up into Valentian with little or no break. They probably lie far away from geosynclinal margins or geanticlines. South-east Ireland, on the other hand, appears to have experienced uplift (probably connected with a resurrected St. George's Channel geanticline) in late Ordovician times, though the evidence is fragmentary, being based on (a) the lack of Ashgill rocks in the area, and (b) the presence of sheared Llandoverian conglomerates above Bala slates north of Courtown.

The same resurrection of the St. George's Channel mass may have been responsible for the surge of coarse greywackes into West-central Ireland (there are no such deposits preserved beneath the Old Red Sandstone in south-east Ireland), though slump conglomerates in the Slieve Bernagh Inlier remind one of the Denbighshire instability at this time. The main Caledonian Storm affected rocks up to Ludlovian age in Central and South-west Ireland, and the Lower Palaeozoic suffered intense buckling into isoclinal, often overturned, folds. The precise dating is not clear though the main orogeny pre-dated the intrusion of the Leinster batholith, and a radiometric date of 386 ± 6 million years is available for this intrusion. In the Dingle Peninsula and in the Comeraghs, unconformities both underlie and overlie the Downtonian Dingle Beds, though the greatest movements here appear to be post-Downtonian. Tremlett has unravelled a very complex succession of tectonic and igneous events during the course of the Caledonian climax in South-east Ireland and it is important to recall the radiometric date for some of the mineralization in the Avoca district of 420 million years. From this episode to the Leinster intrusion is a span of 34 million years, and yet further compression was to occur, as shown by the thrusting of the Carrick klippe of Bray-type rocks on to the metamorphic aureole of the great batholith.

The end of the Silurian did not correspond to the last of the Caledonian fold episodes for there was considerable Caledonian movement, as described in the next chapter, up to the Middle Devonian. The last push along the Moine Thrust was Lower Devonian.

CHAPTER 3

Tectonic Evolution of the British Isles

DEVONIAN - PRESENT

An early Devonian event was the formation of the tectonic depression later to become the "Midland Valley of Scotland" by both down-warping and marginal faulting. The Highland Boundary Fracture-zone probably dates back to at least the Arenig, but the Southern Uplands Fault was probably not initiated before the late Silurian. From the beginning of the Devonian the Midland Valley must have formed a major intramontaine basin (analogous to the Swiss Plain) which received debris from high Caledonian mountains on both margins.

Caledonian earth movements still occurred locally (e.g. in South-west Ireland) and off S. Pembrokeshire as shown by the northward carriage of the Ridgeway Conglomerate. The deposition of the Lower Devonian in South Wales and the Welsh Borderland was followed by appreciable regional uplift and, at least locally, by folding. Rocks of Middle Devonian age are, therefore, not present and in areas like the Clee Hills the gap is considerable. These could be the movements which were responsible for at least the partial removal of Devonian strata over considerable areas of the Midlands. It is just possible that at this time some folding occurred in the west Midlands, e.g. in areas like Trimpley or even the Abberleys. One other effect of this Devonian unrest was the influx of Old Red Sandstone type sediment into the North Devon edge of the mid-European geosyncline.

There are signs also of active rising areas to the south of Britain. Some of the thick spreads into Kerry, in Ireland, appear to come from the south. The spread of Gramscatho-type greywackes into the southern fringes of Cornubia in mid-Devonian times again hints at a southerly source. Renouf (1974) has (his figures 3E, 3F and 3G) suggested a rising source area in the Western English Channel in Lower Devonian time. This "Domnonaea" supplied debris to Brittany on the other side. Some Lower Devonian conglomerates at Roseland (Lizard Peninsula) have debris derived from the same barrier.

Along most of the Highland Boundary Fracture-zone the main movement, resulting in depression of the south-east side by many thousands of feet, was post-Lower/pre-Upper Old Red Sandstone. Ramsay (1962) has shown that this was due to north-west/south-east compression and was, therefore, late-Caledonian rather than proto-Hercynian. At the west end of the fracture-zone, however, in Mayo, there seems to have been little Devonian depression south of the Highland Boundary Fracture-zone; this accounts for the appearance of Lower Palaeozoic rocks and the Connemara schists on its south side.

Post-Lower Devonian folding and faulting were powerful in the Midland Valley; among other results were the formation of a number of north-easterly trending anticlines. That these movements affected at least part of the Southern Uplands can be seen in the Eyemouth district.

Late-Caledonian movement accompanied in some cases by sharp folding also took place along some of the north-easterly faults which cut the Highlands and north-west Ireland. Some of these are later than the post-tectonic granites (probably mainly Lower Devonian). In Donegal (Pitcher *et al.*, 1964) some of these faults are considered to be pre-Carboniferous. On the other hand, the initiation and further development of many may be Hercynian (see below). In the eastern part of the Northern Highlands there is evidence of pre-Middle Devonian movements. Important folding, which can also be classified as late-Caledonian, affected the Middle Old Red Sandstone, but not the Upper, of the Northern Highlands, Orkney and Shetland. Faulting also occurred at this time in the north (including movements along the Great Glen Fault); the trend of these fractures in Caithness, Orkney and Shetland becomes northerly.

Devonian compression forced the folded and metamorphosed rocks of the Caledonian Orogen over the edge of the north-west kratogen, thus forming the Moine Thrust-zone. Slight movements also took place in the foreland, where the Cambro-Ordovician succession is folded and gently faulted. Some of these movements are Devonian, as the thrusts themselves are folded and faulted: others are earlier and may correspond to Lower Palaeozoic fold-phases in the orogen - possible evidence is the absence of Middle/Upper Cambrian in the Durness Limestone.

Lower Carboniferous times were ushered in by an extensive, though shallow, marine transgression into the southern cuvettes of the Old Red Sandstone continent. How far northwards into England this Tournaisian invasion spread is still difficult to estimate. Proto-Hercynian developments of later intense structures were now beginning to take place, as for example along the Neath Disturbance, the Ritec Fault, the Usk Anticline and the Towy Anticlinorium. The trend of the shoreline of St. George's Land in central Pembrokeshire, for example, appears to have been controlled by movements along the locally developed Ritec Fault (Sullivan, 1965).

As Carboniferous times progressed, so the South Wales Coalfield area itself began to become defined in position, particularly with the rise of kratons to the south or south-west. In southernmost Ireland, important earth movements took place near the close of Dinantian times and this Sudetic unrest also halted deposition over large parts of the southern Welsh Borderland. Periodic rises of the Mercian Highlands prevented the southward spread of Dinantian, Namurian and even early Westphalian deposits. During later Westphalian times, crustal instability in the Midland area is typified by such marked breaks as the "Symon Fault" and the removal of earlier Coal Measures over the Nuneaton and other central England charnoid and malvernoid axes.

The age of the intense folding and thrusting in Devon and Cornwall is a complex affair. Several phases of folding appear to have affected the southern side of the Devon Synclinorium and early Namurian uplift must have occurred to the south of Cornubia to provide the spread of clastics into the southern side of the Cornubian trough. This uplift may have been linked with important orogenic phases which were occurring in France and Spain during Namurian times. Late Namurian folding may have been responsible for the different tectonic styles of the southern Cornubian peninsulas. Appreciable uplift and probably orogenesis must have also occurred at this time just south of Ireland in order to spread clastics northwards into the southern Irish coalfields. The influx of Pennant-type debris into the South Wales, Bristol and Kent Coalfields points to the rapid rise of mountains not too far away to the south. These folded mountain chains may even have covered Cornwall and at least southern Devon by mid-Morganian (highest Westphalian) times, though Fitch and Miller (1964) placed the paroxysmal phase of the Hercynian Orogeny as having occurred in the early Stephanian (Asturic phase).

Whenever the exact timing, and it may have reached its climax at slightly differing times in different areas, ultimately the Hercynian "Storm" broke over the British area but various areas responded in different ways to the overall force, believed to be a N-S compression (e.g. Roberts, 1966). Where the cover was thin, older axes or fractures modified the directions of response. Powerful northward thrusting (perhaps even of rocks folded during earlier phases) occurred in Devon and Cornwall and it is very possible that thrust sheets of Devonian or older rocks (since eroded) were thrown over the contorted Culm of Devon. Reading (1965) has drawn attention to the possibility of at least four distinct, but contemporaneous, Upper Carboniferous successions being involved in complex structural units. Some blocks originally deposited to the south might have overridden more northerly ones so that today the original sedimentological pattern is reversed, as in the Alps. This, according to Reading, would be the only way of explaining why turbidites might today occur to the north of *northerly* derived paralic beds.

The Devonian rocks of Exmoor were also thrust northwards over still buried Carboniferous strata, and klippen of Dinantian rocks were thrown northwards on to the Coal Measures of Radstock and Vobster. A great granite batholith was intruded into the roots of the tectonic pile of Cornubia, though it was not to be exposed to erosion for a considerable time to come. Intense mineralisation accompanied the magmatic rise.

Further north, in the southern part of the Welsh Borderland, the compressive forces appear to have been deflected into more E-W directions, building up folds such as the Malvern and Abberley flexures. The anomalous N-S direction of these Hercynian structures is puzzling but could have been produced by an inward movement of the two boundary massifs of St. George's Land and the Mercian-East Anglia Highlands (Owen, 1958). In the Midlands, around the north-eastern and eastern edges of the Welsh Block and in Lancashire, intense fracturing accompanied more gentle folding, the fault movements being of wrench or thrust type at first but, with the entry of Permian times, horst-rift settling movements became dominant resulting in contemporaneous erosion and deposition in localised areas, producing variable sheets of breccia, conglomerate and sand. The reddened character of these deposits is sufficient evidence of the great changes in climate which had occurred.

Radiometric ages of igneous rocks from the Midlands indicate two distinct periods of magmatic intrusion (Fitch and Miller, 1964). One phase (Asturic) of intrusion in Worcestershire, Shropshire and Derbyshire yields dates of about 295 m.y., the second (Saalic) in Staffordshire and Shropshire gives dates of about 265 million years. A later, primary, mineralisation event appears to have occurred at about 225 m.y. (near the base of the Mesozoic).

Stratigraphical breaks show that there was prolonged activity during the Carboniferous in the north of England. In particular, the Alston and Askrigg Blocks moved positively (influencing sedimentation) and there were displacements along their boundary faults and along the fractures which separate them. Marked folding and faulting did not, however, occur until the end of the Carboniferous. This included the development of the Pennine folds and of flanking flextures to the east which pass under unconformable Permian. The south margin of the Alston-Askrigg Block became strongly defined by major displacements (including lateral movements) along the Craven Faults. The west margin was demarcated by upthrows (at this stage) along the Pennine Fault of the Lake District Block to the west. The Carboniferous strata overlying the Lower Palaeozoic core of this block were thrown into fairly gentle folds, well seen in the Cumberland Coalfield.

Proto-Hercynian movements also occurred in the Southern Uplands, resulting, among other effects, in the formation and development of the Dumfries-Sanquhar Basin and controlling Carboniferous sedimentation, both in this basin and in other parts of the region. The main Hercynian folding and faulting in the Southern Uplands affects strata up to the Coal Measures and is clearly pre-Permian: the further development at this time of the Dumfries-Sanquhar Basin can be accepted as a continuation of the Pennine folding.

The Carboniferous of the Midland Valley shows marked lateral variations which were due to tectonic influence. The late Caledonian movements left north-easterly anticlinal ridges which not only separated basins of deposition but also underwent spasmodic uplift. Sedimentation changed suddenly across north-easterly trending faults (probably of late Caledonian initiation) suggesting contemporaneous movement.

Large-scale folding and faulting took place in the Midland Valley at the end of the Carboniferous; reversed faulting and overfolding are, however, rare, the only important sample being the Pentland Fault. The folds and faults show two dominant trends, east and north-east to north by east. The second trend is due to the posthumous effect of the late Caledonian folds.

In the Scottish Highlands and their continuation into Ireland, N-S compression of these rigid blocks resulted in the production of shears; these are chiefly developed, owing to the pre-existing Caledonian grain, as north-easterly transcurrent faults with sinistral displacement. Some of these were probably initiated earlier; this certainly applies to the Highland Boundary and Great Glan Faults, although Hercynian movements also took place. Hercynian shearing also probably occurred in the Southern Uplands.

Some Hercynian folding can also be distinguished in the "Highlands", mainly in Ireland where Lower Carboniferous strata spread far across the Boundary Fracture-zone.

During Rhaetic and Jurassic times, widespread marine conditions returned to considerable tracts of the British area, but prolonged uplift over the East Anglian-London Platform prevented the preservation of any such deposits in this eastern area. Periodic removal probably also occurred over parts of the Welsh Massif and over other positive "shallows" such as in the Mendips, Moreton and Market Weighton regions. On the other hand, appreciable subsidence occurred in areas such as the Wessex Basin, the South Weald and north-east Yorkshire and in present-day off-shore areas. Renewed regional uplift towards the end of the Jurassic Period resulted in the breaking up of the marine tract and these movements gradually grew in intensity as Lower Cretaceous times progressed with (especially in the Weymouth district and the English Channel) the production of folds and faults. Local intensification of these intra-Cretaceous structures may have been, in places, due to the presence of underlying saline layers.

In many of the sea areas around Britain, there is a marked unconformity at the base of the Upper Cretaceous. The Chalk is markedly transgressive upon horizons down to Permo-Triassic and even Palaeozoic in the seas off S.W. England.

Widespread regional subsidence at the beginning of Upper Cretaceous times resulted in the commencement of an important marine transgression, culminating in the deposition of Chalk over perhaps almost the whole British area. This was, however, followed, at the close of that period, by an equally as extensive uplift, accompanied by some warping and some partial (in places, complete) removal of the Mesozoic blanket.

These mid- and late-Cretaceous movements and the succeeding cyclic transgressions and regressions of the Palaeogene sea were fore-runners of the Alpine Orogeny which, by mid-Tertiary times, converted the Tethys Geosyncline into the great folded mountain chains of the Alpides. The southern fringe of the British area felt only the outer ripples of the great crustal unrest which was occurring across southern Europe. The most marked folds in Britain are, therefore, limited to the extreme southern flank of the Hampshire downfold with shallower echelon-type flexuring over the London Basin, Wessex, the Weald, Mendips and the Vale of Glamorgan. Further north, over the Midlands and the southern Pennines, broader warping took place, often sited over earlier (Hercynian or Permo-Triassic) lines, accompanied by considerable fracturing, this faulting too being frequently located along earlier formed weaknesses in the crust. Widespread collapse probably took place at this time in areas such as Cardigan Bay. In S.W. England, Dearman (1963) has demonstrated that Tertiary dextral movements have occurred along NW-SE faults. One such fault is the Sticklepath fracture which cuts the Oligocene deposits of the Bovey Tracey basin. The Mochras Fault has suffered post-Oligocene faulting equal to at least the thickness of the Tremadoc Bay Oligocene sequence (650 m) plus possibly the height, above sea level, of the Cambrian mass of the Harlech Dome today (800 m).

Mesozoic movements, shown above to be important in the sedimentary history of southern and eastern England, also occurred further north and west, although they can be proved in only a few areas owing to the scarcity of post-Triassic/pre-Tertiary outcrops. In Ireland the major break beneath the Upper Cretaceous and non-sequences within the Upper Cretaceous itself are evidence of Mesozoic instability; pre-Upper Cretaceous faulting also took place. Non-sequences also occur in the Hebridean Mesozoic, and the powerful Camasunary Fault in Skye is post-Oxfordian (or perhaps post-Liassic) and pre-Upper Cretaceous. On the opposite coast the Helmsdale Fault was active in Kimmeridge times. In both Scotland and Ireland there was post-Upper Cretaceous/pre-Tertiary basalt warping.

Folding and faulting of the New Red Sandstone of northern England, the Southern Uplands and the Midland Valley is conventionally dated as Tertiary. This may be the case, but there is no stratigraphical proof and some of the movements may, in fact, have been intra-Mesozoic. The most important structure in this category is the Lake District Dome, the uplift of which was accompanied by peripheral folding and faulting affecting in general strata up to the Triassic and up to the Liassic near Carlisle. These presumed Tertiary movements also resulted in the further development of the Pennine Fault, this time with downthrow to the west.

Tertiary folding and faulting on a large scale can be proved in North-east Ireland and in western Scotland where Tertiary lavas and intrusions are present. In Ireland marked folding occurred, and considerable faulting, leading, for example, to differences in level of the base of the Chalk of over 3000 ft (915 m). Datable deposits of Senonian Chalk found by Walsh (1966) between Killarney and Dingle Bay could mean that the Old Red Sandstone ranges of Iveragh and Slieve Mish were uplifted by at least 2500 ft (763 m) during Tertiary times. There could also be Tertiary downwarping here along E-W lines.

In North-east Ireland, wrench faults with dextral strike-slip of up to 1½ miles (2½ km) cut Tertiary igneous complexes. In Scotland the folding is weaker but there are several normal faults with downthrows of over 1000 ft (350 m), some of which probably follow older fractures. Tertiary vulcanism was accompanied not only by the building of high volcanic mountains but by intrusive activity with striking structural by-products such as the North Arran Dome, the calderas of Loch Bà and Slieve Gullion and the peripheral folds of north-east Mull.

George (1965) has demonstrated that appreciable movements occurred in N.W. Britain in post-Cretaceous, post-Eocene and post-Oligocene times. In Northern Ireland, post-Oligocene deformation is evident in the downfold of the Lough Neagh Syncline (with an amplitude of over 1000 m) and in its accompanying faulting.

It is tempting to think beyond the individual ripples of Miocene deformation to a more regional warping of various British areas. Wales may be a regional Miocene upwarp, as also are parts of Ireland and Northern England. Scotland almost certainly is a major upwarp. These upwarps need not have been symmetrical. Intervening areas such as the Irish seas were correspondingly major downwarps. How much a part was played in this regional contrast of broad structure by boundary fractures is not easy to assess. Moreover the rising swells and intervening downwarps (including the major North Sea downwarp) were to continue active in post-Miocene times. As George (1974) has indicated, "both gentle folding .... and hinterland uplift were instrumental in determining the late-Neogene landforms: the exposed land surface of Wales and southern England may then be looked on as mild swells rising out of the Pliocene sea in complement to the synclinal sags, long-sustained, of the southern Irish Sea and the North Sea".

Post-Pliocene movements were of course largely related to Pleistocene glaciation and the succeeding post-glacial recovery. Shotten (1965) has shown that small-scale faulting in Pleistocene deposits in Warwickshire was due to tension induced by the cooling of frozen ground in winter. He interpreted larger-throwing faults in Northamptonshire and Leicestershire in terms of the later Pleistocene reactivation of earlier fractures, as a result of the updoming due to isostatic recovery after the retreat of the ice. There is a close degree of correlation in trend between these Pleistocene fractures and the post-Trassic faults in the Precambrian rocks of Charnwood Forest. Lumsden and Davies (1965) from a study of the buried channel of the River Nith have suggested that there was a Pleistocene renewal of movement of almost 200 ft (60 m) along the Southern Uplands Fault.

Small movements, resulting in minor, but in some cases repeated, earthquakes have occurred much more recently along several British faults, notably the Great Glen, Highland Boundary, west Glasgow, Dent, Berw, Bala, Sticklepath and Malvern fractures.

CHAPTER 4

Precambrian Terrains

4.1 The North-West Highlands and part of the Hebrides

The Precambrian Block (Fig. 4.1) which lies W. of the Caledonian Front in Scotland (The Moine Thrust Zone), is a remnant of a much larger block (sometimes called Eria), part of which is drowned and part of which lies on the other side of the Atlantic; a remnant occurs around Rockall.

The Precambrian Block forms a long outcrop on the mainland of the NW. Highlands extending from Cape Wrath to Loch Alsh, a sparsely-inhabited region deeply indented by fjord-like sea lochs. The Lewisian "basement" largely makes bare, hilly ground with small lochs and peaty hollows. Above, the spectacular relic mountains of Late Precambrian (Torridonian) sandstone rise to 3483 ft (1062 m), and are sometimes capped by Cambrian quartzite. The Precambrian also appears in a number of islands of the Inner Hebrides and forms the long island chain of the Outer Hebrides, including Lewis, the type locality. Here, the gneisses reach a height of 2622 ft (799 m).

The succession is as follows:

	Tertiary		
	Jurassic		
	Permo-Triassic		
	Ordovician		
	Cambrian		
Upper Proterozoic	Upper Torridonian	(Aultbea Group (Applecross Group (Diabaig Group	800 m.y.
	Lower Torridonian	Stoer Group	995 m.y.
Lower Proterozoic	Lewisian	(Laxfordian cycle	1850-1450 m.y.
Archean		(Inverian cycle	2200 m.y.
		(Scourian cycle	2400-2800 m.y.

No regional metamorphism and only moderate folding took place after the Laxfordian cycle. This contrasts with the Sveconorwegian metamorphism of Scandinavia (1000 m.y.) and the strong foreland folding of parts of the Baltic Shield.

Lewisian rocks have been identified as far N. as NW. Shetland and they probably also occur under the sea as part of the West Shetland Shelf. The main Lewisian outcrop extends from North Rona, 40 miles (64 km) NW. of Cape Wrath to Islay. South-west of Scotland Lewisian rocks make up Inistrahull off the N. coast of the Irish Mainland. Pre-Caledonian gneisses in the NW. of County Mayo have given metamorphic dates of roughly Grenville age (Max and Sonet, 1979) but may well be reworked Lewisian. The total length of outcrop in the British Isles is therefore probably 550 miles (880 km).

The Lewisian metamorphics were named after Lewis (Fig. 4.1) in the Outer Hebrides and were from the outset thought to be very ancient. This view was finally proved on a stratigraphical basis when in 1891 the unconformably overlying Torridonian was itself proved to be Precambrian.

No stratigraphical sequence has been, or is likely to be, made out for the Lewisian. Its history has to be seen as a complex series of tectonothermal events worked out from structural studies and radiometric dates (see below).

Interpretation of the radiometric dates for the latest Lewisian (Laxfordian) events is open to debate. However, following these events an interval, which may have been as long as 500 m.y., elapsed before the deposition of the Lower Torridonian, on a land surface with relief of up to 1312 ft (400 m).

The Stoer Group, confined to small outcrops, including the type locality in the Stoer Peninsula, consists of hard, fairly steeply-dipping red sandstones with a mudflow horizon containing volcanic material. Its separation by an unconformity from the Upper Torridonian was not recognised until about 1967. That the unconformity represents a big time-interval has been shown by a radiometric age for the Lower Torridonian of 995 m.y. and for the Upper Torridonian of 800 m.y. This is in accord with the very different palaeomagnetic pole positions reported for the Lower and Upper Torridonian. Further S., strata, including the Sleat Group, within the Thrust Zone (5.1), are also probably at Lower Torridonian age. The Stoer Group, which shows dips of 25° to 30° to the W., was subjected to marked deformation before the deposition of the Upper Torridonian. This unconformably overlies the Stoer Group but rests mainly on the Lewisian (Plate 1) with a strongly diachronous base, frequently marked by breccia or conglomerate of local origin. At some localities, notably Slioch on the E. side of the Loch Maree, (Fig. 4.2B) the Torridonian abuts against hills of Lewisian gneiss which rise up to 2000 ft (610 m) above the general gneiss surface.

The main subdivisions of the Upper Torridonian are:

Aultbea Group
Applecross Group
Diabaig Group

The Diabaig Group consists of red sandstones, red mudstones and grey, sandy shales. It thins out to the N., where the Applecross Group, deposited in several alluvial fans, rests on an ancient pediment of Lewisian gneiss (Williams, 1969). The Group in general consists of red arkoses and pebbly conglomerates, with current-bedding indicating origin from the W. The pebbles include exotic supracrustal metasediments and acid extrusives which can be matched in Greenland. The Aultbea Group consists of sandstones, flags and grey shales; nannofossils (Downie, 1962) from the Upper Torridonian support a late Precambrian age. Red sandstones and conglomerates N. of Stornoway have been mapped as Torridonian but are more likely to be Permo-Triassic (Storetvedt and Steel, 1977).

Successions correlated with varying certainty with subdivisions of the Lower or Upper Torridonian occur in Hebridean islands from Skye to Islay. As nearly all of these, however, occur within the Caledonian frontal thrust-zone, they are dealt with in the next chapter (5.1).

In the North-West Highlands the Cambrian rests unconformably on an almost planar Lewisian/Torridonian surface. The succession of Cambrian to Lower Ordovician strata is as follows:-

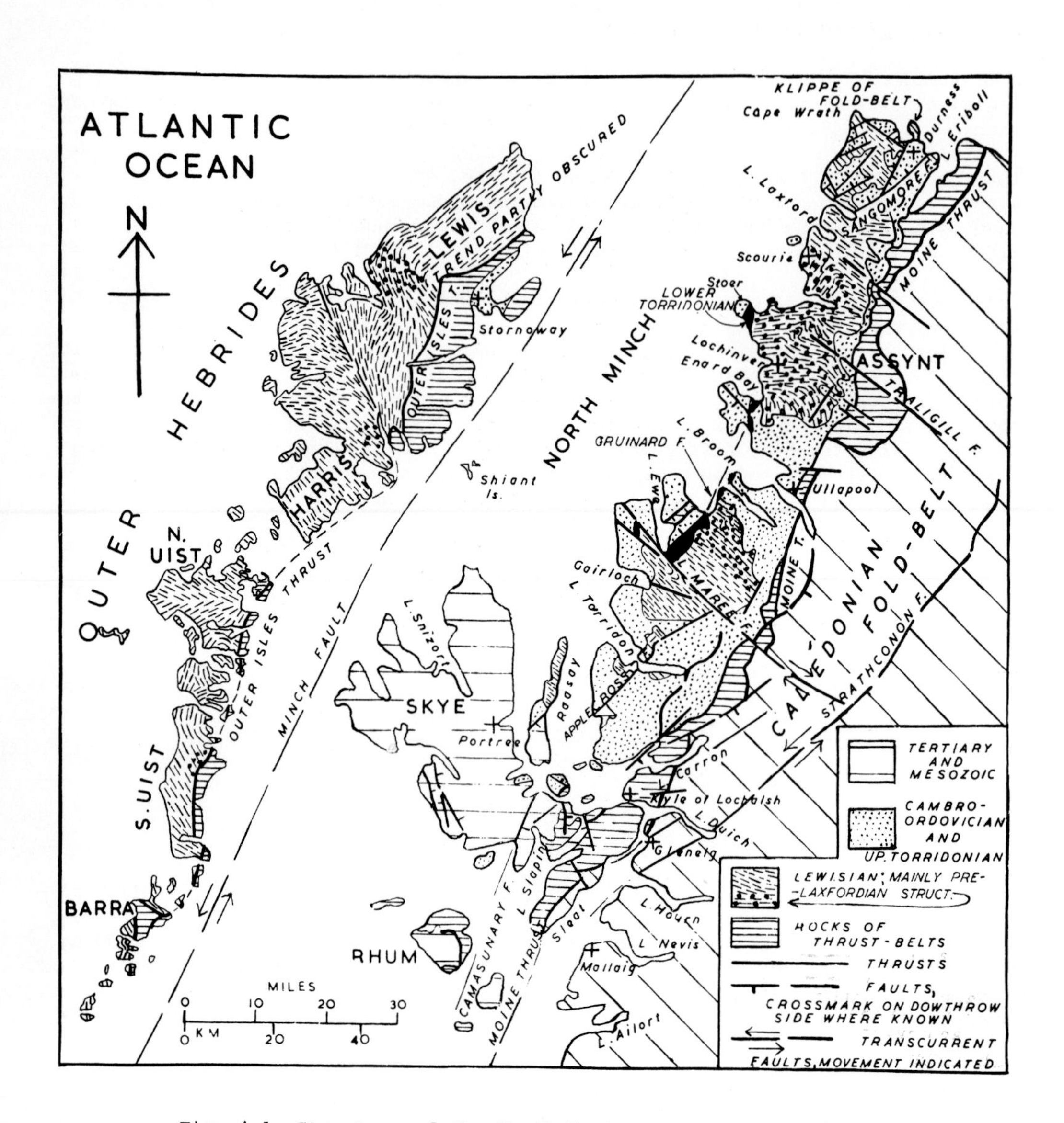

Fig. 4.1 Structure of the North-West Kratogen of Scotland.

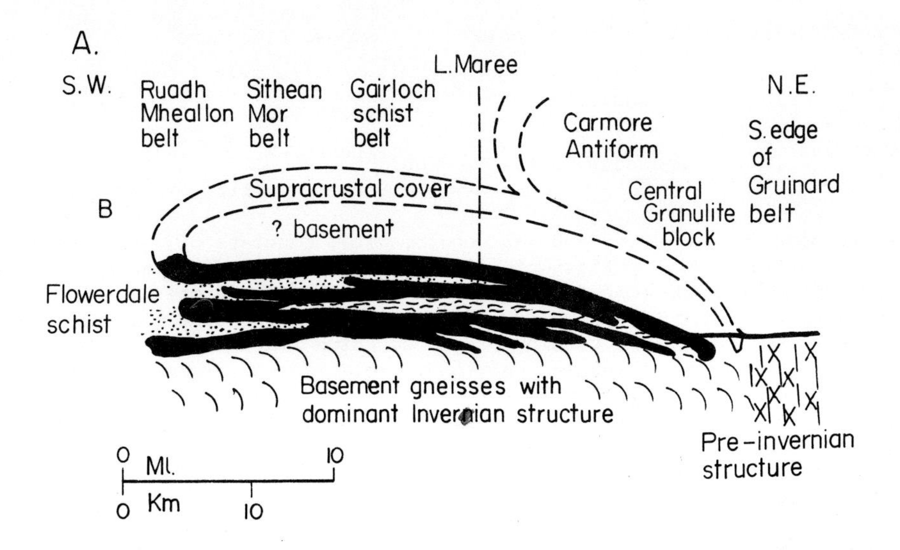

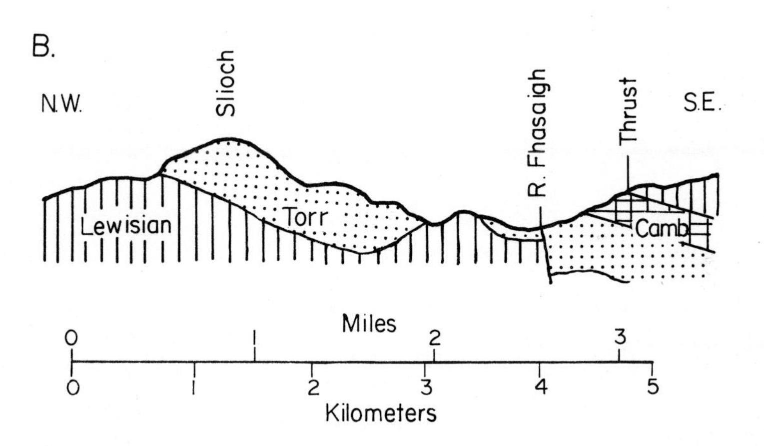

Fig 4.2 Sections of North-West Highlands

A. Reconstruction of the Invernian Structure between Ruadh Mheallon and Gruinard Bay (after Park)

B. Cambrian/Torridonian/Lewisian relationships near Loch Maree.

Cambro-Ordovician-Calcareous Series		Durness Limestone
	Middle Series	(Serpulite Grit (Fucoid Beds
Cambrian	Arenaceous Series	(Pipe-Rock (Basal Quartzite

The Arenaceous Series, above a basal conglomerate, consists of current-bedded quartzite, often of high purity (over 99% SiO_2 in places). The quartzitic Pipe-rock is named after the sand-filled tubes it contains, interpreted as worm-burrows. This is followed by the Fucoid Beds, 12-27 m of dolomitic shales and siliceous dolomites, on the bedding planes of which flattened worm-casts, originally interpreted as seaweed markings, are abundant. They contain the Lower Cambrian trilobite Olenellus. The Serpulite Grit is a gritty quartzite with Salterella. The Durness Limestone, up to 3700 ft (1326 m) thick, is largely dolomitic; the lower sub-divisions contain Lower Cambrian fossils. There must be a disconformity within the formation, for the upper sub-divisions yield an abundant gastropod/cephalopod/trilobite fauna of Upper Cambrian/Lower Ordovician age.

There are a number of patches of red Permo-Triassic rocks in the North-West Highlands and Islands including almost certainly the strata near Stornoway mentioned above. Rocks of the same age are thought to occur on the sea floor of the Minch.

Parts of the Jurassic succession are present, and there are widespread Tertiary lavas in Skye.

Alkaline plutonics, including syenite and felsite-lamprophyric and lamprophyric minor intrusions, all probably of Lower Devonian age, cut formations up to the Durness limestone.

There are abundant Tertiary basic dykes.

Structural events were as follows:

(a)	Pre-Lower Torridonian	Archean and Lower Proterozoic
(b)	Post-Lower Torridonian/ Pre-Upper Torridonian	Upper Proterozoic
(c)	Post-Upper Torridonian/ pre-Cambrian	Upper Proterozoic
	Caledonian, Mercynian, Mesozoic and Tertiary.	

Pre-Lower Torridonian events

No isotopic dates significantly older than 2800 m.y. have so far been recorded; nevertheless, as discussed by Watson (1975, 19), some gneisses may be older than supracrustal rocks affected by the 2800 m.y. metamorphism and may therefore be Katarchean.

The Scourian takes its name from Scourie about 20 miles (32 km) SSW. of Cape Wrath. Structures mainly developed during this cycle are predominant from a few miles N. of the type-locality to Loch Maree. In the Outer Hebrides, Scourian gneisses, (isotopic age 2800 m.y.) form northern Lewis (David, Lisle and Watson 1975) and make up the "grey gneiss" complex (2690 ± 140 m.y. to 2640 ± 120 m.y.)

A. Rodding (Lineation) in Moinian Metasediments. Near Glen Carron, Ross-shire, Scotland (J.G.C.A.).

B. Folded Cambrian Sandstone. East of Caerfai Bay, Dyfed (Pembrokeshire), South Wales (J.G.C.A.).

of the southern part of these islands (Moorbath, Powell and Taylor, 1975). The relationship of the mainland outcrops to those of the Outer Hebrides led Dearnley (1962) to postulate a major NNE. sinistral transcurrent fault under the Minch, the strait E. of the Outer Hebrides, although this has been disputed.

The rocks forming the Scourian Complex and their relationship to the earliest fully authenticated metamorphism (c. 2800 m.y. - the Badcallian metamorphism) may be summarised as follows:-

Rocks	Relationship to Badcallian metamorphism
Supracrustal rocks and intrusive igneous rocks Intermediate/acid banded gneisses and granulites of uncertain origin	affected by gneiss-forming Badcallian metamorphism c. 2800 m.y.
Intrusive igneous rocks Supra-crustal rocks (Loch Maree Group)	Not affected by (?younger than) Badcallian metamorphism).

The Scourian cycle opened with the deposition of the supracrustal rocks, including fine clastic sediments, calcareous cherts and ferruginous sediments and basic volcanics. Basic and ultrabasic intrusions are also present, some layered, (Bowes, Wright and Park, 1964) and a few containing accessory chromite. Anorthosites also occur, mostly in small lenses, although there are a few outcrops up to several kilometres in length in the Hebrides, including a large body in Harris. These may be compared with the Fiskenaesset Complex of West Greenland (Windley 1971), and in fact these Archean supracrustal assemblages as a whole have much in common with the Malene supracrustals and their equivalents in Greenland (Bridgewater and others 1973).

The formation of these supracrustal rocks was followed by the early Scourian (Badcallian) metamorphism which raised the rocks to the granulite or amphibolite facies. Many of the gneisses have a migmatitic aspect. Coeval or near-coeval deformation was complex and caused the obliteration of primary structures other than igneous layering, the inter-leaving of supracrustal rocks with gneisses and repeated folding, producing complicated geometrical forms. These tectonic patterns have been greatly modified in many areas by the later episodes described below, but a north-easterly grain is characteristic of the early Scourian for considerable areas.

Over much of the southern part of the mainland outcrop and most of the Outer Hebrides, the gneisses which appear to retain Badcallian metamorphic minerals are of amphibolite facies; in fact there is no definite evidence that gneiss-forming Badcallian metamorphism in the S. and W. ever reach granulite facies over the whole of the Lewisian Complex.

The metamorphism associated with the early Scourian episodes affected, as far as is known, all the Lewisian outcropping in Scotland but later Scourian episodes were more restricted in their effect. These later Scourian episodes included the intrusion of small igneous bodies of basic to acid and also anorthosite composition which locally cut the banding of the early gneisses, especially in the Outer Hebrides; an example from Barra has yielded an age of 2600 m.y. (Francis et al., 1971). Subsequently, there was a hydrous, retrogressive metamorphism during or preceding a metamorphic episode with which the age-dated Scourian pegmatites (2460 m.y. - Giletti *et al.*, 1961) are genetically related. The pegmatites often contain a high proportion of microcline-perthite along with quartz and plagioclase, dykes of pegmatite up to a few metres wide are widely distributed in the mainland and Outer Hebrides.

The Loch Maree Group of metasediments and metavolcanics includes mica-schists, quartz-schists, calc-silicate rocks, limestones and kyanite-schists; magnetite-banded ironstones also occur, as in so much of the Precambrian, but not in workable quantities. There are also thick sheets of metabasics which are regarded as intrusive into the metasediments.

There has been considerable discussion about the age-relationship of the Loch Maree Group to the gneiss. Several authors have put forward the view that they were deposited as supracrustal rocks after the Badcallian metamorphism. Crane (1978) has shown that they lack the Badcallian fabrics and structures in both the Loch Maree and Gruinard areas. These supracrustal rocks, however, underwent penetrative metamorphism and deformation prior to the emplacement of the dykes of the Inverian cycle.

The duration of the Inverian cycle (Fig. 4.2A) has been subject to varying definitions; it seems best to regard it as following the 2500-2400 m.y. Scourian pegmatites (see above). The early part of this cycle was marked by a series of metamorphic events, often of amphibolite-facies, and deformational events including north-westerly monoclinal folding. Later came the intrusion, mainly in a static environment, of dominantly basic dykes, classic in the study of the Lewisian as they were used by Peach and Horne (1907) to demonstrate that the evolution of the Lewisian was polymetamorphic, although the term was not used at that time. As some of the early studies were in the Scourie district they are often referred to as the Scourie dykes which led to some confusion when the Inverian cycle was established.

In zones which have escaped Laxfordian metamorphism the dykes are very little altered and even ophitic structure is preserved; where the dykes pass into the zones of Laxfordian metamorphism they become progressively amphibolitized and schistose.

Some authors, following the view of Peach and Horne, held that the dykes were intruded as a large swarm over a single time-span. Bowes and Khoury (1965), and Bowes (1975), have shown that early dykes have suffered deformation and metamorphism prior to the emplacement of later dykes, and Hopgood (1971) has identified dykes deformed by folds equated with structures which are cut by dykes at another locality. Park and Cresswell (1972) have stated that there is no evidence of two or more swarms separated by tectonic episodes.

The Laxfordian cycle, named after Loch Laxford in the NW. of the mainland, started about 1850 m.y. ago, and the main phases seem to have been completed some 1600 m.y. ago. Some pegmatites within the complex have however, been dated at 1450 m.y., and a minimum age of 1150 m.y. has been obtained. These more recent dates may indicate considerable delay in the uplift and final cooling of the complex, or else dilution by later events. Lambert and Holland (1972) have pointed out that the Laxfordian dates show a closer correlation with those of comparable rocks in Greenland than with those from rocks in Fennoscandia.

The Laxfordian cycle produced the greater part of the Lewisian structures now seen, largely by the reworking of the products of earlier cycles.

The main events may be summarised as follows:-

Repeated deformation and metamorphism of amphibolite or granulite facies

Metamorphism and migmatisation with the intrusion of granites and pegmatites

Later pegmatites.

Although Laxfordian deformation and metamorphism is thus clearly polyphase, variations in radiometric data suggest that successive phases in different districts may have been diachronous.

For parts of the mainland there was an early phase of granulite facies metamorphism and the development of NW.-trending major folds with steep axial planes. Thrusting and possibly nappe formation have also been proposed.

The later phases of Laxfordian metamorphism on the mainland were of amphibolite-facies and were associated with migmatisation and the intrusion of granite and pegmatite, of most importance in the northern part of the outcrop. The granites and pegmatites contain abundant potash felspar. They are locally associated with dykes of biotite-amphibolite some of which have lamprophyric affinities.

Granulite-facies metamorphism has been suggested by Dearnley (1962) for an early Laxfordian phase in the Outer Hebrides but Coward and others (1969) have put forward the view that early Laxfordian metamorphism in the Outer Hebrides was mainly of amphibolite-facies. Later, Coward (1973) distinguished four Laxfordian phases of deformation in the Outer Hebrides, and Hopgood and Bowes (1972) have described minor structures relating to six episodes.

Granites and pegmatites are also associated with the later Laxfordian phases in the Outer Hebrides and are abundant in Lewis and in Western Harris.

At many localities in the Lewisian thin belts of flinty-crush rock and mylonite occur, frequently with NW. trend. Some may be contemporaneous with the north-westerly Laxfordian folding, others may be later, although they are certainly pre-Torridonian.

A wide belt of mylonite and flinty-crush rock, inclined at low angles to the ESE. skirts the eastern coasts of the Outer Hebrides. Sheared and mylonitized gneisses are intruded by flinty-crush rock; the latter generally occur in thin veins but bands up to 100 ft (30 m) thick have been recorded. The zone is held to mark a thrust zone roughly contemporaneous with the Moine Thrust but may have been localised by a more ancient fracture of late Precambrian age (see below).

Post-Lower Torridonian and pre-Upper Torridonian events

The unconformity between the Lower Torridonian Stoer group, which shows dips to the W. of 25°-30°, and the Upper Torridonian is evidence of considerable movement between the deposition of the two groups. The Outer Hebrides probably underwent uplift as a horst during the same time interval. This uplift may have started some 1000 m.y. ago (Watson 1979) and have taken place at intervals during late Precambrian times, so that the block acted as part of the source of Torridonian/Moine sediments deposited in a basin to the E. The horst was probably bounded to the E. by a fault near the present E. coast of the islands.

Precambrian events of post-Upper Torridonian age

West of the thrust-zone the Upper Torridonian strata have low dips. Nevertheless, the overlap of Cambrian on to Lewisian shows that Precambrian folding took place. In particular, an upfold developed N. of Assynt, where Cambrian rests on Lewisian. In Assynt the Torridonian was inclined to the W. at 10° to 20°, to be restored to the near horizontal by post-Lower Ordovician tilt to the E. of the same amount. (see below). In the occurrence of these late Precambrian movements the Scottish Shield contrasts with the Baltic Shield where the Eocambrian generally passes up conformably into the Cambrian.

Caledonian events

Gentle folding, including the restoration of the Torridonian of Assynt to the near horizontal mentioned above, affected the Scottish foreland. The folding may well have been late-Caledonian as it deforms the border-thrusts, the last movements of which are now known to be Lower Devonian. The folds are certainly post-Lower Ordovician as they affect the Durness Limestone.

Hercynian, Mesozoic and Tertiary events

Numerous faults cut the region, dominant directions being NE. to NNE. and NW. to WNW. In part, at least, they are wrench-faults. Several of the north-westerly faults cut across the thrust-zone (5.1) and continue into the orogen. The greatest of the north-westerly fractures is the Loch Maree Fault, along which dextral transcurrent movement of about 4 miles has taken place. Vertical, as well as lateral, displacement along such faults has occurred, and the combination can be deduced in the thrust-belt where strata and thrust-planes with different, and often opposing, dips are common. Examples are the Loch Carron, Loch Maree and Traligill faults.

Evidence for a Hercynian age for transcurrent displacements in the Highlands is discussed in the next chapter. Many of the faults, however, probably date back much further; the Loch Maree Fault, for example, lies along a zone of weakness initiated in pre-Torridonian times.

Strata of Triassic and Jurassic age are displaced in the North-West Highlands, and in Skye Eocene lavas, showing that faulting took place well into the Tertiary. Some of the movement, however, was certainly Mesozoic, such as the renewed downthrow to the E. along the fault near the E. coast of the Outer Hebrides which became a positive block bounding a basin to the E., under the Minch, in which thick Mesozoic sediments were deposited. The Jurassic succession in Skye also provides evidence of Mesozoic uplift and subsidence. Tertiary dyke intrusion and associated fracture on a NW. to NNW. trend are further indications of Tertiary disturbance of the Precambrian block of the Outer Hebrides.

CHAPTER 5

Caledonian Terrains with Caledonian Metamorphism

5.1 North-West Caledonian Front

Thrust structures separating the north-west kratogen from the highly folded and metamorphosed rocks of the Caledonian Orogen are spectacularly displayed on the faces of the deeply dissected 3000 ft (915 m) mountains near the seaboard of North-west Scotland. Moreover, the district was one of the first in the world where large-scale reversals of normal stratigraphical order, consequent on thrusting, were demonstrated, and it has attracted geologists from many countries for over 80 years. From Loch Eriboll on the N. coast to Assynt, major, but individually discontinuous, thrusts bring westwards autochthonous nappes consisting of Lewisian slices with unconformable Cambrian. From Assynt southwards, Torridonian also appears, and folding becomes significant. Beneath these nappes there is generally a zone of imbricated Cambrian resting on a basal thrust or "sole". Above the autochthonous nappes the Moine Thrust, certainly continuous for over 120 miles (192 km) brings forward an exotic nappe consisting of Moinan metasediments of the fold-belt. Mylonite is developed on a large scale along many of the thrusts (Fig. 5.2). The undisturbed stratigraphical succession is clearly seen in the fore-land.

The superposition of altered or unaltered rocks in North-west Scotland caused early controversy, which became acute when Charles Peach's (1854) discovery of Cambrian fossils in the foreland strata made it clear that Lower Palaeozoic sediments are overlain by highly metamorphosed rocks. Murchison (1860) suggested that the over-lying Moinan metasediments (or Eastern Schists as they were then called) were altered Silurian strata, whereas Nicol (1861) and Calloway (1883) held that rever-sed faulting accounted for the relationship. Lapworth's (1883) investigations in the northern part of the area demonstrated that thrusting had taken place on a large scale, a view finally confirmed through the detailed mapping by the Geol-ogical Survey of the whole region, including the famous Assynt district (Peach and Horne 1884; Peach and others, 1907). The succession involved in the Thrust zone is as follows:-

Durness Limestone	-	Cambro-Ordovician
Serpulite Grit) Fucoid Beds) Pipe-rock) Basal Quartzite)	-	Cambrian
Upper Torridonian part of Moinan (metamorphic)	-	Upper Proterozoic
Lower Torridonian	-	Upper Proterozoic
Lewisian	-	Lower Proterozoic and Archean.

The Moinian is described later in this chapter (5.2); the Upper Torridonian (or Torriden Group of some authors), the Cambrian and the Cambro-Ordovician were des-cribed in the previous chapter. In the more southerly part of the Thrust-zone the

Diabaig Group, characterized by a much thinner succession than was formerly believed to be the case, is underlain by a thick sequence of grey sandstones and shales referred to the Sleat Group and probably in part the equivalent of the Lower Torridonian Stoer Group.

In the Inner Hebridean islands S. of Skye slightly metamorphosed metasediments within the Thrust-zone are accepted as Torridonian, but correlation with the divisions recognised further N. is difficult. The structural elements are broadly as follows:

Moine Nappe - exotic
(probably continuous for 250 miles; 400 km)

Autochthonous Nappes
(individually discontinuous; various local names)

Imbricate Zone - autochthonous (not always present)

Kratogen - autochthonous

The N. coast section of the Caledonian Front is not only of great structural interest but is also of historical significance as it was here that Lapworth carried out his classic research. Although his broad conclusions hold good, Soper and Wilkinson (1975) have shown that here have been four fold deformation sequences, probably wholly of post-Cambrian age.

East of Loch Eriboll (Fig. 5.1A) the Moine Thrust, marked by a thick zone of mylonite, brings forward Moinian metasediments within which there is a slice of probably Lewisian rocks. Under the Moine Thrust there is an autochonous nappe of Cambrian Pipe-rock and Basal Quartzite resting on the Lewisian. This has been pushed westwards along the Arnaboll Thrust over an imbricate zone of Cambrian resting on a sole thrust referred to as the Heilem Nappe by Soper and Wilkinson (1975). This nappe overlies the normal Cambrian succession of the foreland. Further W. step-faults, downthrowing to the W. bring down the Durness Limestone of the Durness Basin. The striking structural feature of this district, however, is that the faulting has also had the effect of bringing down the Moine Nappe to form a klippe, 10 miles (16 km) in advance of the main outcrop of the thrust, thus proving the minimum displacement. The klippe consists both of Moinian granulites and Lewisian-like rocks and rests directly on the Durness Limestone, showing that here the Moine Nappe has overridden the underlying structures.

In the Assynt district (Fig. 5.1B), centering on the village of Inchnadamph, an eroded culmination, on the axis of which the Moine Nappe has been denuded back 7 miles (11 km), provides a wide outcrop of the autochthonous nappes. Above a zone of imbricated Cambrian the Glencoul Thrust brings forward a 1600 ft (487 m) thick slice of Lewisian overlain by Cambrian. Above this the Ben More Thrust carries a similar slice of Lewisian with a cover of both Torridonian and Cambrian. This in its turn is overridden by the Moine Thrust.

Peach and others (1907) took the view that near the centre of the district the Ben More Thrust overrides the Glencoul Thrust. Bailey (1934b) suggested that the Glencoul Thrust continues southwards and that the Ben More Thrust is a comparatively minor plane splitting a large sheet which Sabine (1953), supporting Bailey, termed the Assynt Nappe. At the S. end of the culmination, at the famous Knockan cliff section, the Assynt Nappe is overridden by the Moine Nappe bringing Moinian metasediments onto the Cambrian Limestone of the foreland. The thrusts have themselves been folded, and klippen occur in eroded synclines.

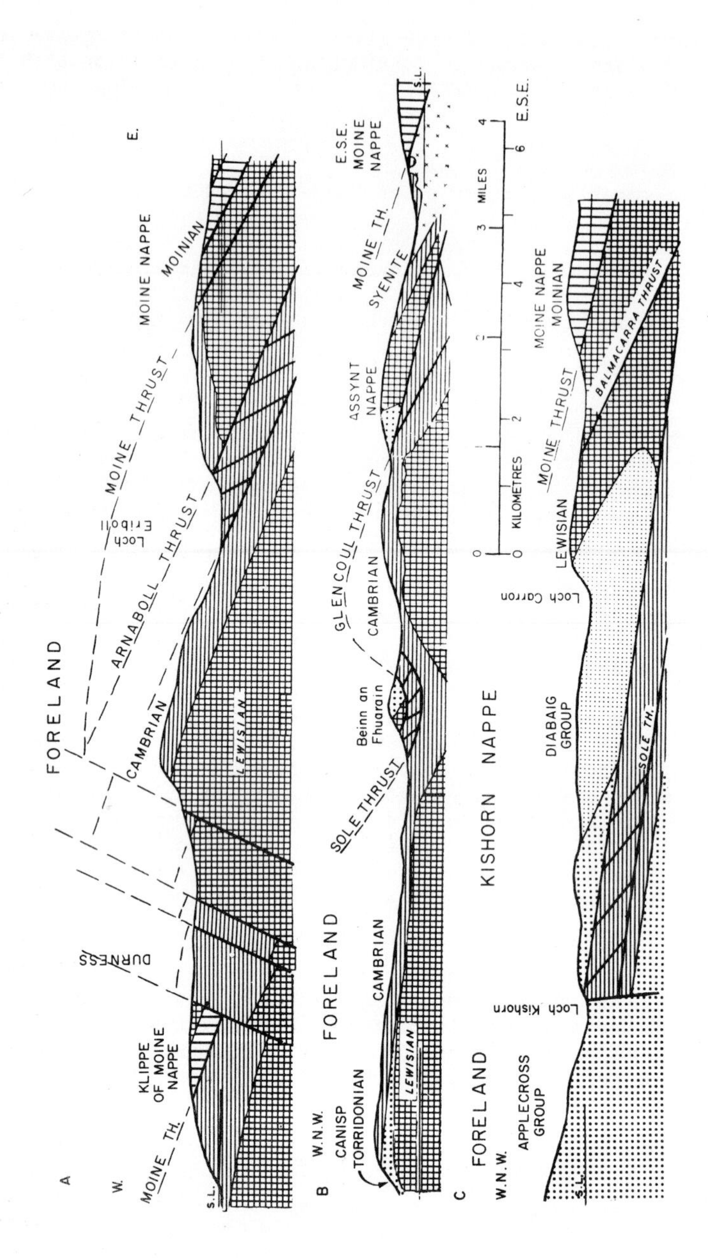

Fig. 5.1 Sections of North-West Caledonian Front
A, Loch Eriboll-Durness; B, Assynt; C, Loch Carron-Loch Kishorn

Spectacular exposures of the Front occur between the S. end of Loch Maree and Loch Alsh. Cambrian Basal Quartzite and Pipe-rock forming 3309 ft (1010 m) Beinn Eighe, are overlain to the E. by Torridonian brought forward by the Kinlochewe Thrust.

The Kinlochewe Nappe in its turn is overridden to the E. by the Moine Thrust. North-east of Beinn Eighe, however, on Meall a'Chiubhais, the Cambrian rocks of the foreland are overlain by a mass of Torridonian forming an outlier or klippe of the Kinlochewe Nappe 3½ miles (5½ km) in advance of the main outcrop. This outlier has a synclinal form, showing that the Kinlochewe Thrust-plane, like those of the Assynt district, has been folded.

Further S. the structural continuation of the Kinlochewe Nappe is known as the Kishorn Nappe (Fig. 5.1C). The main part of this nappe consists of a great recumbent syncline, closing towards the E. and pitching N. Near Loch Kishorn, imbricated Cambrian is faulted down to the W. against Torridonian of the Applecross Peninsula. Above the imbricate zone, E. of the loch, the Kishorn Thrust brings forward the Diabaig and Sleat Groups of the Torridonian overlain still further E. by the Lewisian. This inverted limb of the recumbent syncline continues as far S. as Loch Alsh. Along the N. shores of the latter, E. of Kyle, the Applecross Group, with inverted current bedding, passes under the Diabaig and Sleat Groups. This in its turn is overlain along the Balmacara Thrust by Lewisian, above which comes Moinian and Lewisian of the Moine Nappe. The overfolding is earlier than the thrusting; in fact, eastwards from Kyle, axial-cleavage is replaced by lower-angled thrust-cleavage. Johnson (1960) has put forward the view that at Lochcarron mylonites were developed prior to the clean-cut thrust movements. For the Moine Thrust-zone of Assynt, too, Christie (1960) postulates an early phase of mylonization and a later phase, characterized by cataclasis, during the thrusting.

In Skye, the lower limb of the syncline is seen, consisting of the Sleat, Diabaig and Applecross Groups (with current-Bedding) in normal order. Here the Kishorn Thrust brings the Applecross Group over Durness limestone, thermally metamorphosed and structurally distorted by Tertiary intrusion (Ch. 9).

In the Ord district of Skye, there is an outcrop of Cambrian rocks, from the Basal Quartzite to the Durness Limestone, surrounded by over-thrust Torridonian. This forms a window, the presence of which is due to the anticlinal folding of the Kishorn Thrust and subsequent erosion; the window in fact shows the Imbricate Zone under the Kishorn Nappe. On the E. side of the peninsula the Sleat Group of the Kishorn Nappe is overridden by the Moine Thrust, bringing forward Lewisian. At the S. end of the Sleat Peninsula, the three Tarskavaig Nappes intervene between the Moine and Kishorn Nappes. The Tarskavaig Nappes consist of gritty felspathic sandstones, strongly granulitized, which contain metamorphic biotite and small garnets; the latter are, however, of a manganese variety, and do not indicate the garnet grade. The lowest Tarskavaig Thrust is well exposed on the S. shore of Tarksavaig Bay.

Rhum, S. of Skye, known for its Tertiary intrusions, also contains Torridonian, probably marking the continuation of the Kishorn Nappe. The Lewisian rocks of Coll and Tiree lie W. of the Moine Thrust, but may be within the thrust-zone, evidence being the crushed condition of many of the gneisses and the presence of zones of flinty crush-rock.

The Moine Thrust passes through the narrow strait between the south-west tip of Mull and Iona, for Moinian metasediments occur on the former island and Lewisian and Torridonian on the latter. The Iona rocks lie within the thrust-zone, as do the altered Torridonian strata of Colonsay and Oronsay. The greater part of the Torridonian succession in these two islands consists of sandstones, grits and

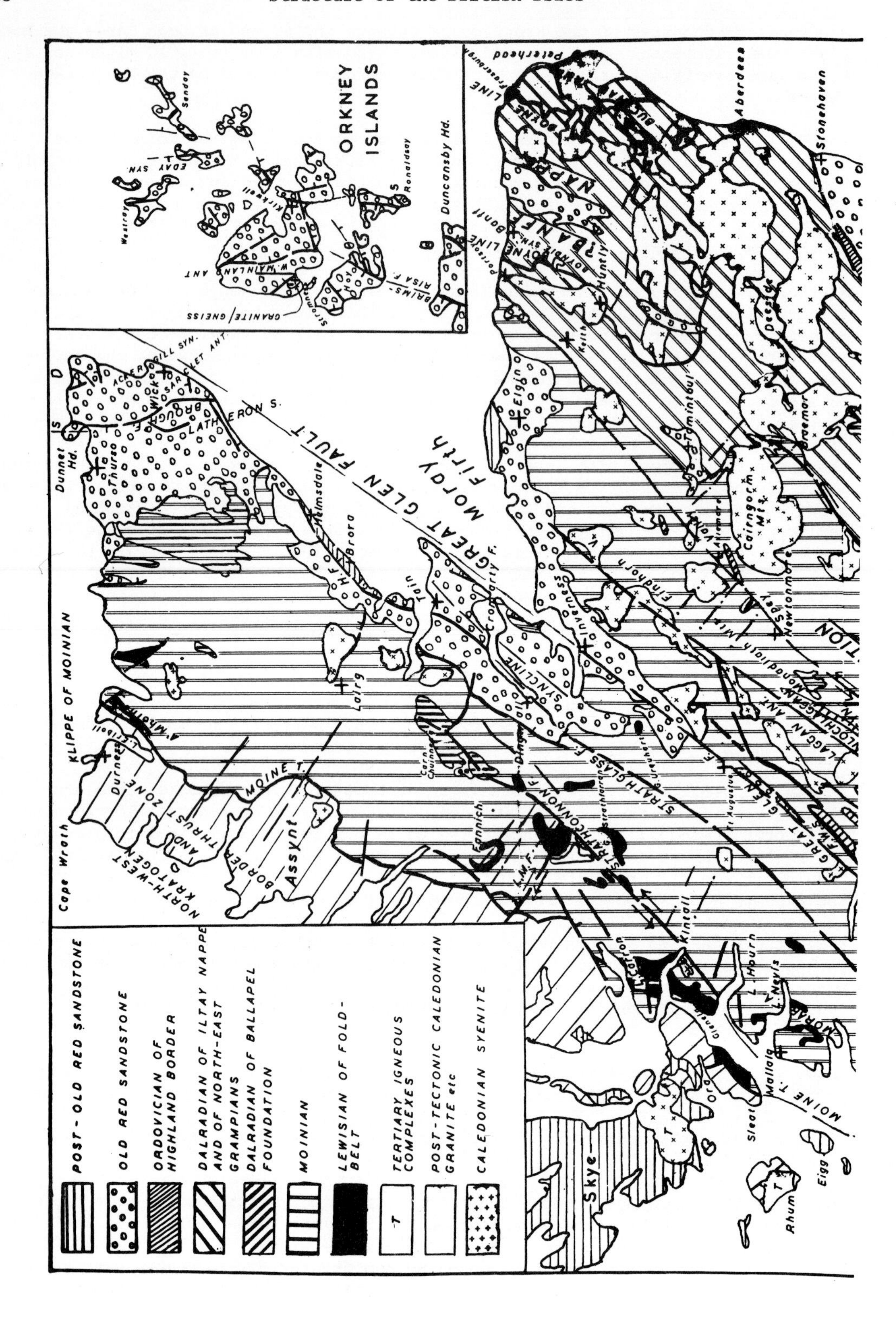
POST - OLD RED SANDSTONE
OLD RED SANDSTONE
ORDOVICIAN OF HIGHLAND BORDER
DALRADIAN OF ILTAY NAPPE AND OF NORTH-EAST GRAMPIANS
DALRADIAN OF BALLAPEL FOUNDATION
MOINIAN
LEWISIAN OF FOLD-BELT
TERTIARY IGNEOUS COMPLEXES
POST-TECTONIC CALEDONIAN GRANITE etc
CALEDONIAN SYENITE
Cape Wrath
KLIPPE OF MOINIAN
Durness
L. Eriboll
NORTH-WEST KRATOGEN
BORDER AND THRUST ZONE
MOINE T.
Assynt
Lairg
Dunnet Hd.
Thurso
Wick
LATHERON S.
Helmsdale
H.F.
Brora
Tain
GREAT GLEN FAULT
Moray Firth
Carn Chuinneag
Fannich
L.M.F.
Dingwall
Cromarty F.
SYNCLINE
STRATHCONNON F.
Strathfarrar
STRATHGLASS F.
G. Urquhart
Inverness
Findhorn
Elgin
Keith
Huntly
Banff
Portsoy
LINE
NAPPE
BANFF
Fraserburgh
Peterhead
Aberdeen
Stonehaven
Deeside
Braemar
Cairngorm Mts
Tomintoul
Spey
Newtonmore
Monadliath Mts
LAGGAN ANT.
Ft Augustus
GREAT GLEN
Kintail
L. Carron
L. Hourn
L. Nevis
Glenelg
Mallaig
Sleat
Skye
Ora
Rhum
Eigg
MOINE T.
ORKNEY ISLANDS
Westray
EDAY SYN.
Sanday
Kirkwall
W. MAINLAND ANT
GRANITE / GNEISS
Stromness
Hoy
BRIMS-RISA F.
S. Ronaldsay
Duncansby Hd.

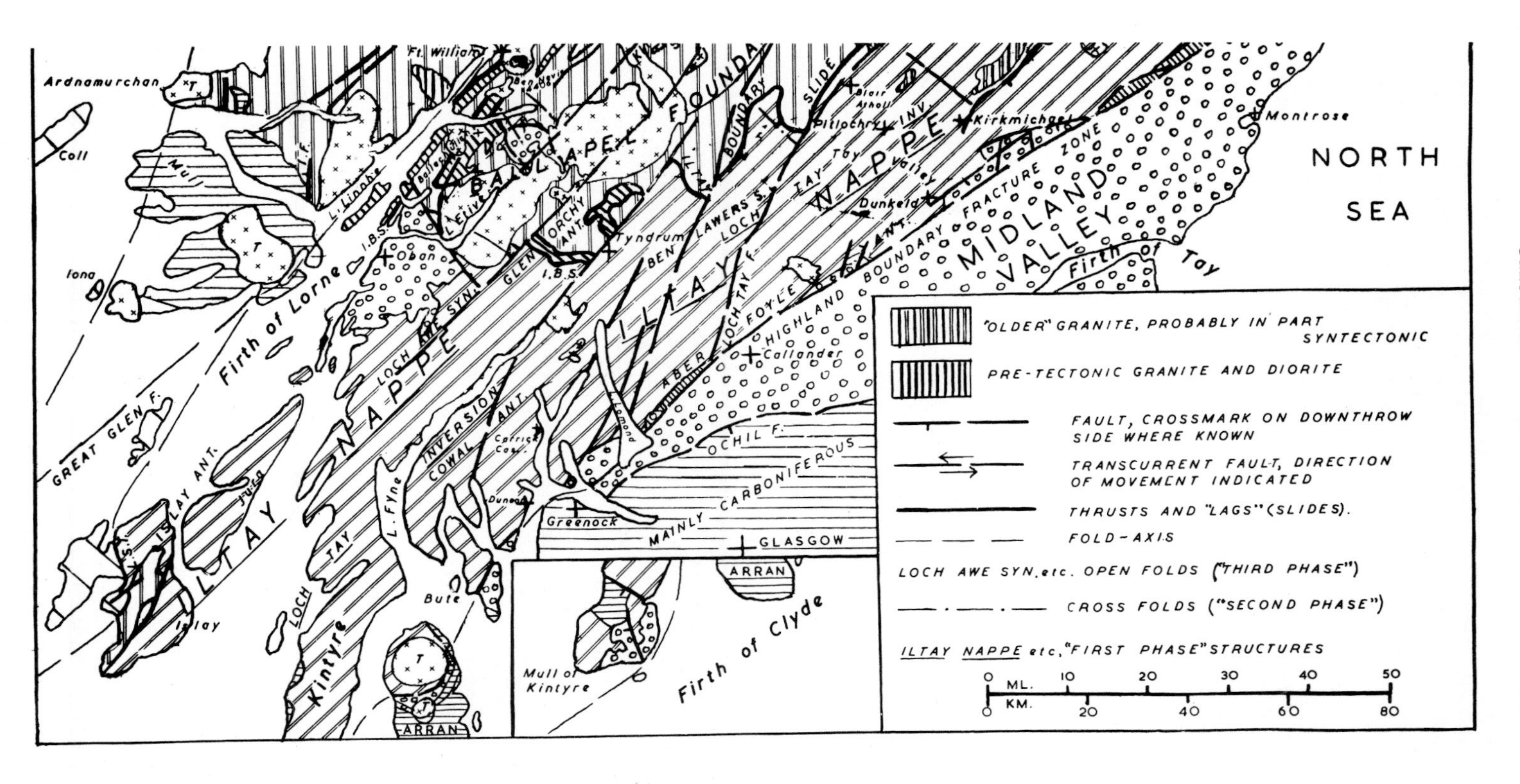

Fig. 5.2 Structure of the Northern and Grampian Highlands of Scotland.
F.W.S. = Fort William Slide; I.B.S. = Iltay Boundary Slide; I.F. = Inninmore Fault; L.M.F. = Loch Maree Fault; L.S.T. = Loch Skerrols Thrust.

phyllites, showing considerable lateral variations, with limestone near the top of the succession. The Torridonian rocks show two cleavages, separated by a period of intrusion of small syenite and diorite masses, and of many lamprophyre dykes (Wright, 1908). The first cleavage dips E. to N.E., the second, which becomes progressively stronger towards the E. of Colonsay, is mostly steep with north-easterly strike.

The Thrust-zone also occurs in the western part of Islay. Lewisian occurs in the south-west peninsula and is overlain, along a moved junction, by Torridonian conglomerate, which is succeeded by a series of grey, green and black slates and phyllites alternating with grey schistose grit and thin sandy limestones, the position of which in the Torridonian succession is doubtful. To the E. they are separated by the Loch Gruinart hollow from the Bowmore grits - red, green and grey arkoses with pebble beds. Some authors held that the Loch Gruinart hollow conceals a fault, and further postulated that the fracture is the continuation of the Great Glen Fault. Other workers (e.g. Kennedy, 1946) deny the presence of the fault: the Bowmore sandstone may in fact be Upper Torridonian and equivalent to the Applecross division.

The relationship of the Bowmore sandstone to the Lower Dalradian rocks making up the eastern half of Islay (5.4) is also controversial. Correlation of the thrust accepted by most workers - the Loch Skerrols Thrust - with the Moine Thrust involved the difficulty that it was out of line with the extrapolation of the Moine Thrust from the north. Kennedy (1946) solved this difficulty by extending the Great Glen Fault to the W. of Colonsay and Islay and using its transcurrent movement to account for this off-setting of the Moine Thrust (Fig. 5.2). The island of Innishtrahull, off the N. coast of Ireland, consists of Lewisian rocks, and it seems probable, therefore, that the continuation of the Moine Thrust passes between the island and the mainland of Ireland. If this is accepted, the Caledonian Front becomes traceable for 250 miles (400 km) in the N.W. of the British Isles.

The last, brittle movements of the thrusts are clearly later than the folding and regional metamorphism of the Moinian metasediments: as regards an upper limit boulders of sheared and mylonitized rocks found in the Middle Old Red Sandstone conglomerates have been claimed to be derived from the thrust-zones. The relationship of the thrusts to the alkaline intrusions is important. These were widely believed to be Ordovician but have now been shown to have a radiometric age of 400 million years. As the thrusting and intrusion seem to overlap, the last movements may, therefore, be Lower Devonian.

For discussion of faults displacing the thrusts see Ch. 4, and for description of the effects of Tertiary intrusions see Ch. 9.

5.2 The Northern Highlands and Orkney

The Northern Highlands (Fig. 5.2) consists mainly of sparsely populated, mountainous country, rising to 3877 ft (1180 m) in Carn Eige (the highest peak in Britain N. of the Caledonian Canal) and deeply incised on the west coast by sea-lochs. Relatively low ground, corresponding to Old Red Sandstone outcrops, occurs near the east coast and in Orkney (Fig. 5.2). The Great Glen Fault, separating the Northern Highlands from the Grampian Highlands, can be traced S.W. from the Moray Firth through Loch Ness and other lochs to Loch Linnhe. The brecciated zone is up to a mile wide; the fault was recognised by Kennedy (1946) as a sinistral transcurrent fracture with a horizontal component of 65 miles (104 km). Comparison of metamorphic levels suggests that there is also a considerable downthrow towards the S.E. The Great Glen Fault passes down the Firth of Lorne (Barber *et al.* 1979) and off the N.W. of Ireland; it may continue as a lineament on the S.E. side of the Porcupine Bank (Riddihough and Max, 1976).

The stratigraphical succession of the Northern Highlands is:-

Tertiary
Cretaceous
Jurassic
Triassic
Carboniferous
Upper Old Red Sandstone
Middle Old Red Sandstone
Lower Old Red Sandstone
Moinian - Upper Proterozoic (mainly)
Lewisian - Lower Proterozoic and Archean

Outcrops E. of the Moine Thrust identified as Lewisian amount to more than 100 square miles. Acid and intermediate banded rocks comparable to the Lewisian orthogneisses of the foreland (Ch. 4) predominate, together with basic gneisses, serpentinites, talc-schists and eclogites. Metasediments also occur in some of the inliers (e.g. Glenelg, Glen Strathfarrar and Glen Urquhart and include marbles and calc-silicate rocks, granulites, ferruginous mica-schists and kyanite-schists.

Late Precambrian and Caledonian events have superimposed structures on the Lewisian rocks within the fold-belt and in this respect they differ from those of the foreland.

The Moinian consists dominantly of altered psammitic, felspathic sediments, often termed granulites; quartzites are rare. The assemblage also contains plentiful mica-schists, but few true limestones or marbles; thin, lenticular beds of calc-silicate rock are, however, fairly common.

The absence of readily recognisable horizons and of an established stratigraphical succession long delayed structure interpretation. In fact, it was not until current-bedding and other "right-way-up" techniques were applied after 1930 that progress was made, an early application being that of Richey and Kennedy (1939) in Morar. The following formations have been recognised, placed by Johnstone and others (1969) in three divisions (Fig. 5.3).

Loch Eil Psammite	Loch Eil Division
Glenfinnan Striped Schist) Lochailort Pelite)	Glenfinnan Division
Upper Morar Psammite) Morar Striped and Pelitic Schist) Lower Morar Psammite) Basal Pelite)	Morar Division

The Morar Division is separated from the Glenfinnan Division by the Sgurr Beag Slide (Fig. 5.3) and the Glenfinnan from the Loch Eil by the Quoich Line, although it is not clear if the latter is a thrust or not. The Morar Division may be the oldest, but its age-relationship to the Glenfinnan is not certain. The Loch Eil Division, however, is almost certainly the youngest.

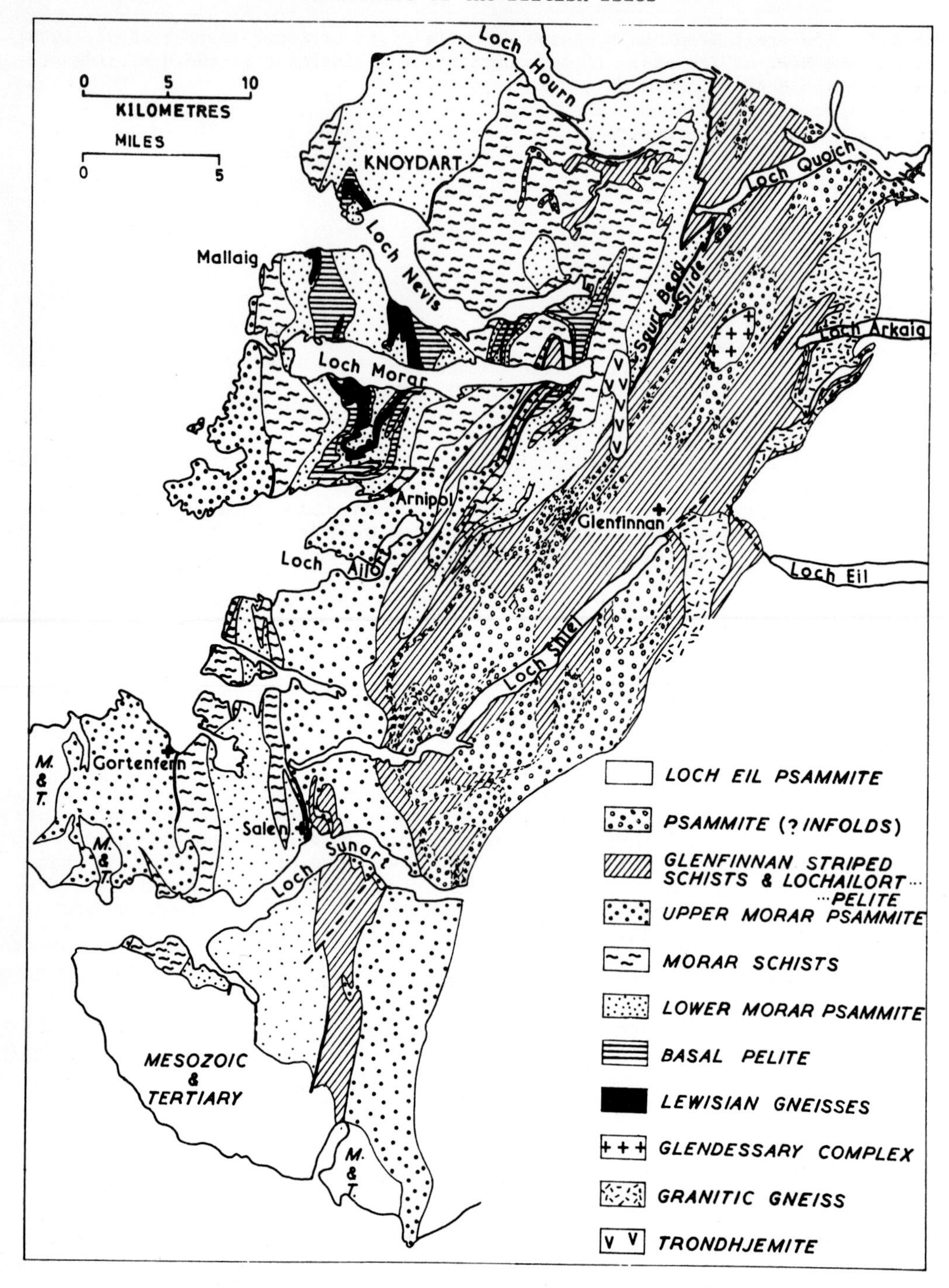

Fig. 5.3 Geological Sketch Map of Western Inverness-Shire, Scotland (after Johnstone *et al.* 1969).

These divisions have been found to hold good throughout much of the south-western part of the Northern Highlands, but cannot be extended with certainty further E. and N. The Sutherland Moinian stratigraphy and structure are difficult to establish partly because of involvement with migmatite complexes.

The Moinian Assemblage may not consist of a single, continuous, stratigraphical succession but may contain an "Older Moinian", affected by Proterozoic events and a "Younger Moinian" of late Precambrian age in which all or nearly all the folding and metamorphism are Caledonian (5.6). The Loch Eil Division would be part of the "Younger Moinian".

The Carn Chuinneag-Inchbae granite gneiss in Rosshire pre-dates the main Moinian fold-phases (5.6) and has an aureole of hornsfelsed sediments which have largely escaped the regional metamorphism.

A number of post-tectonic Caledonian intrusions cut the Moinian. These are mainly calc-alkaline plutons except for the Ben Loyal Syenite in the far north which is similar to the alkaline intrusions of Assynt (Chs. 4 and 5.1). The Old Red Sandstone rests unconformably on the Moinian metamorphics and the granites. In Caithness, and round the Dornoch and Cromarty Firths, Lower, Middle and Upper Old Red Sandstone have been recognised. In Orkney the Middle Old Red Sandstone rests unconformably on Moinian and is followed by the Upper, which contains basic volcanics (Island of Hoy).

Successions are as follows:-

	Orkney	Caithness	Dornoch and Cromarty Firths
Upper Old Red Sandstone	(Hoy Sandstone (Hoy volcanics ((unconformity)	Dunnet Head Sandstone	Grit and sandstone
Middle Old Red Sandstone	(Eday Sandstone (Rousay Flags	John o-Groats Sandstone	
	(Stromness Beds	Caithness Flagstone Series	Sandstone, conglomerate flags, calcareous and bituminous shale and thin lst
		Upper Barren Red Series (unconformity)	Sandstone and conglom. (unconformity)
Lower Old Red Sandstone		(Lower Barren (Red Series	Basement Group

At Inninmore Bay, on the N.E. side of the Sound of Mull 300-600 ft (92-183 m) of the Upper Carboniferous strata rest on the Moinian and underlie Trias and Tertiary lavas. They consist of sandstones and plant-bearing shales with thin lenticles of coal, overlain by pebbly sandstone.

West of Loch Linnhe E. to S.E. dykes and elongated bosses of quartz-dolerate are probably of Permo-Carboniferous age. Camptonite and monchiquite dykes trending in the same direction in the N. of the region, including Orkney, and in the S.W., are also probably Permo-Carboniferous.

Mesozoic strata occur in a number of small outliers in the W. and in fairly extensive downfaulted strips on the E. coast. The Trias consists of sandy and conglomerate strata; the Jurassic ranges from Liassic to Kimmeridgian. The Estuarine Series at Brora contains coal which has been worked. The Kimmeridgian near Helmsdale includes a boulder bed the tectonic significance of which is discussed below. Upper Cretaceous sandstone and silicified chalk occur under the Tertiary in the S.W. of the region; a very pure sandstone on the N.E. side of the Sound of Mull is mined for glass-sand.

Apart from a few small outcrops of thin sediments the Tertiary, seen only in the W. consists of Eocene basalt-lavas (see also Ch. 9). Tertiary basic dykes trending N. to N.W. are abundant from Loch Alsh southwards.

Structures to be considered are:-

- Pre-Moinian structures of Precambrian age affecting Lewisian
- Pre-Devonian structures of later Precambrian and Caledonian age affecting Moinian
- Post-Moinian structures of late-Caledonian age affecting Devonian
- Structures of post-Devonian age.

Pre-Moinian structures of Precambrian age affecting Lewisian

Since the Lewisian rocks were involved in Caledonian movements their pre-Moinian structures have been largely obscured. In places, however, Lewisian structures similar to those of the foreland (Ch. 4) are preserved. Occasionally a band of epidiorite or serpentinite has been mapped cutting the foliation of acid gneisses as do the basic dykes in the Lewisian of the kratogen.

The original interpretation of the Lewisian outcrops as simple stratigraphical inliers can no longer in most cases be sustained. Many of the "inliers" are, in fact, thrust wedges or tectonic inclusions of a Lewisian "basement" within the younger formation. It is true that south of Glenelg a basal Moinian conglomerate has an unmoved, although inverted, junction with Lewisian, but the alternating Lewisian and Moinian strips of this region are mostly in tectonic contact (see also below) and the angular conformity is obliterated.

Barber and May (1976) have identified Scourian metamorphism in the Glenelg Lewisian but there is no evidence of a Laxfordian event. Further S. Clifford (1957) has interpreted a lower group of Lewisian rocks as the basement of a Moinian succession and an upper as part of a nappe (see also below). In the Scardroy and Fannich areas Sutton and Watson, after at one time interpreting the Lewisian rocks as migmatized Moinian (1953 and 1954), later accepted them as thrust wedges which had been folded. One interpretation of the core of the Morar Anticline involves a somewhat similar view. In the case of the Glenstrathfarrar "inlier" Ramsay (1956) has interpreted supposed Moinian basal conglomerate as tectonic. A slice of probable Lewisian which appears to be partly in stratigraphical and partly in tectonic contact with the Moinian occurs not far above the Moine Thrust, east of Loch Eriboll.

In the Armadale-Strathy district of Sutherland grey gneisses with subordinate amphibolite and marble have undergone prolonged high-grade metamorphism which Harrison and Moorhouse (1976) tentatively correlate with the Badcallian (early Scourian) event.

Pre-Devonian structures of later Precambrian and Caledonian age affecting Moinian

Polyphase folding in the Moinian of the Northern Highlands has long been recognised. At one time these phases were thought to be all Caledonian *(sensu lato)*. However, radiometric evidence (5.6) has shown that parts of the Moinian have undergone Precambrian thermal and tectonic events.

The problem of the stratigraphical age, or ages, of the Moinian has to be considered in relation to the ages of the Moinian/Dalradian in the Grampian Highlands and of corresponding rocks in Ireland. This problem is therefore considered later (5.6), and in the present section only the structures will be considered.

It has not been possible to establish firm tectonic and metamorphic sequences applicable to the Northern Highlands as a whole nor, in fact, can one be sure that the earliest and latest events of one district are of the same age as those of another.

Tectonic regimes in the Moinian will therefore be illustrated from a few districts where the structural sequences are reasonably well authenticated.

Near Glenelg, immediately E. of the Moine Thrust, Sutton and Watson (1958) recognised three fold-phases prior to the formation (considered to be Lower Devonian) of the Border Thrust-zone, namely isoclinal folding (D1) leading to the interleaving of Lewisian and Moinian, folding on easterly-dipping axial planes (D2) and folding under more rigid conditions on axial planes striking N.E. (D3). Further S. the three highest groups of the Morar succession (see above) were first established by Richey and Kennedy (1939) in the Morar Antiform (Fig. 5.4B). At least three major deformations are now recognised (Poole and Spring, 1974). The earliest folds resulted in the isoclinal interleaving of Lewisian and Moinian (the Lewisian in part is referred to by Richey and Kennedy as sub-Moinan). The second phase produced a major westerly-closing recumbent antiform, the Knoydart Slide. The open Morar Antiform (Fig. 5.4) belongs to the third phase and has been traced as far S. as Ardnamurchan.

Further E., in the mountain known as The Saddle, S. of Glen Shiel, three fold phases have been recognised by Simony (1973), the first involving the tectonic emplacement of Lewisian basement sheets into the Moinian.

North of Glen Shiel, in Kintail (Fig. 5.4A) Clifford (1957) postulated the formation of a Kintail Nappe (with a slice of Lewisian wedged between overlying Moinian and the underlying thrust) followed by folding on N.E. axes, and also the development of W.N.W. cross-folds.

All the Moinian involved in the structures so far considered belongs to the Morar Succession, which to the E. is separated from the Glenfinnan succession by the Sguir Beag Slide (Fig. 5.3) (Tanner, 1970), traceable for over 12½ miles (20 km) S. from the head of Loch Hourn. This slide has brought up Lewisian in the Scardroy area.

The tectonics of the Glenfinnan division are complex and are characterised by a general steep disposition of the metasediments in long, attenuated flexures (probably D2), folded by younger (D3) structures which may be co-axial with D2 or may be oblique.

In the southern part of the outcrop of the Glenfinnan division Brown and others (1970) have recognised four deformation episodes. This outcrop contains the Ardgour granitic gneiss, of migmatitic origin, which probably pre-dates the D2 folds.

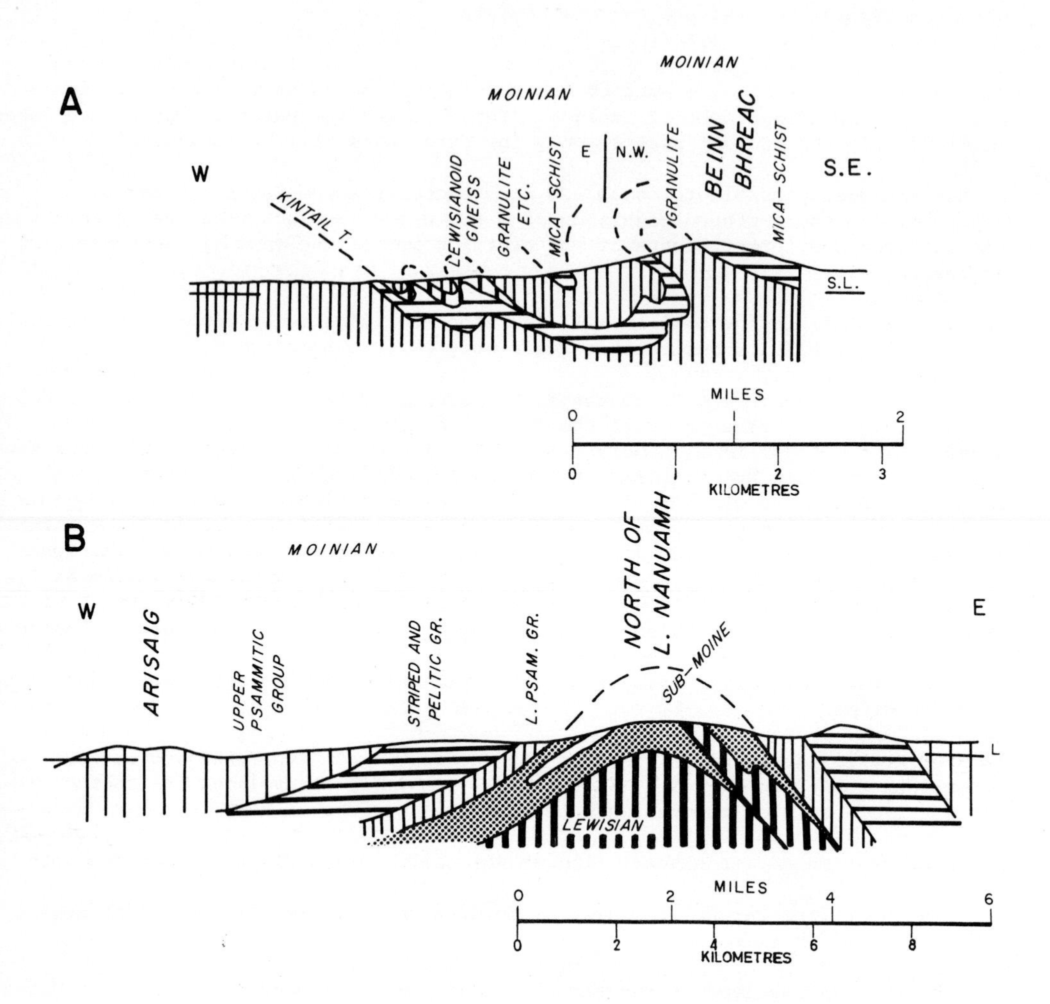

Fig. 5.4 Sections of Northern Scottish Highlands

A. Section near Beinn Bhreac, Kintail, Ross-shire (after Clifford 1957).

B. Section of southern end of Morar district showing the Morar Anticline (after Kennedy 1955).

Age - determination (see 5.6) of the gneiss are of significance in considering the age of the Moinian.

The gneiss lies immediately on the W. side of the Loch Quoich line, which separates the Glenfinnan division from the tectonically different Loch Eil division, and the emplacement of the gneiss may thus have been tectonically controlled.

The Loch Eil division consists essentially of one formation, the Loch Eil Psammite, but there are infolds of Glenfinnan rocks. The Loch Eil division has been termed the "Flat Belt" but there are steeply-inclined zones, suggesting a large-scale basin and dome regime.

Northwards from Invernessshire it is difficult to correlate either the stratigraphy or structures with those just described. In the Fannich district Sutton and Watson (1954) have recognised N.-S. folds overturned towards the W., with contemporaneous N.W. flexures, affecting a Moinian metasedimentary succession, complicated by what were later accepted as Lewisian tectonic wedges.

Further E., a long outcrop of striped and pelitic schists S.E. of the Strathconon Fault marks the continuation of the Glenfinnan division.

The Carn Chuinneag augen-gneiss was intruded into the Moinian sediments, prior to their folding and metamorphism; in the contact aureole which resisted the regional metamorphism sedimentary structures are preserved. North-west folds in the envelope were held to pre-date the intrusion but are now thought to have been an accompaniment to its emplacement. Further N. still in Sutherland, both the stratigraphy and structure are difficult to elucidate. Long narrow bodies of Lewisian-like rocks in the far N. may have been interleaved with Moinian by early folds.

The Moinian metamorphism ranges up to the sillimanite grade or granulite facies and is clearly polyphase, extending, according to Dalziel and Brown (1965), from before the D2 folds into the D3 phase. Retrograde metamorphism is evident in places. The sillimanite grade is largely associated with a belt of migmatization which with some gaps extends from the Ardgour augen-gneiss, mentioned above, to large migmatite complexes in Sutherland. The Ardgour augen-gneiss, and others, predate the D2 folds but it would be unwise to assume that all the migmatites are of the same age; in fact pegmatites of widely different dates have been recognised.

Post-Moinian structures of late-Caledonian age affecting Devonian

In Orkney (Fig. 5.2) the thick Middle Old Red Sandstone succession, resting unconformably on Moinian, is strongly folded; dips of up to 30° are common, and in places the beds are inclined as much as 60°. Most of the important folds, such as the West Mainland Anticline and the Eday Syncline, have a N.-S. trend but north-westerly folds occur in the S.E. of the islands and a north-westerly anticline in Sanday. The main folding was pre-Upper Old Red Sandstone, as this formation, which forms most of Hoy, is almost flat. In Caithness the much faulted Sarclet Pericline has a N.W. major axis; it is flanked to the N. by the Ackergill Syncline and to the S. by the Latheron Syncline. The Dunnet Head sandstone, belonging to the upper division, is faulted against Middle Old Red Sandstone and forms a syncline with a northerly pitch. In this county folding of the lower part of the Barren Red Series took place before the unconformable deposition of the upper part, and it is possible that the lower part is Lower Old Red Sandstone.

The Old Red Sandstone of Orkney and Caithness is also cut by a number of faults, some of which are reversed. By far the greater number have a general northerly trend; there is another group trending from a little N. of E. to N.E., and a less numerous set striking N.W. Some of the faults may be pre-Upper Old Red Sandstone as this higher series is less fractured. Some, however, have a post-Upper Old Red Sandstone component, such as the northerly Brims-Risa Fault which lets down this formation in the S.E. of Hoy and cuts off the north-easterly Scapa Flow step-faults, all dropping the Middle Old Red Sandstone strata down to the S.E. The Brims-Risa Fault may be continued in Caithness by the Brough Fault which at its N. end brings the Middle Old Red Sandstone over the Upper Old Red Sandstone of Dunnet Head to the W, at about 45°. Along this, and several other northerly faults, there is marked minor folding or rucking, suggesting posthumous movement along pre-existing basement fractures.

The Middle Old Red Sandstone of the Dornoch and Cromarty Firths lies in a major north-easterly syncline with subsidiary folds on the same axes. A basement (probably Lower Old Red Sandstone) is unconformably covered by fossiliferous Middle Old Red Sandstone and N. of Tain is thrust over Moinian by pre-Middle Old Red Sandstone movements (Westoll, 1964). At the N. end of the syncline there is an outcrop of Upper Old Red Sandstone the relationship of which to the middle series is not clear.

Structures of post-Devonian age

North-east to N.N.E. and N.W. to W.N.W. faults are significant structural features of the Northern Highlands. It is probable that the major displacements along these faults were Hercynian, but some, e.g. the Loch Maree Fault, were active in the Precambrian, and others may have been initiated as Caledonian structures. These faults have important transcurrent components, generally sinistral in the case of the N.E. faults, and dextral in the case of the N.W. Of the N.E. faults, apart from the Great Glen Fault forming the S.E. margin of the region, the most important are the Strathconon Fault, which has been traced for over 70 miles (112 km), with horizontal displacement of up to 3 miles, 5 km), and the Strathvaich Fault further E. Both of these are marked by zones of brecciated rock up to half-a-mile (0.8 km) or more wide in places. Of the N.W. faults the most striking is that of Loch Maree.

Johnstone and Wright (1951) have shown that camptonite dykes (probably Permian) are later than the main Great Glen movements. Some faults have important Tertiary components. The north-easterly Helmsdale Fault lets down a coastal strip of Trias/ Upper Jurassic strata against Moinian, the Helmsdale Granite and Middle Old Red Sandstone; it may continue south-westwards through Dingwall and Strath Glass. The last movements in the Helmsdale district were post-Kimmeridge: Bailey and Weir (1933) have shown that it was an active submarine fault-scarp in Kimmeridge times, from which were derived huge angular blocks of Old Red Sandstone, now included in breccias interbedded with normal marine Kimmeridge sediments. The latter also contain fossil fauna and flora of littoral type, which are thought to have been swept from shallow water on the north-western or upthrow side of the scarp by tunamis energized by earthquakes along the fault. The Mesozoic strata adjacent to the fault dip away from it but further away show gentle folding along W. to W.N.W. axes, with a general northerly ascending sequence.

North and south of the mouth of the Cromarty Firth, Jurassic strata are let down against Middle Old Red Sandstone by north-easterly faults which are close to the undersea prolongation of the Great Glen Fault. The Mesozoic sediments show steep south-easterly dips away from the boundary fractures.

In the S.W. part of the region, north-easterly, northerly and north-westerly faults displace strata up to the Tertiary lavas. One of the most striking faults, scenically, is the N.-S. Inninmore Fault, on the N. side of the Sound of Mull, which has a downthrow to the W. of at least 1000 ft (305 m) bringing down a discontinuous succession ranging from Coal Measures to Tertiary lavas against Moinian. The Mesozoic and Tertiary rocks are only very gently folded, except in the vicinity of faults and as a result of Tertiary igneous intrusion (Ch. 9). Under the Firth of Lorne splay faults from the Great Glen Fault with normal throw control the size and nature of a basin containing Mesozoic and probably Tertiary (Barber and others, 1979).

5.3 Shetland

The Shetland Islands (Fig. 5.5) are the highest parts of a mostly submerged, irregular Palaeozoic and older block separating two large submarine sedimentary basins. The deeply-indented rocky coast is fringed in places by cliffs over 500 ft (152 m) high. Inland, hills rise to 1475 ft (450 m) above peaty moorlands. The islands lie about 103 miles (165 km) N.E. of the Scottish mainland and about 210 miles (336 km) from the nearest point in Norway.

The formations present are:

Upper Old Red Sandstone
Middle Old Red Sandstone
Lower Old Red Sandstone
Shetland Metamorphic Rocks, including units of Dalradian, Moinian and (?) Lewisian ages

There are marked contrasts in the geology E. and W. respectively of the Walls Boundary Fault, which is probably the continuation of the Great Glen Fault (5.2).

Acid and hornblendic gneisses of Lewisian aspect form the N.W. corner of the Mainland. These could be part of the Caledonian foreland or else a Northern Highland-type inlier (5.2). The rest of the metamorphic rocks W. of the Walls Boundary Thrust consist of impure quartzite, hornblendic gneiss and muscovite-schist followed by green schists and calcareous rocks. These metasediments are unlike those E. of the Fault. Although radiometric dates give a Caledonian age it has not been possible to equate them with any members of the Moinian/Dalradian sequence.

The metamorphics of the Mainland E. of the Walls Boundary Fault form a long succession starting with psammites (granulites) like those of the Moinian; similar rocks make up Yell. On the mainland the psammites are followed by quartzites, pelites, metalimestones, grits and altered spilitic lavas which can be tentatively correlated with the Dalradian of the Grampians (5.4). There are two belts of migmatites, granites and pegmatites. Along part of the E. coast of the Mainland gneisses, semipelites and gritty limestones, tectonically separated from the metasediments to the W., are termed the Quarff Succession (Fig. 5.5).

In Unst and Fetlar serpentinites, metagabbros, etc., are involved in a Nappe Pile. Post-tectonic Caledonian granites occur in both the E. and W. Mainland.

The Walls Sandstone, of Lower/Middle Old Red Sandstone age, outcrops in the W. and contains basalt, andesite and rhyolite lavas as well as ashes. Upper and high Middle Old Red Sandstone strata occur near Lerwick. The sediments of Foula, the most westerly island, have been shown from spores to be Middle Devonian (Donovon and others, 1978).

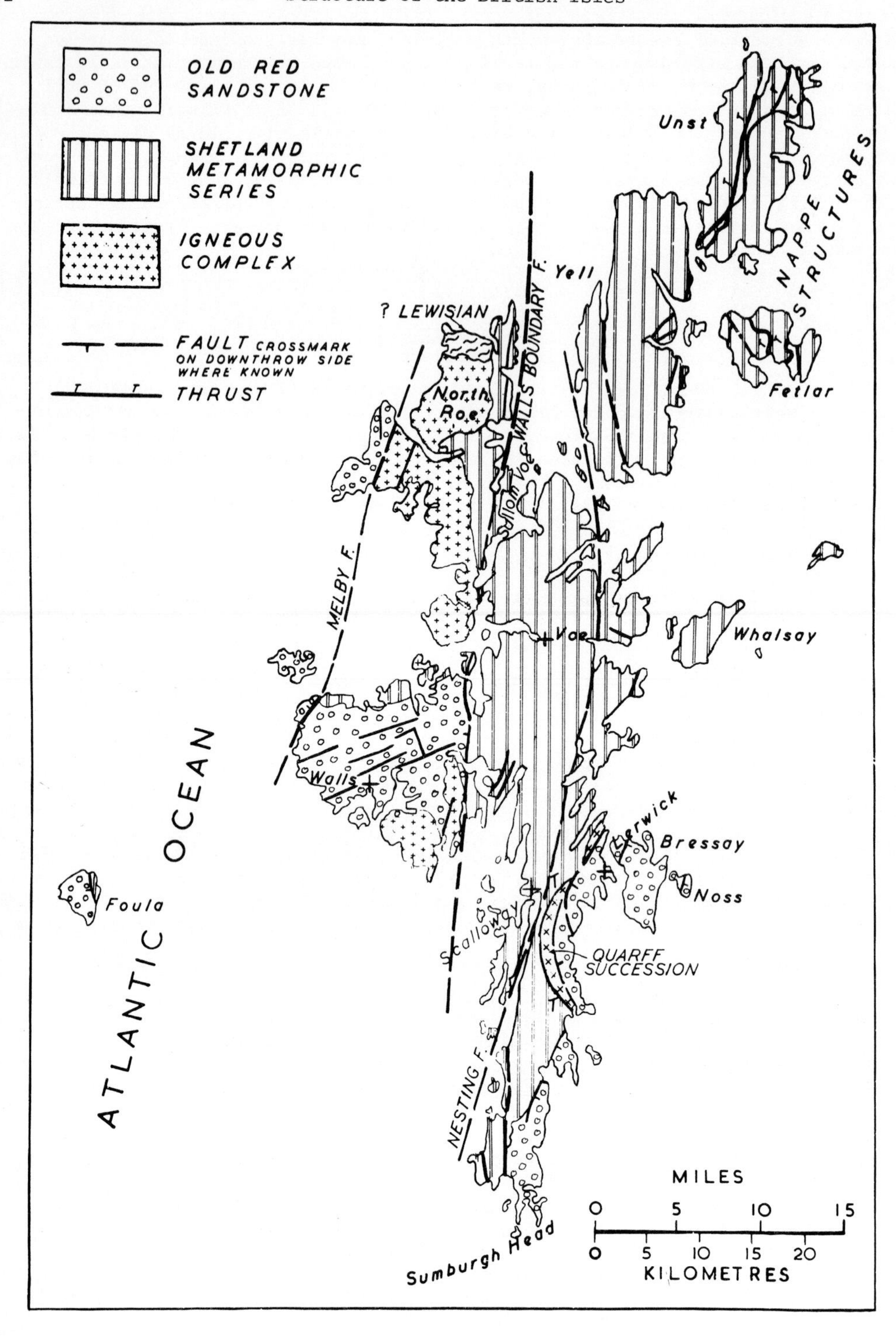

Fig. 5.5 Structural Map of Shetland.

The main structural events were:-

Pre-Caledonian
Caledonian - pre-Devonian
Late or Post-Devonian

Pre-Caledonian Structures

The gneisses of probably Lewisian age in the N.W. of the Mainland were affected by several periods of metamorphism and deformation before the deposition of the rocks to the E. (Pringle, 1970), from which they are separated by the Webster-Keolka Shear Zone.

Caledonian - Pre-Devonian Structures

The metamorphics to the E. of the Shear Zone belong to several geographically separated groups, only tentatively correlated with each other. All these rocks have undergone at least two phases of folding with metamorphism, radiometrically dated as Caledonian, accompanying or closely following the first phase. The southerly-dipping foliation in the Walls Peninsula must be due to late Devonian movements (see below) as it parallels the southerly bedding dip of the overlying Old Red Sandstone (Mykura, 1976, 23).

East of the Walls Boundary Fault the tectonic fabric of the metamorphic rocks (striking mostly N. by E. but bending N.E. in the N. of the Mainland) was produced by only one phase of intense deformation, termed the Main Deformation, of Caledonian age. According to Flinn (1967) this was followed by porphyroblast metamorphism and migmatisation, etc. Refolding of earlier structures has, moreover, taken place. However, May (1970) concluded that in the Scalloway district the migmatisation accompanied the main movements and that the latter post-dated an earlier and first phase of metamorphism. Near the middle of the E. coast of the Mainland the Quarff succession (see above) forms a nappe which was pushed westwards along a "melange" or schuppen-zone.

In Unst and Fetlar two major nappes of basic and ultrabasic rocks are separated from each other, and from the metamorphic rocks to the W., by thrust-planes and schuppen-zones. The zone separating the two nappes contains pebbles derived from the erosion of the nappes, suggesting nappe-emplacement at a high tectonic level (Mykura, 1976, 7).

Late or Post-Devonian Structures

Major N. by E. faults include, from E. to W., the Nesting Fault, the Walls Boundary Fault and the Melby Fault. The Walls Fault is marked by intense shearing and shattering. In the Walls Peninsula an E.N.E. set is also important. Large dextral displacement, claimed for the Walls Fault, complicates its correlation with the Great Glen Fault; similar movement has been suggested for other faults.

The Lower/Middle Old Red Sandstone of the Walls Peninsula has undergone intense folding along early E.N.E. axes and later N.N.E. to N. axes. The Old Red Sandstone of the East Mainland is gently folded along mainly N. axes.

5.4 The Grampian Highlands (Scotland and Ireland)

Apart from the area around and N. of Aberdeen the Grampian Highlands are noted for their rugged mountain scenery. Ben Nevis (4406 ft; 1345 m), within a Devonian ring-complex, is the highest point in the British Isles, and there are numerous other mountains, mostly of metasediments, rising to above 3000 ft (915 m). Structurally, the Grampian Highlands extend into Ireland where the mountains rise to nearly 2500 ft (760 m) in the N.W. (Figs. 5.2 and 5.9).

Exposures are excellent on the mountain ridges and on the sides of the deeply-glaciated valleys. Glacial drift, however, hides much of the solid geology in N.E. Scotland, and in Ireland, apart from the N.W. The deeply-indented coast provides long sections in rock, and in W. Ireland cliffs drop over 1500 ft (460 m) to the Atlantic.

The Great Glen Fault (5.2) forms the N.W. boundary. The Highland Boundary Fault, marking the S.E. margin, extends for 380 miles (608 km) from Stonehaven on the North Sea to Clare Island off the Atlantic coast of Ireland. This is a complex fracture-zone, along which movements, not all affecting the total length, have taken place from Arenig (or perhaps earlier) times to Tertiary times.

In Scotland, where it is almost continuously exposed, and is often marked by ophiolites or by carbonated derivatives, it runs in an almost uniform bearing of 5.57°W from the North Sea to the upper Firth of Clyde. Under the lower Firth of Clyde and through the islands of Bute and Arran however, its course averages 5.38°W. This swing, it is suggested on geophysical evidence, by McLean and Duggan (1978, 105) is only apparent and due to cross-faulting. However, the swing under the Firth is parallel to the regional bend southwards of the Caledonian strike seen in the Dalradian to the N.W., which is demonstrably not due to cross-faulting. In Ireland considerable sections are hidden by Tertiary lavas in the E. and by Dinantian in the centre. Towards the W. its course is more doubtful. A fault along the S. shore of Clew Bay is regarded by some authors as the Boundary Fault. However, another fault-zone, not far to the S., separating Dalradian to the N. from Silurian to the S., contains serpentinite and seems more likely to be the main fracture-zone. The offset of the fault-zone from the undoubted Highland Boundary on Clare Island (also with serpentinite) could be due to the N.W. Doo Loch faults (5.5) or their *en echelon* equivalent under the sea. It may be that on approaching the continental margin the Fracture-zone not only becomes westerly but splits up.

The formations present in the Grampian Highlands are:-

Pliocene
Tertiary volcanics
Liassic

Triassic) New Red Sandstone
Permian)

Old Red Sandstone
Lower Ordovician
Dalradian Metamorphic Assemblage or Supergroup (partly Cambrian)
Moinian Metamorphic Assemblage) Precambrian
?Lewisian (Mayo only))

It is only in the N.W. of County Mayo, Ireland, that any rocks are seen in the Grampian Highlands which may be a basement to the Moinian (Fig. 5.8). Here gneisses have been correlated with the Lewisian (Sutton 1972), although radiometric dates (Max and Soret 1979) give a younger age.

The Moinian Metamorphic Assemblage forms a large part of the Grampian Highlands in Scotland and considerable outcrops in western Ireland. It consists of metasediments, most of which are meta-psammites, traditionally known as granulites, and metapelites, generally similar to those of the Northern Highlands (5.2). As in the latter region they are best regarded as making up an assemblage of metamorphic rocks for both in Scotland and Ireland it appears unlikely that the sediments belong to one continuous succession. In the Scottish Grampians Piasecki and Van Breeman (1979) distinguish an older Central Highland Division and a younger Grampian Division. The latter in places passes upwards into the Dalradian without structural or stratigraphical break; it includes the metasediments previously termed the Central Highland Granulites or Central Highland Psammitic Group, and also pelites, some quartzites and thin calc-silicate lenses.

The Dalradian metasediments are much more varied than those of the Moinian. Consequently many district successions, sometimes interpreted in the wrong stratigraphical order, were worked out at an early stage in research. The use of "way-up" techniques has made it possible to establish a sequence which is broadly accepted by most geologists. Correlation tables by Anderson (1965, Table I) and by Harris and Pitcher (1975, Fig. 12) give most older locality names. The two latter authors have selected certain locality names for subdivisions of the broad sequence, but, on the other hand, lithological terms have the merit of providing descriptions. Both are, therefore, set out in the summary table below which also shows the top of the Moinian.

		Sub-groups and (groups)		
CAMBRIAN	U.Psammitic Gr. U.Pelitic and Calcareous Gr.	(Southern Highland)		UPPER
DALRADIAN	L.Psammitic Gr.	Crinan	Argyll	MIDDLE
	L.Pelitic and Calcareous Gr.	Easdale		
	Carbonaceous Gr.			
	Quartzitic Gr.	Islay		
	Calcareous Gr.	Blair Atholl Ballachulish	Appin	LOWER
TRANSITION FROM DALRADIAN TO MOINIAN	Pelitic and Quartzitic Transition Gr.			
MOINIAN	Grampian Division			

Where there is no tectonic contact between Moinian and Dalradian there is a variable interbedded succession of pelites and quartzites between the Psammitic Group and undoubtedly Lower Dalradian limestones. Some authors have placed this Transition Group in the Lower Dalradian. There seems, however, to be some advantage in using the first limestone horizon in this continuous succession as a base, especially as the Eilde Quartzite, sometimes taken as the lowest Dalradian formation, is not always present.

The Ballachulish Sub-group reaches its fullest development in the nappe-complex known as the Ballapel Foundation (see below). The Blair Atholl Sub-group forms extensive outcrops in the overlying Iltay Nappe. Nevertheless in the Blair Atholl district itself (Smith and Harris 1976) and in the Schichallion district (Treagus and King 1978) relatively thin successions, stratigraphically below the Blair Atholl Limestone, have been recognised as the equivalent of the Ballachulish sub-group.

On the other hand the Kinlochlaggan Limestone (Anderson 1947), because of its association with a sparsely-developed glacogenic boulder bed, seems likely to be equivalent to the Blair Atholl Limestone, although it rests directly on the Transition Group. In the North-East Grampians the Sandend Limestone is an equivalent of the Blair Atholl Limestone. These various carbonate horizons would make a logical stratigraphical base, probably diachronic, of the Dalradian.

The broad lithology of the Dalradian subdivisions can be judged from their descriptive names. In Ireland a Middle/Upper Dalradian succession, very similar to that of the S.W. Highlands of Scotland, is present. Further S.W. in Ireland the Middle/Upper Dalradian divisions are not as clearly defined, but what has been termed the Kilmacrenan succession is almost identical with the Lower Dalradian of the Iltay Nappe in Scotland (see below). North-west of the large Donegal Granite the Creeslough and Fintown successions resemble the successions of the Ballachulish district.

An important horizon at the base of the Quartzitic Group is a glacogene with large boulders, often of nordmarkite, with interbedded upper and lower contacts, which has been traced from Aberdeenshire to the Atlantic coast of Ireland (see also 5.6).

Near Loch Awe in Scotland and Strabane in Ireland, abundant basic lavas, some with pillow-structure, occur throughout much of the Middle Dalradian outcrop. The basic volcanics may have been associated with the Cambrian opening of the Iapetus Ocean (Graham 1976).

Dating of the fold and metamorphic phases is discussed below (5.6). Stratigraphically, most of the Moinian and part of the Dalradian are late Upper Proterozoic. The Upper Dalradian, on the other hand, is Cambrian as pagetid trilobites of late Lower Cambrian age occur near Callander, Perthshire, in the Leny Limestone, which is part of the Upper Psammitic Group. As acritarchs (Downie and others, 1971) of Lower Cambrian type occur down to the Loch Tay/Tayvallich limestone horizon at the base of the Upper Dalradian the Precambrian-Cambrian boundary may be about here.

Narrow strips of basic lavas, black shales, cherts and other sediments occur along the Highland Border, mostly separated from Upper Dalradian by faults but in places, notably Arran, following on without structural or metamorphic break. Fossils, some doubtfully diagnostic, indicate Upper Cambrian/Lower Ordovician ages.

Many post-tectonic Caledonian intrusions of granitic and related rocks penetrate the schists. They include large masses, probably of batholithic form, and ring-complexes such as that of Ben Nevis (for structure see below). Large numbers of hypabyssal intrusions also occur, including north-easterly dyke swarms. Some thick, easterly, quartz-dolerite dykes of presumed Permo-Carboniferous age are present. In the W. of Scotland and in Ireland Tertiary basic dykes are numerous (Ch. 9). Considerable outcrops of Lower Old Red Sandstone sediments and volcanics unconformably overlie the metamorphic rocks, notably in Lorne (near Oban) in Aberdeenshire and immediately N. of the Highland Border. Middle and Upper Old Red Sandstone conglomerates and sandstones cover a large area around Inverness and S. of the Moray Firth. Upper Old Red Sandstone sediments rest directly on the Dalradian in the Loch Lomond/Firth of Clyde district, where they are overlain by a small outcrop of Lower Carboniferous strata. Both Lower and Upper Carboniferous overlie the Dalradian at the S. end of Kintyre. A patch of Upper Carboniferous occurs along the Pass of Brander Fault E. of Oban.

The Grampian Highlands in Ireland differ from those of Scotland in the presence of extensive outcrops of Lower Carboniferous strata, including thick limestones. Throughout much of the region sedimentation seems to have started in the Visean but in the Omagh district a considerable thickness of sandstones and sandy limestones below C_2S_1 beds are Tournaisian. In the extreme N.E., at Ballycastle, there is a Lower Carboniferous succession, including basalt lavas, similar to that of the Midland Valley of Scotland, followed by sandstones, shales and coals belonging to the Upper Carboniferous.

On the W. coast of Kintyre red sandstones, formerly assigned to the Upper Old Red Sandstone, are now considered to be of Permian age. Triassic strata, with reptiles, unconformably overlie the Old Red Sandstone S. of the Moray Firth. Gravels in Aberdeenshire are believed to be Pliocene.

In N.E. Ireland Triassic red conglomerates, sandstones and marls occur round the rim of the Antrim Tertiary basalt lavas. These are followed in places by Rhaetic and Lower Liassic shales and thin limestones, followed by Upper Cretaceous white limestone, a hard chalk.

Structural events were:

Precambrian
Caledonian (Pre-Lower Devonian)
Structures connected with Lower Devonian intrusions
Late Caledonian (Pre-Upper Devonian)
Probably Hercynian
Probably Mesozoic and Tertiary

Precambrian structures

In the N.W. of County Mayo, Ireland, gneisses which may be Lewisian (Sutton, 1972) show Precambrian structures modified by intense flattening and reconstitution.

In what is probably the "Older Moinian", named the Central Highland Division by Piasecki and van Breeman (1979), three generations of structures, each associated with high-grade metamorphism, have been recognised. The first is a regional schistosity, the second isoclinal folding of this schistosity and the third large recumbent folds. These structures are held to pre-date the Grampian Group (see above) and to be separated from the group by a slide.

The possibility that some of the structures of the Grampian Division itself may be late Precambrian rather than Caledonian is discussed below (5.6).

Caledonian (Pre-Lower Devonian) structures

It is generally accepted that most, if not all, of the Dalradian was affected by three major fold-phases (Johnson, 1969), although some authors have postulated more; for example, Harris and others (1976) recognise four phases in the Tay Nappe and Roberts (1974) no less than eight in the S.W. Scottish Highlands.

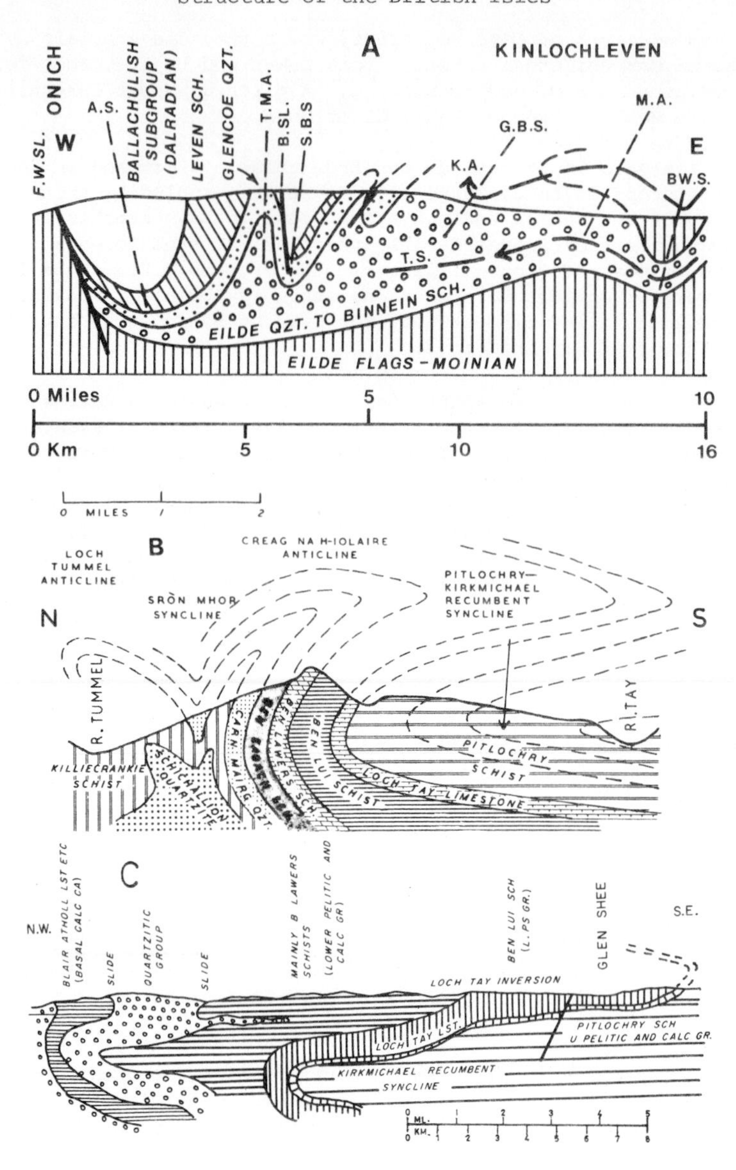

Fig. 5.6 Fold-Styles in Grampian Highlands

A. Lochaber (after Hickman). Axial traces of recumbent folds - long dashes. Axial traces of secondary (upright) folds - short dashes. T.S., Treig Syncline-K.A., Kinlochleven Anticline; F.W.Sl., Fort William Slide; A.S., Appin Syncline; Bw.S., Blackwater Synform; G.B.S., Garbh Bheinn Synform; M.A., Mamore Antiform; S.B.S., Stob Ban Sybform; T.M.A., Tom Meadhoin Anticline.

B. Central Perthshire (after Harris)

C. Eastern Perthshire (after Bailey)

The nappe hypothesis for the Grampian Highlands originally put forward by Bailey has been greatly modified. Nevertheless, it is accepted that a large part of the Grampians, both in Scotland and Ireland, is made up of the generally north-easterly striking Iltay Nappe-Complex. This structurally overlies the Ballapel Foundation seem mainly to the N.W. of the Iltay Nappe both in Scotland and Ireland. A Banff Nappe has been postulated in the N.E. Grampians but some authors regard this as part of the Iltay Nappe (Fig. 5.2).

The Iltay Nappe-Complex is largely characterized by north-easterly trending recumbent folds (Dl) accompanied by sliding. These close towards the S.E. in the S. of the Nappe-complex and towards the N.W. in the N., the two parts being separated by a belt of steeply inclined structures (Fig. 5.6). The Dl folds are modified by fairly open cross-folds (D2) and folds (D3), trending N.E. Later (D4) structures sometimes take the form of minor folds and buckling.

These fold-phases represent distinct and separable orogenic pulses. Where, however, more phases have been described, it must be asked, as in other regions, what is the nature of true polyphase folding and whether minor structures of relatively limited distribution may not be due to local structural and geological inhomogeneity.

From near Kirkmichael (Fig. 5.6) to Kintyre the lower limb of a major recumbent anticline forms the "Loch Tay Inversion" affects mainly Middle and Upper Dalradian metasediments.

The Upper limb is preserved in the Loch Awe Syncline (probably a D3 fold) where the metasediments (see Allison, 1940) and volcanics occur in normal order. The recumbent fold was held by Clough (1897) to close at Carrick Castle.

Towards the Highland Border the structures become steep, and the metamorphic grade rapidly falls off. The fact that these steep structures are virtually parallel to the Highland Boundary Fracture-zone is hardly likely to be coincidental and supports the view (expressed above) that the Fracture-zone was initiated in Arenig (or perhaps earlier) times.

The Loch Tay recumbent fold may close in the Highland Border steep belt although differing interpretations of the evidence are possible (cf. Anderson 1947, Shackleton 1958a and Roberts 1974).

The lower limb of the recumbent fold forms the upper limb of another recumbent structure, the Ben Lui Syncline. This has been held to be a Dl structure but Roberts and Treagus (1975) argued that it is a D2 structure similar to others they recognise in the Dalmally district.

The north-easterly (Dl) structures are accompanied by numerous homoaxial minor structures and b-lineations. In Perthshire, the north-easterly structures (Dl) have been shown to be modified (Rast, 1958; Sturt, 1961; Harris, 1963) by N.W. "cross folds" (D2), also accompanied by minor structures formed at the same time as, or later than, the (Fl) folds. Such "cross-folds" are held by Rast to account for the marked "twist" of strike in the Schichallion district. In the Pitlochry/Blakr Atholl area Harris (1963) distinguishes a further set (F3) of open folds whose b-axes trend E.N.E.

To the N.W. of the steep belt the overfolds, directed N.W., are shorter and are separated by the Iltay Boundary Slide from the folds of the Ballapel Foundation (see below).

South-easterly plunging structures are a marked feature near the Cairngorm Mountains where there is a structural swing southwards. Further N.E. the folds may be autochthonous or parautochthonous and may have been relatively held back, possibly by the resistance of the large pre-tectonic basic intrusions which occur here, while those to the S.W. were driven more powerfully south-eastwards accompanied by more intense overfolding and by sliding. It is perhaps significant that the Iltay Boundary Slide has not been recognized N.E. of the southward swing.

Still further N.E. in Banffshire, Johnson (1962) has postulated four fold-phases. The first resulted in the formation of isoclines and also probably of major recumbent folds and of a slide underlying the Banff Nappe (if the existence of this structure be accepted). The second folds are usually tight and at an acute angle to the N.N.E. striking primary schistosity which is axial planar to the first folds. The third phase includes the major Boyndie Syncline (a huge monocline (Sutton and Watson, 1956) and the complementary Buchan Anticline (Read and Farquhar, 1956). The fourth phase-folds are of small to medium dimensions, assymmetrical or monoclinal in shape with axial planes striking N.N.E.

Only 13 miles (21 km) of sea separates the Dalradian of Kintyre from that of the Antrim Inlier in Ireland (Fig. 5.7) where Upper and Middle Dalradian, seen also in the Sperrin Mountains further S.W., form the continuation of the Iltay Nappe. It is probable that the "Loch Tay Inversion" is present, but firm evidence is lacking (Wilson, 1972).

From Antrim and Malin Head the Iltay Nappe can be traced south-westwards through Donegal to the Atlantic coast of N.W. Mayo.

West of and south of Lough Foyle as far as Strabane, a late syncline forms the structural continuation of the Loch Awe Syncline and like the latter contains an uninverted Upper Dalradian succession including basic pillow-lavas.

Around Lough Derg Moinian psammites ramified by migmatites (Anderson, 1948) are brought up in a dome which is probably a D3 structure (Fig. 5.10). Earlier folds have been recognised by Borradaile (1974). To the N. the Dalradian strata of central Donegal are considered (Pitcher and others, 1971) to occupy a tight early fold - the Ballyboffey anticline. This is held to be separated from the Moinian psammites of Lough Derg by a thrust, but the evidence is doubtful; the thrust, if present, may be further N.

In N.W. Donegal the Iltay Nappe contains the Kilmarcrenan Succession similar to that of the Scottish structure. Apart from complications due to cross-folding (see below) the strike is south-westerly to near Ardara where it swings westerly. Overfolding and thrusting towards the N.W. and N. are common. Thus in the Portnoo area, Akaad (1956) has recognized four such thrusts, and in the northern part of the Slieve League Peninsula, Anderson (1954) has mapped an E.-W. thrust, which, however, fades out westwards (Fig. 5.7).

The thrust forms part of the southern margin of the large Slieve-Tooey Syncline containing the quartzite of the same name (=Quartzitic Group). At the tip of the peninsula, two smaller synclines with cores of the same quartzite have a north-westerly strike. This is probably due to cross-folding.

In the Upper Dalradian rocks at the S.W. end of the Ox Mountains, Currall (1963) has distinguished four movement phases. The last phase involved thrusting from the N.W.; the lower thrust brought the Dalradian rocks on top of the migmatites and metasediments, of probable Moinian age, forming the main part of the Ox Mountains. In the latter it seems possible to distinguish from the minor folds three movement phases.

In the large outcrop of Dalradian (and perhaps some Moinian) metasediments N. of Clew Bay (Trendall and Elwell, 1963) poor exposure and doubt regarding stratigraphical correlation makes structural interpretation uncertain. The strata are probably disposed in a syncline with flanking anticlines overturned in the S. towards the W. to S.W.; refolding of these folds has taken place.

South of Clew Bay the Upper Dalradian Westport Grits form a narrow outcrop N. of the Highland Boundary Fracture-zone. The dominant structural trend is northwesterly;inverted graded bedding makes it clear that a recumbent fold is present.

The Ballapel Foundation, which contains only Lower Dalradian, forms a nappe-complex separated from the Iltay Nappe by a Boundary Slide. It is characterised by the full development of the Ballachulish sub-group. Structurally the Ballapel Foundation was interpreted by Bailey (1934a) as consisting of north-easterly recumbent synclines closing towards the S.E. (implying over-movement to the N.W. with slides (lags) affecting the unreversed limbs of the recumbent folds. The latter are themselves refolded. The stratigraphical discordances, accounted for by Bailey by slides, may be due to facies variation (Hickman, 1975); the lowest Dalradian rocks rest along a sedimentary contact on Moinian. Bowes and Wright (1973) state that two early phases of deformation preceded the north-easterly folds but Roberts (1976) holds that the only phases of any importance were those postulated by Bailey. Hickman (1978) making a number of modifications in Bailey's hypothesis, has shown that recumbent (D1) folds, refolded by upright (D2) folds (Fig. 5.6) extend S.W. to the island of Lismore which consists of thick limestone equivalent to the Islay Limestone.

In the N.W. of Ireland there are two Dalradian sequences, the Creeslough and Fintown successions, which belong to the Ballapel Foundation. They are bounded to the S.E. by a slide which is partly obscured by, and partly lies not far S.E. of the Donegal Granite.

North-west of the granite the Creeslough Succession is itself cut by three slides superimposed on which are north-easterly folds partly overturned towards the N.W. the most important of which are the Gweebarra Anticline and the Creeslough Syncline. These are affected by north-westerly cross-folds. The Fintown succession forms a narrow outcrop S.E. of the granite and is overlain along a slide by part of the Iltay Nappe.

Metamorphic zoning on the basis of certain index minerals was established by Barrow (1893) in the S.E. Grampians and has served as a model for similar work in many other metamorphic terrains. Andalusite - and cordieite-bearing schists occur in the N.E.; this has been termed the Buchan-type of metamorphism from the Buchan district and is probably due to elevation of the regional temperature at a comparatively shallow depth.

Two early phases of metamorphism (M1 and M2), up to the garnet grade (or amphibolite facies), appear to be associated with D1 and D2 folds. Dolerite intrusion followed, and a phase, or phases, of high temperature metamorphism in a static environment leading to the formation of later garnet, kyanite and sillimanite in both Scotland (S.E. Grampians, S. Inverness-shire) and Ireland (Lough Derg).

Retrogressive metamorphism, probably associated with D3 and later fold-phases, has been widely recognised.

Structures connected with Lower Devonian intrusions

Within the classic cauldron subsidence of Glen Coe (Clough, Maufe and Bailey, 1909) a roughly circular pile of volcanic rocks resting on Lower Old Red Sandstone sediments, overlying Dalradian and Moinian metasediments, has subsided, to the accompaniment of ring-dyke intrusion, by more than 3000 ft (915 m). At the same time the volcanics and their foundation buckled into basin form (Fig. 5.8). The formation of the Ben Nevis Complex involved several phases of ring-dyke intrusion to the accompaniment of the foundering of a central block of Lower Old Red volcanics. The amount of subsidence cannot be measured but Dalradian schists outside the complex are in the garnet grade whereas those inside, beneath the volcanics, although hornfelsed, are at a lower regional grade suggesting a very considerable drop from a higher metamorphic level. That the Ben Nevis and Glen Coe volcanics, originally in areas of subsidence amounting to several thousand feet, now form high mountains is a striking example of the dependence of British scenery on differential erosion rather than on structure.

Both the Ben Nevis Complex and the Etive Complex, S. of Glen Coe, are the focus of great north-easterly dyke-swarms. The dilation of the crust at right angles to these was a structural event of some significance, amounting in the case of the Etive swarm to about a mile.

The formation of the mile-broad Outer granite of Ben Nevis seems to have involved more than the ring-dyke mechanism. On its northerly side, normally south-westerly striking metasediments swing into an almost westerly strike suggesting outward push by magmatic pressure. This phenomenon is even more striking near the north-westerly orientated Strath Ossian quartz-diorite (an anomalous alignment for a Caledonian mass) on the margins of which the metasediments are swung almost through a right angle.

In Ireland the intrusion of the Main Donegal granite (Pitcher and Read, 1960) has caused marked folding of its envelope.

Late Caledonian (Pre-Upper Devonian) structures

The main movement along the Highland Boundary Fracture-zone was of post-Lower, pre-Upper Old Red Sandstone age. The Lower Old Red Sandstone outliers N. of the fracture-zone (Allan 1940) are folded along north-easterly axes and are cut by normal or possibly reversed faults, parallel to, and at about 30° to, the Highland Boundary Fault. These folds and faults are likely to have been of Middle Old Red Sandstone age and if this is so their presence in the Grampian Highlands would support the view that the last folds affecting the metamorphic rocks may be as late as Middle Devonian.

The Middle Old Red Sandstone, S. of the Moray Firth, is fairly gently folded along N.E. axes. As this formation is unconformably overlain by the Upper Old Red Sandstone, the folding must be, in part at least, of pre-Upper Devonian age.

Probable Hercynian structures

An important structural feature of the Grampian Highlands (and also the Northern Highlands, (5.2) is the presence of many major and minor north-easterly to north-north-easterly faults. In the case of some, powerful transcurrent movement in a sinistral sense can be demonstrated, for example the Ericht-Laidon Fault (up to 4½ miles, 7.2 km), the Tyndrum Fault (up to 3 miles, 5 km) and the Loch Tay Fault (up to 4 miles, 6.4 km). Many of these faults may have been initiated during the

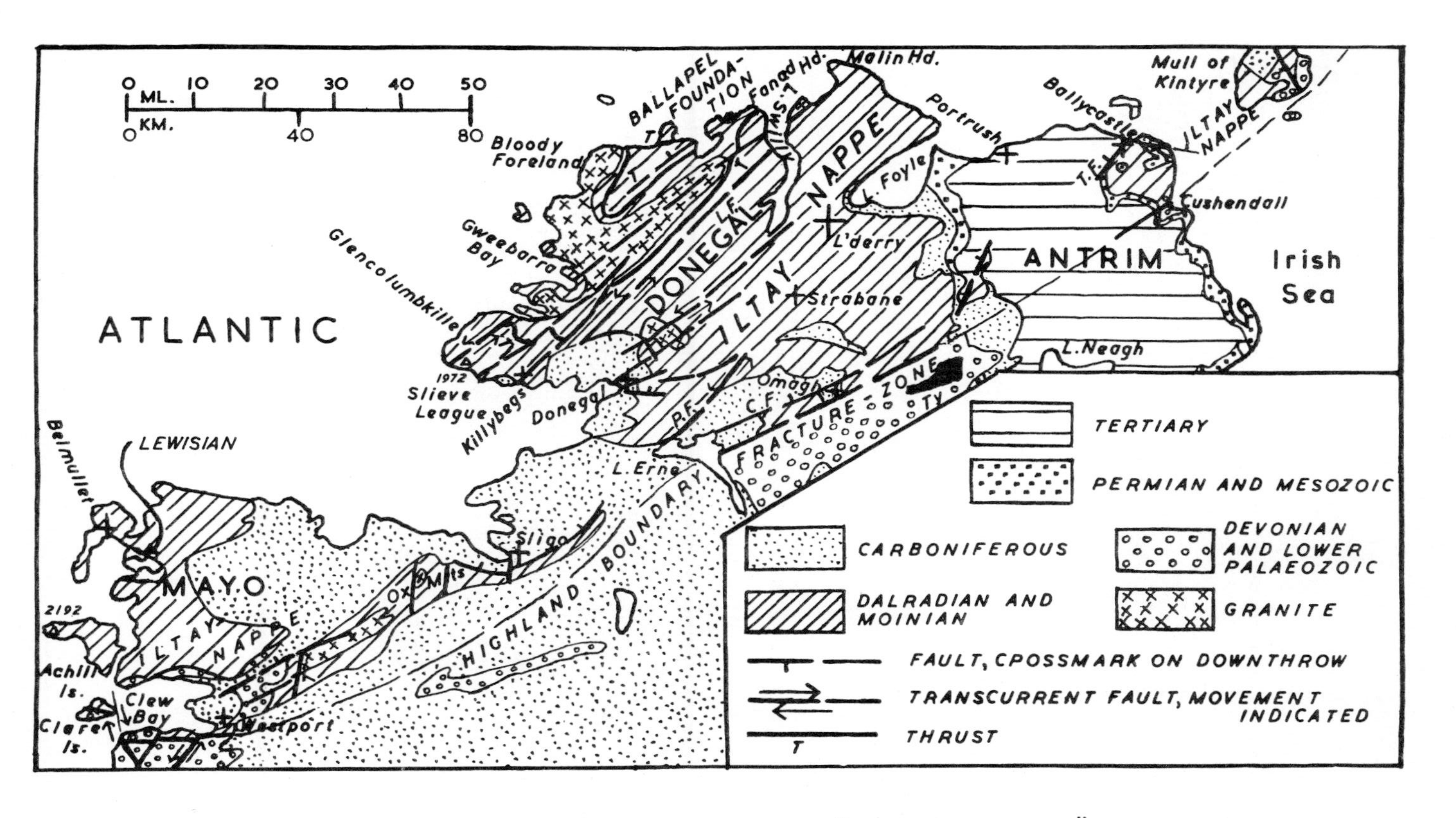

Fig. 5.7 Structural Map of the Irish "Grampian Highlands".
C. F. = Cool Fault; L.F. = Leannan Fault;
P.F. = Pettigo Fault; T.P. = Tow Fault;
T.y. = Tyrone Inlier.

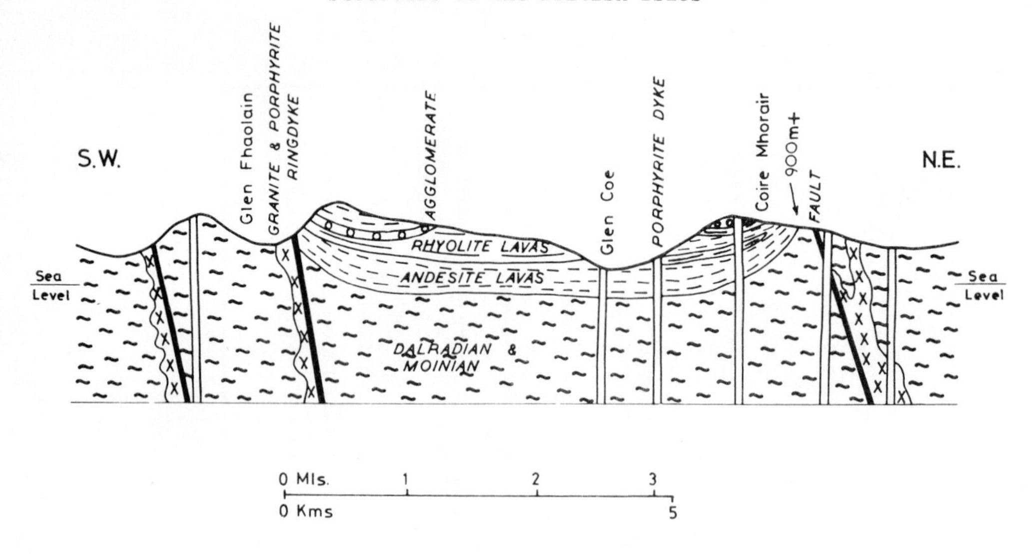

Fig. 5.8 Section of Lower Devonian Cauldron - Subsidence of Glen Coe.

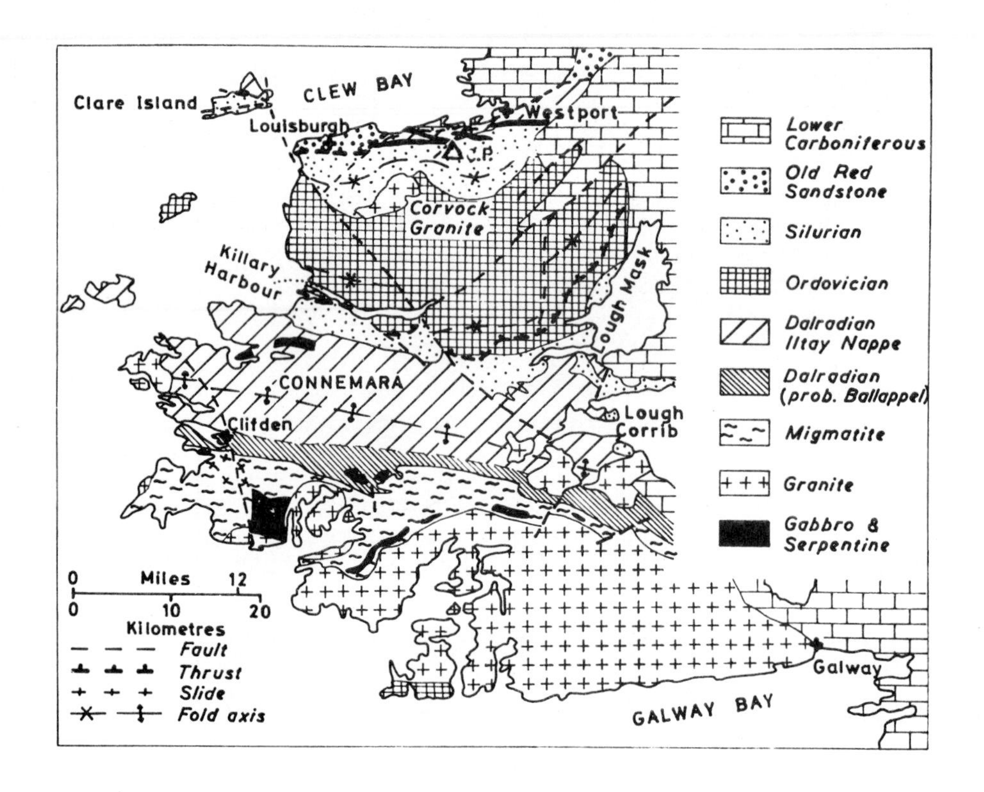

Fig. 5.9 Structural Map of Connemara and its Flanks
C.P. = Croagh Patrick.

Lower Palaeozoic and also have moved in the Lower Devonian, but important components of movement are certainly later than the post-tectonic (probably Lower Devonian) granites and some, near the Great Glen, displace Middle Old Red Sandstone, and others, near the Firth of Clyde, the Upper Old Red Sandstone. It appears likely, therefore, that much of the faulting was Hercynian, in response to N.-S. compression. The predominant north-easterly grain would favour development in this direction. Complementary north-westerly fractures are in fact much less common, but there are a number of minor north-westerly faults. A major north-westerly fault is that which determines the Pass of Brander near Oban, although in this case the obvious displacement is downthrow towards the S.W. of over 3000 ft (915 m).

In Ireland, the Lower Carboniferous rocks resting on the metasediments are thrown into gentle anticlines and synclines, important examples being the Omagh and Donegal synclines. Many of these folds have a north-easterly trend and their development as Hercynian folds is probably due to the influence of the underlying Caledonian structure. Folds W. of the Lough Derg Anticline and in the Clew Bay area run E.-W., again in accord with the underlying structures.

Many of the north-easterly faults which traverse the Moinian and Dalradian rocks can be shown to have important post-Visean components, partly at least transcurrent. In some cases, however, the main movements seem to have been pre-Carboniferous. One of the most important is the Leannan Fault (Pitcher and others, 1964); in fact correlation with the Great Glen Fault has been suggested, although it seems more likely to be a splay or parallel fault to the latter. At its N. end the fault displaces red rocks, generally accepted as Lower Old Red Sandstone; the relationships are such that earlier movement cannot be excluded. At its S. end, N.E. of Killybegs, both the Leannan Fault and others parallel to it either fail to pass into the Carboniferous or do not appear as important structures in that formation. Sinistral transcurrent movement of as much as 25 miles (40 km) is possible for the Leannan Fault, although not definitely proved. Sinistral transcurrent faults with movement of up to 3 miles (5 km) cut the Barnesmore granite of probably Lower Devonian age, and although these extend into the Carboniferous, it has been suggested that the Post-Visean component was small (Pitcher and others, 1964, p.252).

Turning now to faults where there is an important post-Visean component we may first note fractures associated with the Omagh Syncline, e.g. the Fettigo Fault, bounding the Carboniferous on the W., the Cool Fault which runs N. of Omagh and continues through the Dalradian for some 25 miles (40 km) and the Castle Archdale section of the Highland Boundary which has resulted in apparent reversal of throw of the older fracture (Simpson, 1955). The Donegal Syncline and the Moinian/Dalradian to the W. are traversed by a number of north-easterly wrench-faults. On the other hand, for some of the north-easterly faults, a vertical component is more important; the North Ox Mountains Fault drops the Carboniferous rocks of the Sligo Syncline some thousands of feet to the N. (George, 1953).

A curving fault downthrowing southwards forms the northern margin of the Donegal Syncline and displaces north-easterly fractures. North-easterly faults are present at the S.W. end of the Ox Mountains (Currall, 1963, p.164; Currall and Taylor, 1965); they were probably initiated in the Devonian and extended well into the Carboniferous. In N.E. Ireland the Great Gaw Fault, striking E.-W. through the Ballycastle Coalfield, drops the strata to the north by some 1300 ft (395 m).

Probable Mesozoic and Tertiary structures

Owing to the scarcity of post-Carboniferous formations in much of the Scottish Grampians and in the W. of the Irish Grampians it is difficult to demonstrate Mesozoic or Tertiary movements. The Triassic S. of the Moray Firth appears to be faulted (although drift makes this difficult to prove) and the probably Permian rocks of Kintyre are cut by a few faults. The existence of north-westerly Tertiary dyke-swarms in Scotland and Ireland indicates a Tertiary tensional phase which may have reactivated older fractures.

Tertiary movements (some probably initiated in the Mesozoic) can be more readily demonstrated in N.E. Ireland. They include the accentuation of the anticline along the Highland Border and the development of downfolds to N.W. and S.E. Faulting took place along the N.W. and S.E. margins of the Highland Border ridge and also at right angles. The scale of the deformation may be judged from the fact that combined folding and faulting has led to differerences in level of the Chalk of over 3000 ft (915 m).

5.5 Connemara and its flanks

The region between the Highland Boundary Fault just S. of Clew Bay and Galway Bay with Connemara in its centre is the only part of the British Isles where undoubtedly Dalradian rocks come to the surface of the Highland Boundary and also the only part of the British Caledonides where strata as young as the Silurian (Wenlock) have undergone regional metamorphism. It is a sparsely inhabited and mountainous district rising to over 2600 ft (792 m). To the E. the region is bounded by unconformable Lower Carboniferous (Fig. 5.9) and to the W. by the deeply-indented Atlantic coast.

Formations present are as follows:

Silurian
Ordovician
Dalradian Metamorphic Assemblage, or
Supergroup (partly Precambrian, partly Cambrian).

The Dalradian rocks have long been known as the Connemara Schists. In addition to mica-schists and migmatites, they comprise quartzites and metalimestones. They have been shown to resemble lithologically and stratigraphically the Lower/Middle Dalradian. They include, as in Donegal andScotland, a boulder bed of glacogene type with a carbonate horizon (containing the well-known Connemara Marble) below and a quartzite above.

Epidiorites and other basic and ultrabasic rocks are present which pre-dated at least some of the metamorphism.

The Ordovician (McKerrow and Campbell, 1960); was deposited to a thickness of over 35,000 ft (10,700 m) in the South Mayo Trough, N. of a rising cordillera of Dalradian. The succession starts with the Sheefry Grit (Arenig) consisting of black shalesor slates with cherts, spilites and basic tuffs followed by greywackes, shales or lenticular conglomerates. The Sheefry Grit is succeeded by the Glenummera Slate (?Llanvirn) consisting of slates with subsidiary graded sandstones and lenses of conglomerate. Above this comes the Mweelrea Grit, made up of a thick succession of coarse greywackes and conglomerates with green shales and thick beds of welded tuffs or ignimbrites.

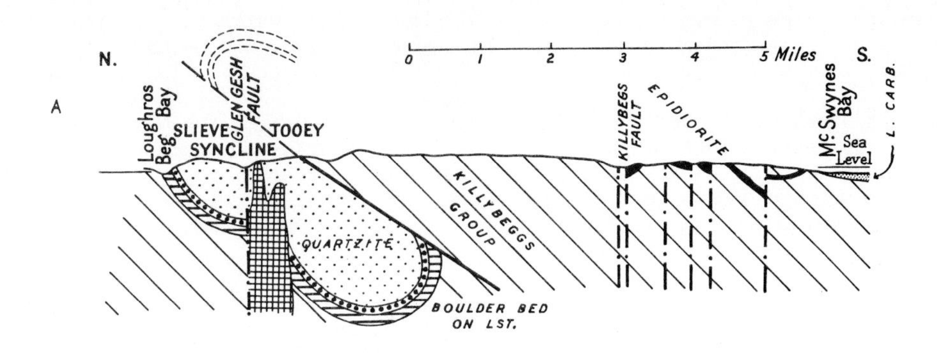

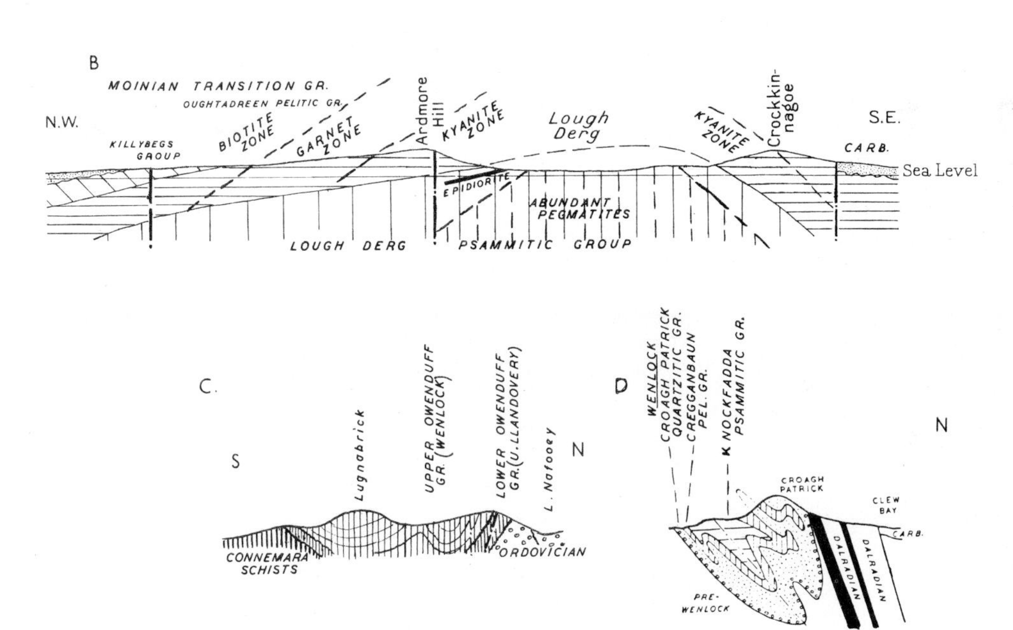

Fig. 5.10 Sections of "Grampian Highlands" in Ireland and of Connemara and its Flanks.

A. Across Slieve League Peninsula (after Anderson).

B. Across Lough Derg, Donegal (after Anderson).

C. Across north flank of Connemara (after McKerrow and Campbell).

D. Across Croagh Patrick, South Mayo (after Anderson).

Silurian strata, S. of Killary Harbour, rest unconformably on Arenig rocks to the N. and on the Connemara schists to the S. The Lower Owenduff Group (Upper Llandovery) consists of a basal conglomerate followed by grits, calcareous flags and purple shale. The Upper Owenduff Group (Wenlock) contains graded greywackes, siltstones and shales with, in places, a basal conglomerate which S. of Killary Harbour rests directly on Connemara Schists. This group is succeeded by the Salrock Group of sandstones, shales and slates.

Between the Ordovician and the serpentinite-bearing fault to the N., regarded (5.4) as the main branch of the Highland Boundary Zone, a Silurian succession of a different character occurs (Anderson, 1960; Bickle and others, 1972). The Croagh Patrick Quartzite with a basal conglomerate is followed by pelites and calcareous pelites succeeded by psammites. In spite of epizonal metamorphism, distorted fossils are preserved and prove a Wenlock age.

In places on the S. margin the Wenlock rests unconformably on a thin Llandovery succession, and to the W. on Clare Island it overlies Dalradian and possibly older rocks (Philips, 1973, 1974).

In the S. of the region the large Galway Granite, dated at 365 m.y., was intruded post-tectonically. On its S.W. margin hornfelsed basic pillow-lavas, tuffs and sediments with Dalradian clasts (the South Connemara Series) are assigned to the Ordovician.

In the N.W. the Corvock Granite post-dates the regional metamorphism of the Wenlock.

Some basic Tertiary basic dykes are present.

Structural events were:

Caledonian	(Late Cambrian to late Ordovician (Post-Wenlock (Pre-Carboniferous)
Hercynian and Tertiary	(Post-Carboniferous (

Late Cambrian to late Ordovician structures

Nearly all the Dalradian metasediments are of a facies similar to that of the Iltay Nappe (5.4) but a strip in the S., eastwards from Clifden, contains a facies more like that of the Ballapel Foundation. This may be separated by a slide from the Iltay type rocks further N. which are themselves divided up by slides.

Four fold-phases have been recognised in the Connemara Dalradian (Badley, 1976; Yardley, 1976), and Barrovian metamorphism ranges up to sillimanite grade and migmatites.

A D1 deformation, evidence for which rests mainly on fabric studies, was followed by upright folds (D2) which were later flattened. Northward-facing nappes (D3) were then formed. Open D4 folds developed after the rocks had largely cooled and include the most obvious structures such as the Connemara Antiform. The time-relationship of the D4 folds to the overthrusting of high-grade rocks over low-grade schists to the S. is not clear.

As Arenig strata with metamorphic clasts come against the schists it has been claimed that all the metamorphism was pre-Arenig. However, as the contacts are invariable large faults this conclusion is by no means certain.

The clasts could have come from older metamorphics or they could have been derived from a part of the Connemara pile which had undergone one of the earlier metamorphisms and have been uplifted at an initial stage of the formation of the Connemara Cordillera and South Mayo Trough. From a date of 510 m.y. obtained by Pidgeon (1969) for metamorphism in the Connemara schists, and from other evidence, the early folding and metamorphism would appear to be due to late Cambrian (Sardic) events, but folding and metamorphism continued into the Ordovician; Phillips (1973) suggests a mid-Ordovician age. Metamorphism of this age could have faded out rapidly northwards and not affected the Ordovician, just as the undoubted post-Wenlock metamorphism S. of Clew Bay (see below) diminishes rapidly southwards. D4 in the Dalradian could be contemporaneous with part of the folding in the Ordovician.

In the Ordovician the main structure is the Mweelrea Syncline, which is flanked to the N. and S. by anticlinal zones. There is, however, some doubt as to the relative importance of pre-Silurian and post-Silurian (see below) folding in determining the final form of these folds (McKerrow and Campbell, 1960, 44).

Structures of Post-Wenlock (Pre-Carboniferous) age

There is evidence of intra-Silurian movement, for Lower Wenlock conglomerates rest unconformably in places on Upper Llandovery. However, the main folding is post-Wenlock. Whether it was late-Silurian or Devonian remains in doubt. The folds trend E.-W. in most of the region, but in the E. swing into the N.E. Caledonian strike.

To the N. lies the Croagh Patrick Syncline (Fig. 5.10), overturned towards the S. so that the Silurian succession on its northern limb is inverted at angles as low as 30° and folded into an antiform (Anderson 1960). The folding was accompanied by epizonal regional metamorphism; this is strongest in the N. and fades out (apart from cleavage) in the Ordovician strata S. of the Silurian. The main folding and metamorphism was followed by N.W. cross-folding. The post-tectonic Gorvock Granite has imposed strong thermal metamorphism on the regional.

Grey and red sandstone and shales N. of the Croagh Patrick Silurian but separated from it by the Highland Boundary Fault have been interpreted as Silurian of a different facies but are likely to be Old Red Sandstone.

The Silurian strata sandwiched between the Ordovician and the Connemara Schists form several wide synclines, with steep northerly limbs and intervening narrow anticlines.

The Salrock Fault is an important fracture which belongs to the post-Wenlock compressional phase. This has an easterly course roughly following Killary Harbour and is here a low-angled overthrust with Ordovician pushed upwards on the N. side. Further E. it becomes a high-angled over-thrust and, like the folds, swings N.E. This fault has been regarded as the continuation of the Southern Uplands Fault.

The whole region is cut by a considerable number of north-westerly and north-easterly faults; transcurrent components of movement can be demonstrated in many cases; some of the faults cut across the Boundary Fault into the Dalradian to the N. It is difficult to distinguish pre-Carboniferous from post-Carboniferous movements, especially as Caledonian compression in the west was N.-S. and therefore could result in the same type of shearing as Hercynian compression. However, as several of the faults are earlier than the Salrock Fault and as differential folding can be demonstrated across others, some of the movement at least was pre-Carboniferous. The most important north-westerly fractures are the Maam-Doo Lough Faults which were initiated in the Ordovician. They have a dextral displacement of 3.5 miles (5.5 km) and 1.7 (2.7 km) miles at Doo Lough and displace the Salrock Fault. They probably continue north-westwards to cut across the western end of the Croagh Patrick Syncline (Anderson, 1960) and to displace the Highland Boundary Fracture-zone. In the competent Mweelrea Grits this zone is marked by two clearly defined dextral tear faults but in the less competent slates, etc., further S. by twisting of the strike and drag-folds.

Hercynian and Tertiary Structures

Post-Carboniferous Limestone movements along north-easterly and north-westerly faults can be demonstrated N.W. of Lough Mask and S.W. of Lough Corrib. It is probable that transcurrent displacement, due to N.-S. Hercynian compression, took place along many of the other N.W.-S.E. faults, as in other regions. Tertiary normal faulting along several of these fractures may also have occurred.

5.6 Ages of Moinian and Dalradian rocks and structures - discussion

Stratigraphical and palaeontological evidence

The presence of Lower/Middle Cambrian trilobites near the top of the succession shows that the Upper Dalradian is of Cambrian age. The base of the Cambrian, often doubtful in non-metamorphic terrains, is even more difficult to identify in a meta-sedimentary sequence. Nevertheless, evidence from acritarchs suggests that the Cambrian extends downwards into the Middle Dalradian. The lower part of the Dalradian is therefore also Precambrian, i.e. Eocambrian or Vendian/Riphean according to varying European usages.

Since the Moinian of the Grampian Highlands (5.4), in places, passes upwards through the Transition Group into the Dalradian - the upper part at any rate of the Moinian Assemblage must also be Upper Proterozoic.

Indirect stratigraphical evidence supports the above views. The glacogenic boulder-bed in the Dalradian of the Grampians and of Connemara belongs to the late Precambrian glacogenic sequence discovered in many parts of the world, although the number of beds varies. In Norway the bed (or beds) has a carbonate horizon below and thick quartzite above; the latter lies not far below strata with Lower Cambrian fossils.

For the Moinian there is no palaeontological evidence and no direct stratigraphical evidence other than unconformably overlying Old Red Sandstone. The Moinian meta-sediments of the Northern Highlands have been regarded by many geologists as the equivalent within the fold-belt of the Upper Torridonian. Since this is known to be late Precambrian - supported by the finding of Upper Riphean acritarchs - the Moinian would therefore be Upper Proterozoic. This would be in agreement with correlation of parts of the Moinian of the Northern and Grampian Highlands across the Great Glen Fault.

Radiometric evidence (discussed below) shows however, that this is too simplistic a view, although this evidence does not exclude such a correlation for parts of the Moinian Assemblage.

On lithological grounds the metasediments of Shetland (5.3) are Moinian/ Dalradian.

In Connemara it has been claimed that the base of the Arenig is an upper limit for the metamorphism of the Dalradian. The contacts are, however, faults (for discussion see 5.5).

Radiometric and structural evidence

Radiometric determinations of Moinian/Dalradian ages have solved some problems but raised others. Some dates are not only at variance with each other but cannot be reconciled with the stratigraphical and structural evidence. Apart from experimental and interpretative errors, the problems of "fast" or "slow" cooling arises.

However, for the Dalradian the stratigraphical, structural and radiometric data are in fair agreement. There are many Dalradian dates around 475 m.y., supporting a view that the D1, D2, and D3 folds and associated metamorphism are mid-Ordovician. This agrees with the structural observation that D1 and D2 deform Arenig along the Highland Border. The Ben Vuroch Granite in Perthshire, which postdates D2, gives 514 m.y. The difficulties are illustrated by the fact that this constrains Bradbury and others (1976) to imply that the 500 m.y. Ordovician time-base is wrong. On the other hand a 510 m.y. date obtained by Pidgeon (1969) in Connemara leads to a suggestion that metamorphism began in late Cambrian (Sardic) times although continuing into the mid-Ordovician.

There are a number of Dalradian (and Moinian) dates around 420 m.y. This could be evidence for metamorphism and folding (perhaps some D3, D4 and subsequent fold-elements) of Silurian age; there is undoubted post-Wenlock folding and epizonal metamorphism in South Mayo (5.5). In fact, some of the higher or older folds could be Devonian as Lower Old Red Sandstone overlying the metamorphics is folded.

For the Moinian of the Northern Highlands (5.2) there are more question marks. There is no doubt that folding and metamorphism of Caledonian age take place but it has still to be discovered to what extent Precambrian tectonothermal events played a part and how much of the Moinian Assemblage consists of pre-Upper Proterozoic sediments.

Dates between 1000 and 1050 m.y. have been obtained for the Ardgour Gneiss (Brook *et al*, 1977). If these dates are correct (and some doubt exists, especially for the Morar pelite) this would show that the pre-Torridonian Moinian sediments shared the Grenville metamorphism of North America - a perfectly tenable hypothesis on plate-tectonic theory.

Other younger, but still pre-Cambrian, dates are those around 730 m.y. (van Breemen and others, 1974) for pegmatites from Morar and other localities. There is evidence that the pegmatites predate or are syntectonic with the local D1 folds. If these are in fact the first folds a problem arises as the metasediments could not in this case have been folded in Grenville times. The relationship of the pegmatites to later folds is not certain, and these folds could be Caledonian. It is also still doubtful whether the Morarian pegmatites mark a major orogenic phase or whether their intrusion and accompanying folding took place locally and at the base of a thick sedimentary pile without markedly interrupting general Moinian sedimentation. Van Breemen and his co-authors (1974) also recognise younger pegmatites dated at around 450 m.y. which corresponds to many similar determinations of the age of Moinian metasediments.

The Carn Chuinneag granite-gneiss, dated at 530 m.y., predates all but the Dl folds in the Moinian, and even these may be contemporaneous (Harper 1970). The folding here is therefore Caledonian.

All the fold-episodes in western Sutherland are held to be later than the initiation of the Moine Thrust and to affect Durness limestone of Ordovician age. This could be reconciled with the Ardgour evidence by assuming that Dl folds in Sutherland are contemporaneous with the younger folds in the S.W.

If all the sediments of the Moinian Assemblage existed in pre-Grenville times most might have been expected to show some effects of that orogeny. A more acceptable view is that there is an undetected, or even undetectable (owing to later events) break between an "Older Moinian" and an Upper Proterozoic "Younger Moinian" equivalent to the 800 m.y. old Upper Torridonian. The 730 m.y. pegmatites do not belie this hypothesis. Nevertheless, if they mark a true orogeny there must be an undetected break within the "Younger Moinian" in the Northern, and probably also in the Grampian, Highlands.

The "Younger Moinian" may include the Loch Eil Division, much of Moinian of Ross-shire, of Sutherland, of the Scottish Grampians and of Ireland. An inlier in the Grantown area is probably Grenvillian and, if the Great Glen Fault is pushed back, this outcrop is much closer to the Ardgour Moinian than at present. "Older Moinian" may also occur in Ireland.

Broad conclusions on age

Much of the Moinian and all the Dalradian form a succession ranging from the Upper Proterozoic (about 800 m.y.) to the Lower/Middle Cambrian. Metamorphism and folding in several episodes took place in the Lower Palaeozoic with a peak in the mid-Ordovician (the Grampian Phase); metamorphism in Connemara probably started in the Cambrian. Folding and metamorphism continued into the Silurian, and folding, thrusting and faulting took place in the Devonian. Parts of the Moinian Assemblage were subject to a 700-750 m.y. old thermal event but to what extend this was an orogenic event causing an undetected unconformity remains to be ascertained.

The Moinian Assemblage probably contains older late Lower Proterozoic sediments affected by Grenville events; the boundary between this part of the Assemblage and the younger metasediments is still to be discovered.

CHAPTER 6

Caledonian Terrains without Caledonian Metamorphism

6.1 The Welsh Block

It will be convenient to consider the North Wales Coalfield (moulded by the Hercynian and later earth movements) and the Longmynd of Shropshire within the framework of this chapter.

The effects of the Caledonian Orogeny (in its widest stratigraphical sense) are seen over much of North, Central and South-west Wales, as far south as the curved line that marks the base of the Devonian System from Pembrokeshire to the Welsh Borders. This Welsh Block has an extensive basement of Precambrian sedimentary, igneous and metamorphic rocks which comes to the surface in Anglesey, Caernarvonshire and Pembrokeshire. Over a very large part of the Block these ancient rocks are covered by a thick Lower Palaeozoic blanket involving all three systems and representing therefore a span of 200 million years. Upper Palaeozoic (Devonian and Carboniferous) strata form the northern, eastern and southern rims of the Block. Triassic rocks are limited in North Wales to the Vale of Clwyd and the eastern fringe of the North Wales Carboniferous.

Large parts of the region have a rugged relief and extensive tracts lie above 1000 ft (305 m) O.D. (see, for example, Brown, 1960, fig. 44). Several mountain units occur. The highest, Snowdonia, contains all of the fourteen peaks in Wales which exceed 3000 ft (915 m) in height, and is noted also for its spectacular, glacially deepened passes. Other prominent ranges include the Rhinogs, Diphwys, Arenig Fawr, Cader Idris and Aran in North-west Wales, Plynlimon in Central Wales, the Denbigh Moors, Clwydian Range and the Berwyn Hills in north-eastern Wales. On the eastern fringe, Radnor Forest and Clun Forest push the north-eastward trending hill-prongs of the Longmynd, Wenlock Edge and Wrekin into the southern side of the Cheshire Plain.

The details of the drainage pattern of Wales, which is at present dominated by easterly flowing (Dee, Tanat, Vyrnwy, Severn), south-easterly flowing (Wye) and south-westerly flowing streams (Dyfi, Teifi, Towy), has suggested to a number of workers (O. T. Jones, 1952: Brown, 1960) that the ancestral river pattern, formed possibly on a Chalk cover, radiated eastwards and southwards from a focus in North-west Wales, but has been subsequently extensively modified during the superimposition of the drainage on to the completed structural pattern of the Palaeozoic rocks.

The succession in the area is as follows:

Triassic (with possibly some thin Permian)
  (unconformity)
Carboniferous
  (unconformity)
Old Red Sandstone
  (unconformity in places)
Silurian
  (unconformity in places)
Ordovician
  (unconformity)

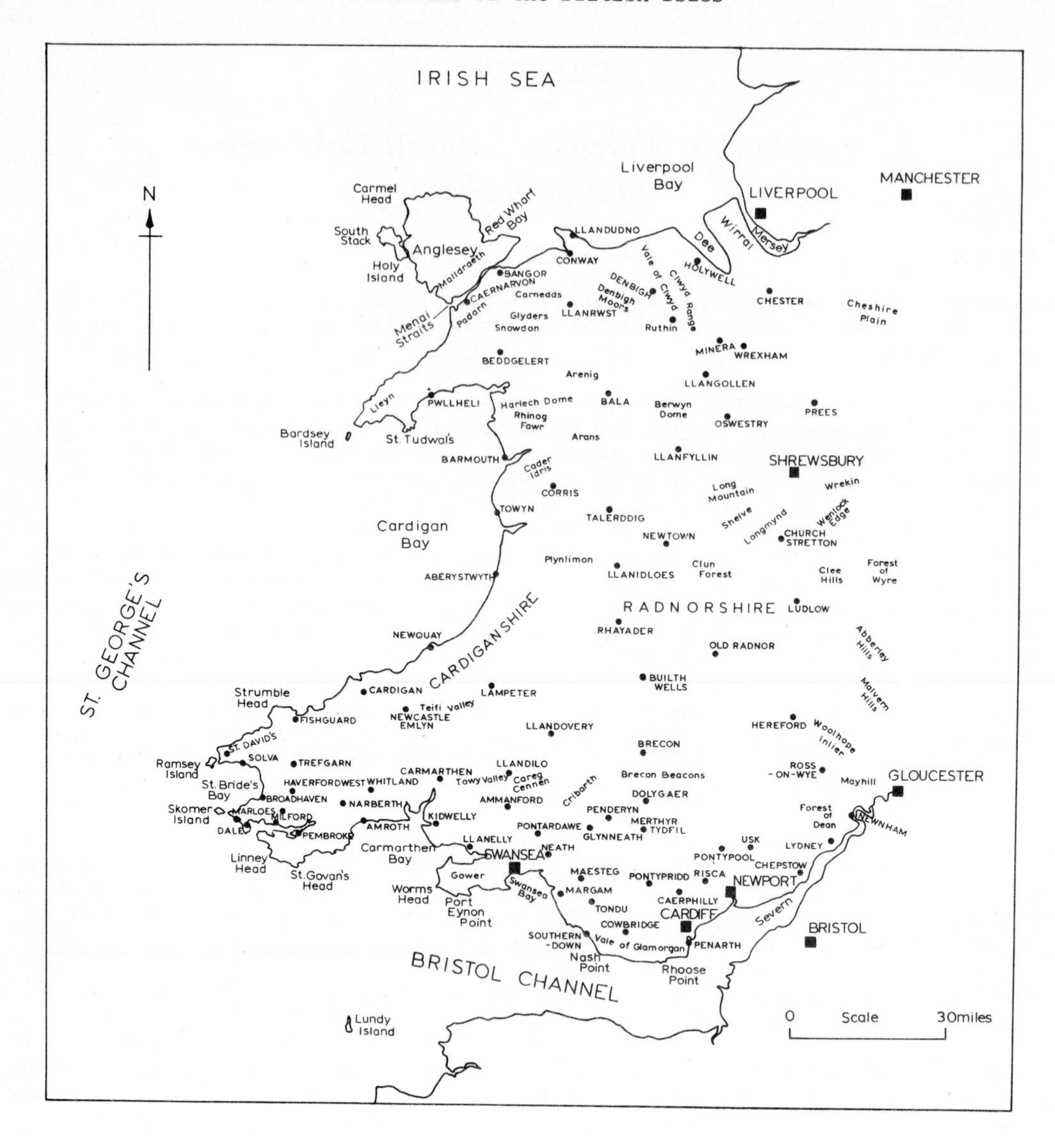

Fig. 6.1. Locality Map For Wales

Cambrian
(unconformity)
Precambrian

The region has been affected by several periods of earth movement, particularly in Late Precambrian, mid-Palaeozoic and late-Palaeozoic times. Of these, the Caledonian Orogeny left the greatest imprint and is still reflected in the present topographical grain of the Welsh Block. Nevertheless, the whole period of Precambrian to Triassic times in North and Central Wales is one of continued unrest (note the uncomformities listed above), and it is therefore convenient to describe the geological structures in their order of evolution.

Precambrian

Precambrian rocks are exposed in Anglesey and South-west Lleyn (Mona Complex), the Bangor-Caernarvon and Padarn ridges (Arvonian volcanics), the Longmynd and Stretton hills (Uriconian volcanics and sedimentary Longmyndian), the St. David's, and Roch-Trefgarn areas (Pembrokeshire). Inliers of Longmyndian rocks occur in Radnorshire.

The Precambrian rocks of Anglesey and S.W. Lleyn have been termed the Mona Complex or the Monian. Greenly (1919, 1920) divided the Complex into three main units: 1, the Gneisses; 2, the Bedded Succession; 3, the Coedana Granite and its related hornfels. The old basement of gneisses was limited in its surface outcrop. Greenly subdivided the Bedded Succession (claimed to be 6000 m thick, and comprising quartzites, schistose greywackes, schists, sodic spilites with pillow structure, soda-rich acid lavas and pyroclastic tuffs) into six units, all to be then involved in appreciable folding and metamorphism (the latter increasing south-eastwards across Anglesey). Final Precambrian phases, according to Greenly's version involved the intrusion of the Coedana Granite, hornfelsing the gneisses and the more metamorphosed "Penmynydd Zone of Metamorphism".

Greenly's interpretation of the order of sequence within the Bedded Succession was to be later completely changed by Shackleton (1954), who also interpreted Greenly's "Gneisses" as merely more metamorphosed and migmatised Bedded Succession. Shackleton, using "right-way up" techniques, placed the Fydlyn Group at the top of his Bedded Succession with the South Stack rocks and their interbedded Holyhead Quartzite at the base of the sequence (see fig. 6.3a, column A). To Shackleton, the Coedana Granite represented the culmination of the metamorphic processes. This granite has now been dated at 615-633 m.y. by Moorbath and Shackleton (1966).

This seemed to be the solution of the Moinian "tangle", but the whole issue has now been raised once again by the recent views of Barber and Max (1979). They have resurrected the Gneisses as once more the early basement of Anglesey. Moreover they say that in the northern part of Anglesey, the Bedded Succession can be separated into 3 *structural* units (see fig. 6.3a, column B). Their South Stack Unit of Holy Island is overlain by *the more highly deformed* (see fig. 6.3b) New Harbour Unit (the contact between the two units being interpreted as a major thrust plane). Their third (and upper) unit, the Cemlyn Unit includes the Skerries and Gwna groups of Greenly and Shackleton, together with the famous Gwna "melanges" (subaqueous olistostromes) and also the Ordovician and Llandovery. They reduce the intensity of the sub-Ordovician unconformity and think the Fydlyn Group could be correlated with the Caradocian volcanics of Parys Mountain. They point out further that the time gap between the deposition of the Gwna Group and the Ordovician is considerably reduced by fossil finds, especially through the demonstration by Wood and Nicholls (1973) that limestone blocks in the Gwna melage contain Vendian to Cambrian stromatolites (Muir *et al*, 1979). Graphitic material from the melange *matrix* has yielded palynomorphs of probable lower Cambrian age.

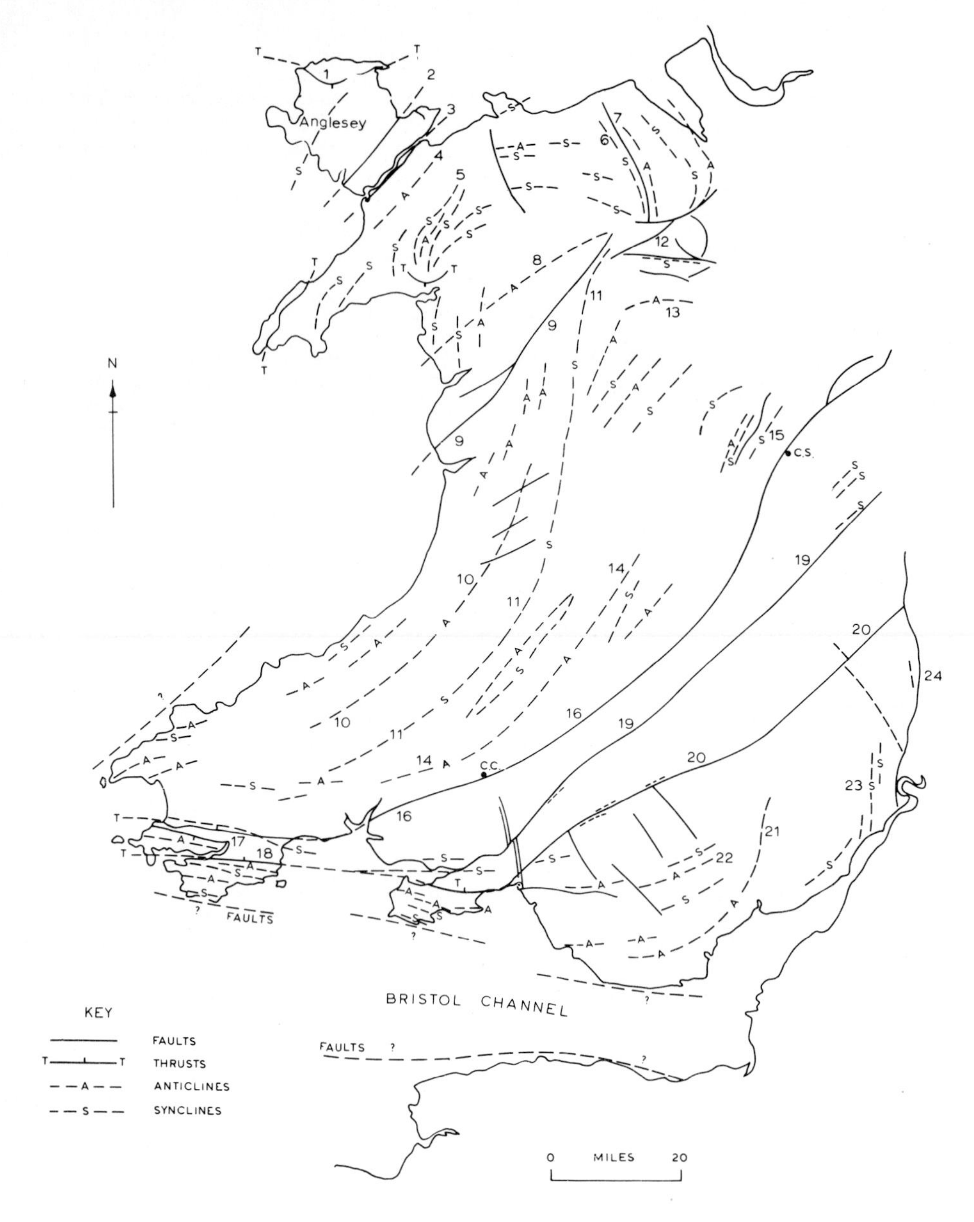

Fig. 6.2. Tectonic Map For Wales. Key: 1. Carmel Head Thrust; 2. Berw Fault; 3. Dinorwic Fault; 4. Padarn Anticline; 5. Snowdon Syncline; 6. Vale of Clwyd Syncline; 7. Vale of Clwyd Fault; 8. Derwen Anticline; 9. Bala Fault; 10. Teifi Anticlinorium; 11. Central Wales Synclinorium; 12. Bryneglwys Fault; 13. Berwyn Dome; 14. Towy Anticlinorium; 15. Longmynd Syncline; 16. Church Stretton-Careg Cennen Disturbance; 17. Benton Fault; 18. Ritec Fault; 19. Swansea Valley Disturbance; 20. Neath Disturbance; 21. Usk Anticline; 22. Pontypridd Anticline; 23. Forest of Dean Syncline; 24. Malvern Fault.

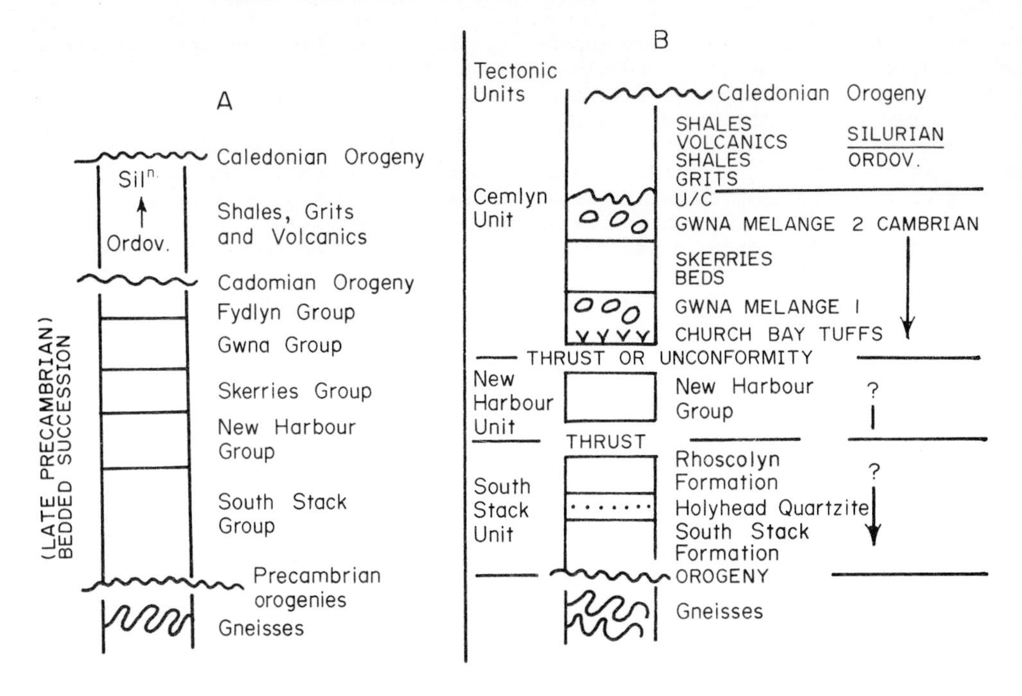

Fig. 6.3(a) Stratigraphical and Structural Sequences in the Mona Complex.
A. Greenly, Shackleton, Baker, etc.
B. Barber and Max (1979).
(Reproduced by permission of the Geological Society).

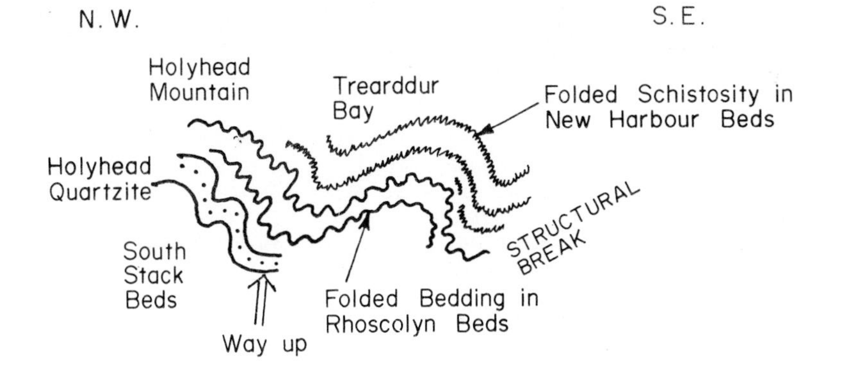

Fig. 6.3(b) Interpretation of the Structure of Holy Island
(after Barber and Max, 1979).
(Reproduced by permission of the Geological Society).

The implications of the above are important. It introduces a considerable Cambrian element into the sedimentary (and plate tectonic) history of Anglesey and of Lleyn As Barber and Max point out, there could then have been no "Irish Sea land mass" in the Anglesey area in the Lower Palaeozoic. Moreover, major structures in the Complex previously attributed to a late Precambrian orogeny are of Caledonian age. Barber and Max's views also introduce a late (Lower Palaeozoic) element into the subduction story for Anglesey. Views about the pattern of subduction on this southern side of "Iapetus" were already varied and controversial (see Thorpe, 1974; Owen, 1976, fig. 20. Anderton *et al* (1979) have also given their own interpretation of the late Proterozoic plate evolution of Southern Britain (their fig. 6.3, part reproduced here as Fig. 6.4). A Cambrian age for the Gwna olistostromes must now make them postdate Coedana intrusion, Longmyndian folding, blueschist metamorphism, etc.

The existence of an ancient basement of gneisses (as once more postulated by Max and Barber) has however been opposed by Beckinsale and Thorpe (1979) who state that new Rb-Sr whole rock isochron data for gneisses from the Holland Arms area indicate a late Precambrian metamorphic episode at 595±12 m.y. and that similar data for the Coedana Granite gives 603±34 m.y. Beckinsale and Thorpe believe that these gneisses are no more than metamorphosed equivalents of part of the Bedded Succession.

In Shropshire, the major Precambrian structure is a gigantic overfolded syncline (fig. 6.5) composed of a thick pile (up to 10,000 m thick) of Longmyndian grits, flags and slates, flanked by the underlying Uriconian volcanics (rhyolitic lavas and pyroclastics, with basalts and intrusions of granophyre and dolerite). James (1956) detected that these Shropshire Precambrian rocks were compressed into a deep NNE.-SSW. trending, overturned syncline, the axial plane dipping westwards at about 70°, with the axis lying within the outcrop of the red Bridges Group (of the Wentnorian or Western Longmyndian). Shackleton believed that this spectacular fold is a local feature, due to a consistently weak crustal zone being always present just west of the Church Stretton Fault (1958). A glance at fig. 6.5 shows that whereas the Precambrian rocks are compressed into this deep syncline, the overlying Lower Palaeozoics are broadly flexured upwards into a major anticline (cut by major fractures, of course, such as the Pontesford-Linley zone, on the west side of the Longmynd, and the Church Stretton zone to the east.

Finally, before leaving the Shropshire Precambrian, one wonders about the floor on which the Uriconian volcanics were deposited. Much older basement could possibly be represented in the small outcrops of garnetiferous Primrose Hill Gneiss and Rushton Schist. Both Baker (1973) and Thorpe (1974) believe this to be the case, but Wright (1969) holds that these 2 areas represent dynamically metamorposed zones of Uriconian volcanics and Longmyndian sediments, respectively. As for the age of the Uriconian, Fitch *et al* (1969) have obtained K-Ar ages of 677±72 m.y. and 632±2 m.y. for rhyolites from the Wrekin. The Longmyndian appears to belong to the very end of the Precambrian and Bath (1974) believes that this thick pile was deformed soon after its deposition.

Cambrian

A thick (up to 15,000 ft, or 4575 m) apparently continuous succession of comparatively shallow-water deposits in Merionethshire, compared with thinner, incomplete successions in north Caernavonshire and in Pembrokeshire implies a down-warping trough flanked by intermittently upwarping landmasses. This interpretation is upheld by the large number of turbidite greywackes in the Merionethshire Succession. Intermittent movements of a marginal shelf area are (probably accentuated near the Pontesford-Church Stretton belts) also exemplified in breaks in the succession in Shropshire, especially at the base and top of the Middle Cambrian.

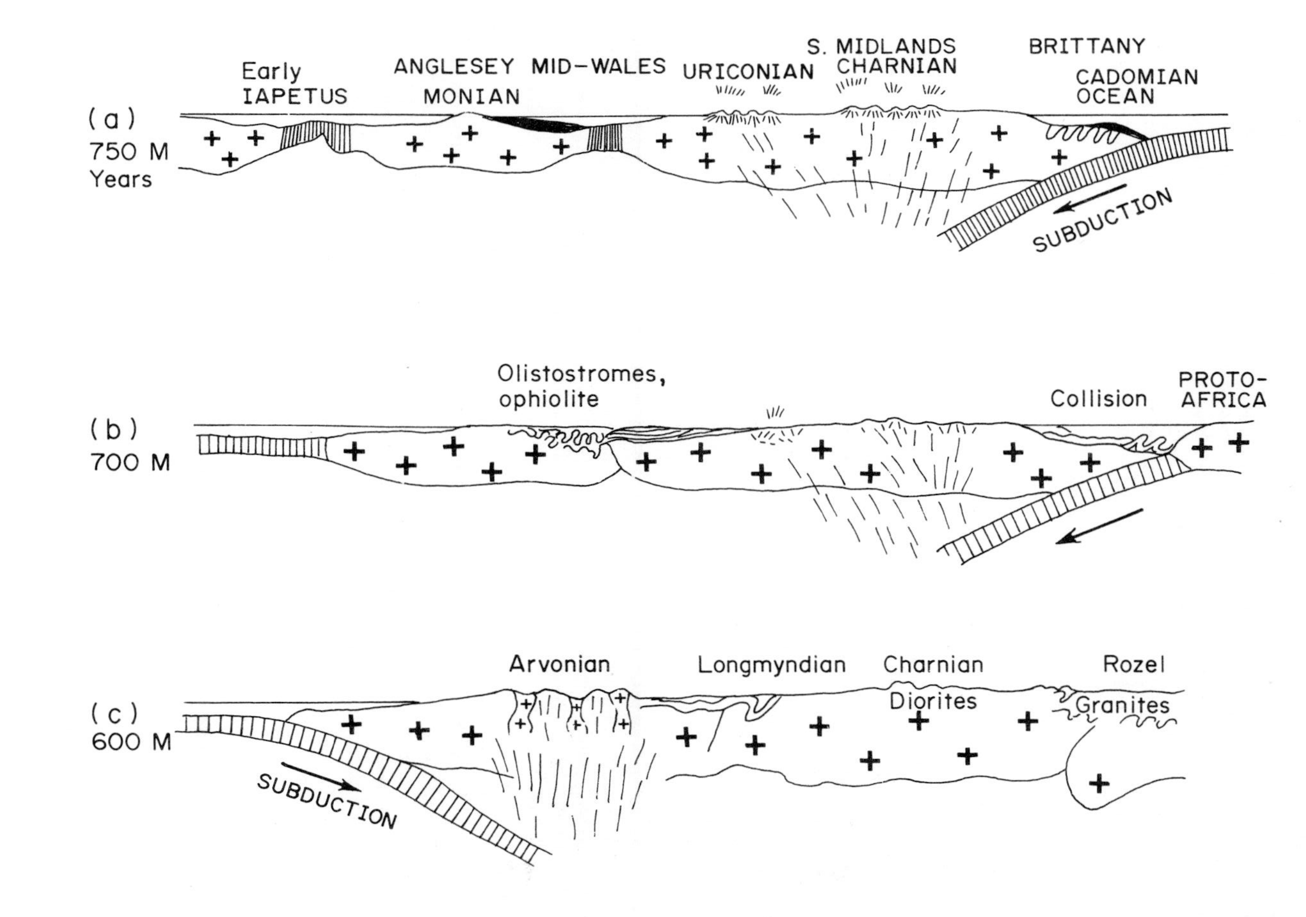

Fig. 6.4. Hypothetical Evolution of the Southern Continental Margin in Late Precambrian Times (based on Anderton *et al*, fig. 6.3).

Ordovician

The most widespread and intense unconformity, however, occurs at the base of the Arenig. Shackleton (1958a) and George (1961a, Fig. 11) have shown the Caledonoid trend of the pre-Arenig uplift in north-west Wales, and according to Shackleton the pre-Arenig dips were of the order of 10°. In Anglesey and South-west Lleyn the Arenig rests on the Mona Complex. The varying intensity of the pre-Arenig gap (Plate 3) between the *localities* of Tremadoc and Arenig signifies that the Snowdon Synclinorium and Harlech Dome were being foreshadowed even at this early stage. In Pembrokeshire, pre-Arenig erosion removed the beds of Tremadoc age, but in Shropshire the amount of this uplift cannot be estimated, because of the eastward-increasing Caradoc overstep across the Longmynd.

Arenig deposits in the Welsh Geosyncline begin with shallow-water conglomerates (very thick in Anglesey), grits or quartzites but gradually give way upwards to flags or shales in the higher Arenig and in the Llanvirn stages. Vulcanicity, part subaerial and part submarine, was intense in the Cader Idris, Aran, Arenig, Manod and Moelwyn areas of North Wales, and also in the Fishguard, Tregarn and Builth areas further south. In Shropshire, vulcanicity was more limited, occurring mostly in the Llanvirn of Shelve. Llandeilian deposits are probably confined in North Wales to the Berwyn Dome (where they comprise 2000 ft (610 m) of shelly facies silts, sands and calcareous sediments, interspersed with acid tuffs and flows). In South Wales, and again in Shelve, the Llandeilian consists predominantly of calcareous flags, vulcanicity being restricted.

Mid-Ordovician uplift and erosion was thus relatively extensive and the (localized) intensity of these movements is shown by the rapid Bala overstep on the Pontesford margin of the Longmynd and again near Tremadoc where a local anticline, over 2000 ft (610 m) high, was removed before Caradocian deposition began. In North Anglesey, along what is now a faulted slice in the Carmel Head Thrust-zone, Basal Caradocian muds were invaded by exotic blocks of mid-Ordovician limestone, probably through violent earthquake shocks. These Bala sediments rest unconformably on the Mona Complex within this remnant klippe and, as George (1963a) has pointed out, nothing remains of over 20,000 ft (6100m) of the earlier Ordovician and Cambrian rocks of the Welsh mainland. Shackleton (1958) sees this great local break as possibly an early upthrust of the Carmel Head Nappe.

On the mainland, Bala deposition was relatively quiet in the south, but was interrupted by violent vulcanicity in Caernarvonshire. The eruptions extruded volcanics ranging from basic to acid (rhyolites predominating) but probably all deritatives of a basaltic magna. Many of the Snowdonian rhyolites have now been interpreted as ignimbrites and these silicic volcanics of Snowdonia then came to be considered as subaerial in origin. More recently however Howells *et al* (1973) have thought that some flows are products of submarine emissions.

Shackleton draws attention to the parallelism in trend of the volcanic belt and the Caledonian structures from Conway to Lleyn. Somewhat earlier Caradocian volcanic activity occurred in the Corris district. In general, the Bala sediments of North Wales were of basin facies in the west and north but of more marginal or shelf type, with intercalated limestones, further east. In South Wales the Towy Anticline appears to have frequently separated the two major Ordovician facies, and in Ashgillian times formed an effective barrier with no deposition to the east. Breaks in Bala sedimentation occur in the Bala district, the west Berwyns and along the Towy Anticline, whilst in the Beddgelert area, mid-Caradocian faulting influenced the position of volcanic piles (Rast, 1961).

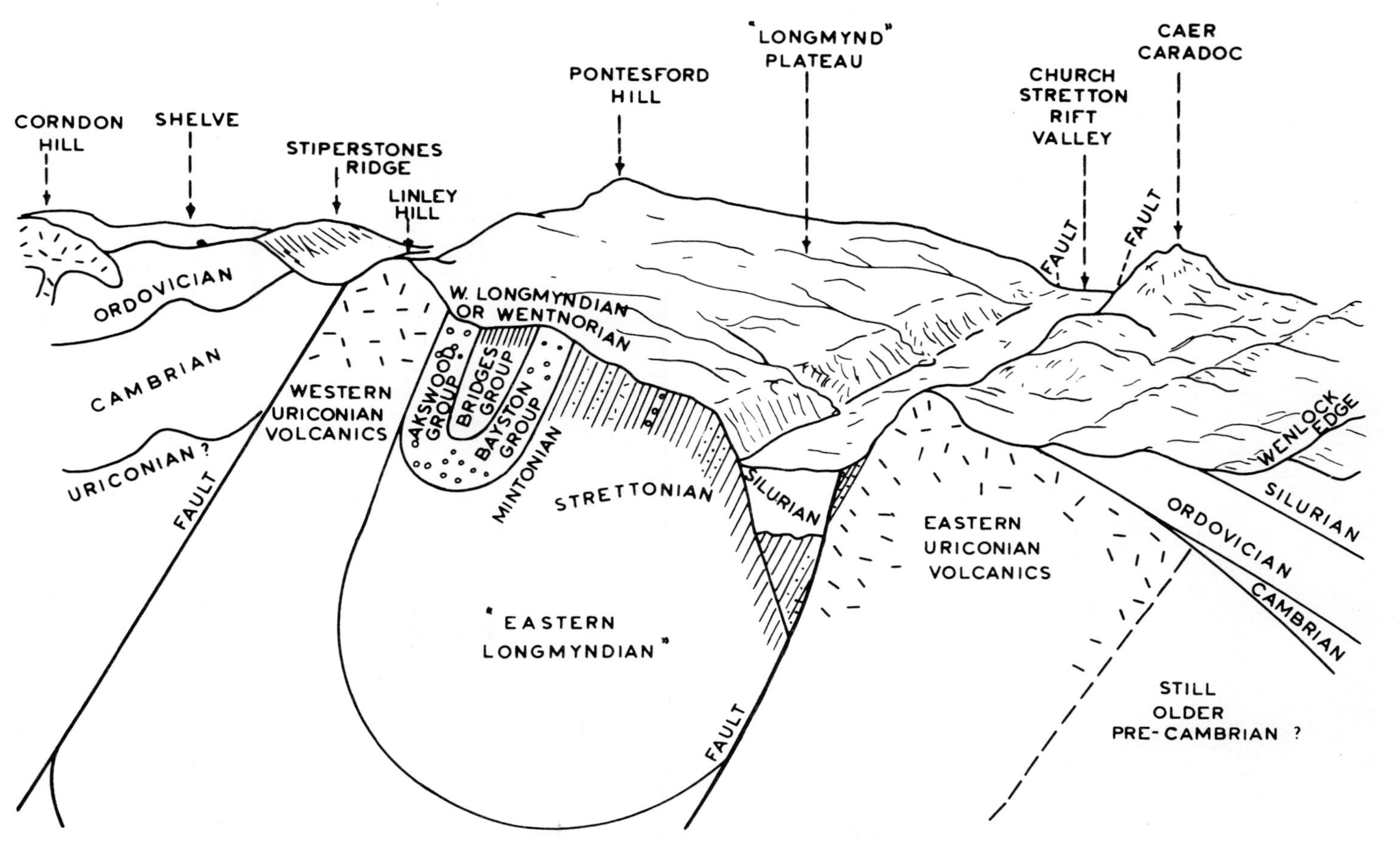

Fig. 6.5. The Structure of the Longmynd, Shropshire (Salop).

A. Ordovician Unconformable on Cambrian, Trwyn Llech-y-doll, St. Tudwal's North-West Wales. (J.G.C.A.)

B. Unconformity of Upper Old Red Sandstone on Silurian Siccar Point, Berwickshire, Scotland. (Geological Survey photograph).

Silurian

At the beginning of the Silurian Period, yet another phase of unrest occurred, these movements reaching considerable intensity on the eastern border of the Welsh Trough but probably dying away inwards. In the Shelve district of Shropshire, NE.-SW. folds with wavelengths of several miles, and limb dips of up to 60°, were produced and strong-wrench faulting occurred (Whittard, 1952). Arching over the Longmynd removed the Bala (and locally any older) beds and the spectacular unconformity of Upper Llandovery on Longmyndian, as at Plowden, resulted. Large-scale movements took place elsewhere along the caledonoid Longmynd-Old Radnor line, removing Ordovician and Cambrian rocks, truncating folds and causing thrusting with large displacement. George (1963a) has hinted at the presence of a pre-Valentian continuation of the Pontesford Fault in the ground west of Old Radnor.

Movements along the Towy Anticline were widespread throughout Llandovery and Lower-Middle Wenlock times (Williams, 1953) with folding, thrusting and even wrench-faulting. The flexuring was not always *along* the major Caledonian structure, being often at right angles, as, for example, the Mandinam Trough (see also George, 1963a, Fig. 12). Bassett (1963) has drawn attention to the views of Davies (1933) and others regarding the cross-folded restriction of Valentian lenticles and rudites along and north-west of the Towy Axis. Boswell (1949) sees the stronger and caledonoid folding of the Ordovician of Caernarvonshire as "Taconic" in contrast to the gentler and later E.-W. folding of the Silurian rocks of North-east Wales. Shackleton (1958a), however, rejects this view and believes that early Silurian movements, even if submarine, were not important in North-west Wales. In Pembrokeshire, pre-Upper Valentian upthrusting occurred along the Benton Fault, rapidly removing what may have amounted to 15,000 ft (4574 m) of Cambro-Ordovician beds after producing an elevated landscape of high relief (George, 1963a, fig. 13). One could say that the climax of the Caledonian Orogeny had already been reached here, as in Shropshire, in early-Silurian times.

That Lower Silurian movements occurred in other areas is suggested by the influx of coarser turbidites from the south-west or south-south-west into the Newquay-Aberystwyth area and slightly later into areas further north-east. In Wenlock and Lower Ludlow times coarse greywackes were carried from the west into the Denbighshire Moors and from the south along the west side of the Berwyn Dome, i.e. along (according to Cummins, 1957) the axes of tectonically controlled troughs now represented by the Snowdon and Central Wales Synclinoria and separated by a submarine ridge, now the Merioneth Dome. The source rocks are believed to be like the Pebidian volcanics of Pembrokeshire. They may, of course, have occupied a large land area occupying the present position of Cardigan Bay.

Devonian

Devonian rocks are restricted to the extreme south-eastern margin of the Welsh Block, where they reach 4000 ft (1220 m) in thickness in Shropshire, and to one outcrop in northern Anglesey, where 1300 ft (396 m) of conglomerates, sandstones and marls occur. It is conventional to think that the Caledonian Orogeny reached its climax in North Wales before the deposition of these redbeds, though McKerrow (1962) suggests this climax could have been at any time between the Upper Ludlow and the Upper Devonian. O. T. Jones (1956) is more forthright and claims that a cover of Old Red Sandstone at least 6000 ft (1830 m) thick extended across a large part of North Wales and "that the characteristic folding of the Lower Palaeozoic rocks of Wales were impressed upon them not at the close of the Silurian but after the deposition of most if not all of the Lower Old Red Sandstone". He claims that the fracture cleavage in the Anglesey Old Red Sandstone was produced with an original Devonian cover of over 12,000 ft (3660 m), and also invokes a thick cover (7000 ft or 2135m) of Salopian rocks over north-west Wales.

Whatever the precise timing of the climax of the Caledonian Orogeny in North Wales, the results are today obvious (Fig. 6.2) and have been clearly outlined by Shackleton (1958a). Some of the major structures produced include from north-west to south-east:

The Carmel Head Thrust
The Padarn Anticline
The Snowdon Synclinorium
The Harlech Dome
The Berwyn Dome
The Central Wales Synclinorium
The Teifi Anticlinorium
The Towy Anticlinorium

The Carmel Head Thrust has a curved E.-W. surface outcrop of 11 miles (17 km) across north Anglesey. The thrust-zone has an average dip northwards of 25° though this varies from 15° to 60° (Anderson, 1951). Striae on the sole thrust at Carmel Head trend S.10°E. (Greenly, 1919). Greenly claimed the movement to be post-Silurian and estimated the forward movement as being at least 20 miles (32 m). In support of this major displacement is the striking difference between the Lower Palaeozoic strata of the Carmel Head Nappe (especially the *gracilis* beds) and those south of it. As Shackleton (1958a) points out, the Precambrian floor was sliced or ruptured rather than folded, and the Palaeozoic sediments firmly welded to the underlying Mona Complex. In other words no decollement took place at the sub-Palaeozoic surface. These fault-cut slices in the basement can, however, in north Anglesey often be traced upwards into shear zones and then quickly into smoothly curved folds in the overlying Ordovician. Anderson (1951) saw the Carmel Head Thrust as a possible southern counterpart of the Moine Thrust. They could well be of roughly contemporaneous origin, though a mid-Devonian age could be as acceptable as a late-Silurian one.

The Padarn Anticline is perhaps better referred to as an anticlinal horst because uplift probably occurred as the incompetent Cambrian mudstones of the south-eastern flank were being squeezed and packed against the older core. The site of the intense cleavage and fault-packed minor anticlinal warps in the famous slate belt was no doubt originally determined by the feather-edge disappearance of the Cambrian north-westwards beneath the transgressing Arenig. Though Shackleton does not believe in "structural sanctuaries", nevertheless it seems that the Padarn mass afforded some protection to the relatively uncleaved Ordovician argillites along the Menai Straits. Within the slate belt, however, the volume was reduced by over 10 per cent, distances in the direction of maximum pressure (NW.-SE.) were halved, and vertical dimensions were almost doubled (Sorby, 1908).

The Snowdon Synclinorium. This complex downfold owes its position to being intermediate between the tough controlling foundational influences of Anglesey and Padarn to the north and the tough Cambrian grits and higher volcanics of Merionethshire to the south. It is a composite structure with a number of *en echelon* downfolds, including the Snowdon Syncline itself, the Arddu, Dolwyddelan and Idwal folds. The Dolwyddelan Syncline trends ENE.-WSW. in the east but bends to about N.40°E. when traced south-westwards. It is a broad, though assymmetrical, structure with limbs dipping at rarely more than 65°. A marked cleavage fan occurs within the synclinorium, the central, vertical cleavage occurring on the northern slopes of Snowdon. The Snowdon Syncline curves under Moel Hebog and bends to a NNW.-SSE. trend towards Tremadoc. Many of these folds plunge southwards. Well out on Lleyn, the Llanor Syncline has an overturned north-western limb near Bodfaen (Shackleton, 1958a).

The Harlech Dome. The swing to a more N.-S. trend of the folds in southern Snowdonia is maintained into the area behind Harlech. The so-called Harlech Dome is a large pericline caused by a combination of N.-S. folds, e.g. the Dolwen Anticline, and a later WSW.-ENE. upwarp - the Derwen Anticline which extends through the Arenigs to just south of Llanbedr on the west coast. This latter structure, according to Shackleton, sweeps right across the deflection marked out by the cleavage and smaller folds and is presumably a later structure. The chief N.-S. fold is the Dolwen Pericline, extending from Trawsfynydd to Bontddu. The rather simple folding of the thick Cambrian grits suggests (Shackleton, 1958a) that they were free to deform without interference from the basement thus implying that the geosynclinal pile must have been thick in this area.

The Berwyn Dome is also complex with structures trending NNE.-SSW. in the west, NE.-SW. in the south and almost WNW.-ESE. in the north-east. The western element, the Berwyn Anticline has an overturned and sheared eastern limb but a long, gently inclined western limb with superimposed smaller-amplitude folds. Shackleton has indicated (1958a, plate XIX) that the reserved limb could pass down into steep reversed faults in the basement. On the southern flanks of the Dome, one of the NE.-SW. anticlines near Llanfyllin is crossed by E.-W. corrugations (O. T. Jones, 1954).

The Central Wales Synclinorium includes a number of elements arranged *en echelon* such as the Tarannon and Llanderfel synclines. The main effect of course is to preserve the narrow Silurian outcrop between the Berwyn Dome and the Bala district. Southwards, as more and more minor folds appear, a broad pitch occurs and the Silurian area swells markedly. These minor folds trend about NNE.-SSW., have northward or southward plunge values of about 10° and the upfolds frequently bring the Ordovician to the surface. O. T. Jones claims that the major Central Wales downfold can be traced as far as the border of Pembrokeshire. In the same very broad sense the north-eastern continuation would be the easterly trending complex Llangollen Synclinorium. Intense minor folding occurs on the northern flank of this downfold, and there is a slight southerly overturning of the southward-facing limbs.

The Teifi and Towy Anticlinoria occur on either side of the southward projection of the Central Wales downfold and bring Ordovician rocks up into the valleys of these two major rivers, more especially in the case of the Towy. The latter structure extends from Carmarthen to beyond Rhayadr, with parallel folds in the vicinity of the Builth igneous complex. The Towy folding is fairly sharp, but the Teifi counterpart is a broader but more frilled structure. This western upfold probably continues as the St. David's Anticline. The occurrence of the Precambrian masses of Pembrokeshire along lines corresponding to the trend of these anticlinal axes has been noted by Pringle and George. These hard basement masses could, like the Padarn ridge, have induced complex upfolding in the overlying Lower Palaeozoics.

When traced from Central Wales into Carmarthenshire and Pembrokeshire, these Caledonian folds swing into a more E.-W. trend (Fig. 6.2) almost at right angles to the folds in the Harlech Dome. Later (Hercynian) deformation in south Pembrokeshire could have appreciably "bent back" what were originally south-westward trending flexures of Caledonian age. It is less easy to explain the equally striking bend that occurs in the trend of the Caledonian structures of Denbighshire and the Llangollen area (Fig. 6.2). The main structure of the Denbighshire Moors is, according to Boswell (1949), a broad east-north-east pitching syncline whose axis runs from Nebo to Denbigh. The structure is, however, quite complex and the severe faulting masks the overall fold pattern. There are many (mostly E.-W.) minor folds - particularly in the areas of Ludlow rocks - with small amplitudes and comparatively long wavelengths. The main downfold axis seems in fact to trend in a more WNW.-ESE. direction well to the south of Denbigh. This makes it begin to bend more into sympathy with the probable, pre-Carboniferous, anticlinal structure under the Clwyd Rift-Valley and again with the broad east-south-east trend of the gentle folds in the adjacent Clwyd Range.

Not only do the folds turn into this more easterly trend in north-east Wales but the broad cleavage pattern does so also (see Shackleton, 1958a, plate xvii). A number of possible explanations come to mind though each one poses difficulties. In the first place the marked S-form of the Caledonian folds in Wales may reflect (a) the form of a subsiding geosynclinal trough and (b) more basically the variation in sub-crustal convective pattern. Secondly, the swing of the north-eastern structures could be due to the presence of a Midland kratogenic block (see Ch. 7.3), the folds beginning to turn around at its northern end (in both north-east Wales and in the Pennines) in a way reminiscent of the Armorican structures in the Dublin area (see Ch. 7.6).

Another possibility is that the change of trend into North-east Wales was controlled by a hard core of older rocks somewhere to the north, in, for instance, what is now Liverpool Bay. This barrier could have approached close to the isoclinally folded and thrust Benarth and neighbouring Conway areas. The mass could even have been a north-eastward projection at depth of the Padarn mass, and, as Shackleton (1958a) suggests, the Benarth belt could be a continuation of that at Nantlle and Bethesda. Hints of the presence of this "northern" mass beyond Denbighshire and Flintshire are given by O. T. Jones (1956, pp. 327-8) for Caradocian and even Llandoverian times. Jones wrote: "it will be noted that in the extreme north of Wales the (Ordovician) isopachytes (plate xvii) trend nearly E.-W., suggesting that a land area may have lain not far to the north". George (1961a) believed there was "a relative southward movement of the surface rocks in north-east Wales to complement the northward underdrive of the deeper and more rigid western structures". Boswell (1949) claimed that the synclinal cleavage fan in the Denbigh Moors became assymetrical during a late Caledonian phase as a result of a superficial overdrive from the north-north-west.

One other possibility is that hinted at by Boswell (1949), i.e. that the more E.-W. trending folds of north-east Wales are later than the ("Taconic?") NNE.-SSW. elements of Snowdonia, etc. O. T. Jones (1954), from a study of the structures implied two successive stress directions, one from an east to east-south-east direction and the other from a more southerly direction. He pointed also to the presence of the E.-W. corrugations, crossing NE.-SW. folds, on the southern edge of the Berwyn Dome and across the N.-S. trending western margin of the Denbigh Moors.

Key areas in the testing of this two-phase explanation are (i) the Conway Valley at the Ordovician-Silurian boundary and (ii) the Central Wales Syncline, again along the junction of the two systems. The first area is unfortunately complicated by faulting,though, as Shackleton suggests, "more detailed structural work is particularly needed to establish the relation between the structures east of the Conway Valley in the Silurian and west of it in the Ordovician". The second critical area includes the Talerddig district which was mapped by Bassett (1955) who discovered both dominant NNE.-SSW. folds and E.-W. flexures (not merely culminations and depressions) in the region. There are also two, similarly aligned, sets of cleavage in the region. Bassett claims that the possibility of the different fold trends being the products of different stages in one orogeny has yet to be proved, but stresses that the *gradation* from the NNE.-SSW. into the E.-W. cleavage indicates that any refolding took place while the material was still in a plastic state. This is a crucial point, and tends to support the virtual contemporaneity of the arcuate structures in North and North-east Wales, i.e. they are all of Caledonian age.

Faults

The fault pattern in the Welsh Block, especially in North Wales, is equally complex and individual elements are often difficult to date precisely. Recrudescence of movement has almost certainly taken place, especially on the large north-east tending faults, much of the movement along these fractures being of post-Carboniferous age.

Strike faults and slides, closely connected with the folding, abound in the Lower Palaeozoic rocks, especially in belts of incompetent strata such as the Cambrian slates. Besides the Carmel Head example, thrusts occur in Snowdonia (Clogwyn Du'r Arddu) and near Tremadoc. Wrench-faults are particularly abundant in the Denbigh Moors (see Smith *et al.*, (1965), an area "shattered into a mosaic of fault-blocks" (Shackleton, 1958a). They also occur numerously in the Harlech Dome. They are relatively rare in the intervening Snowdonian area. In the Denbigh Moors the trend of the main fractures changes from north and north-north-west in the Llanrwst area to north-east and east-north-east in the Ruthin area further east. This veer supports the impression gained of an overall curved downfold axis swinging into a more south-easterly trend in the eastern portion of the Moors. In the Harlech Dome and in Anglesey NNW.-SSE. and NE.-SW. fractures are numerous. Many of the Merionethshire examples were injected by a dolerite magma. A huge bulk of basic magma was in fact injected into the folded Lower Palaeozoic cover, being forced up vertical fissures and spreading on reaching the shales and layered volcanics of Upper Cambrian and especially Ordovician age to form a great sill complex (Shackleton, 1958a). The only acid intrusion of Caledonian age in north Wales is the Tan-y-Grisiau granite, whose contact metamorphosed zone is post-cleavage. Shackleton believes the granite is younger than the dolerites.

Upper Palaeozoic

Whether or not a considerable thickness of Old Red Sandstone was deposited over North Wales, nevertheless uplift and erosion must have occurred in Devonian to early Carboniferous times and the next transgression was a late-Visean one, depositing, after conglomerates, and sandstones, a considerable thickness of limestone - up to 3000 ft (915 m) thick - on the western fringe of the Flintshire Coalfield. George has suggested that the coincidence in position of thinner banked successions with anticlinal cores in the Lower Palaeozoic rocks (e.g. the Cyrn-y-Brain Anticline) could be a sign of continued, if gentle, arching during Carboniferous sedimentation (1961, Fig. 25).

Millstone grit deposition appears to have been basin-type in Flintshire but of more inshore arenaceous character in Denbighshire, and Wood (in Trueman, 1954) has noted the early growth of the (Hercynian) Horseshoe Anticline in defining the boundary between these two facies in Namurian times. The Coal Measures of North Wales are confined to the North Wales Coalfield and the small Malldraeth area in Anglesey. The succession comprises a lower Productive Series and an upper Barren Series of marls, conglomerates, breccias and thin freshwater limestones. Renewed movements caused slight doming in areas like Flintshire, producing uplift and erosion of the higher Productive Measures before deposition of the (overlapping) Red Measures. Similar upwarps with contemporaneous erosion occurred in the southern parts of the North Wales Basin, whilst the Longmynd, too, appears to have either suffered periodic uplifts, or to have continually stood out as an upland mass, during much of Westphalian times, the northern end of the Longmynd not being crossed until Keele Bed times.

These movements were the precursors of the great Hercynian Orogeny which was to follow towards the close of the Carboniferous Period. The earth waves, producing tense buckling in areas further south, broke against what was by now a fairly rigid and resistant Welsh Massif, though crossed in part by Upper Palaeozoic sediments, more particularly on its northern, eastern and southern fringes. Folding was therefore fairly gentle but the brittle "foundation" (in places now up to Silurian levels) gave way frequently along fractures, some old, some new, the sites of some of the latter perhaps being influenced by older foundational structures. The impulse was from the south or south-south-west, but owing to (a) the general hard core of the Lower Palaeozoics and Archaean, (b) the presence of irregularities on the upper surface of this foundation and (c) of Caledonoid structures in this basement, lateral diversions of stress occurred widely. These stress variations produced peculiarly aligned folds such as the sub-charnoid flexures of the Flint Coalfield, Clwyd Syncline and Clwyd Range Anticline (Wedd's Crescentic Anticlines) and much wrench-faulting, especially in the Llangollen-Wrexham area (George, 1961a, Fig. 31) and probably also along the Bala Fault. The ensuing rotation resulted in the production of curved fractures such as the Aqueduct and Minera faults.

The Bala Fault can be traced from the west coast near Towyn (it has now been traced even further west in Cardigan Bay, see Ch. 10) through Bala Lake and thence towards the core of the Derwen Anticline. In this north-easternmost part of its course, the fracture runs into the strike of steeply dipping Lower Palaeozoic strata and becomes difficult to detect, though the dissipation of the movement through the bedding surfaces probably has much to do with its apparent disappearance. The plane of the fault is never seen but the straight course implies a near vertical hade. It has a number of branches, particularly in the Dolgelly district, and in the Bala area (Bassett *et al* (1966) have demonstrated the presence of a number of nearly N.-S. splays on both sides of the main fault. Their work has also shown that movements started along the main belt at least as far back as mid-Ordovician times, a pre-Ashgill unconformity being more intensely developed on the northern side of the fault. Moreover, the nature of the movements has differed with time. The greatest difficulty, as with most large faults, is to demonstrate the amount of relative lateral shift along the fracture. Jehu (1926) suggested a sinistral shift of about 2 miles on the basis of the displacement of the Aran volcanic front and of fold axes near Towyn. Bassett *et al* point out a similar apparent sinistral shift of the faults on either side of the main fracture near Bala but interpret them as second- and third-order shears associated with the main fault. Shackleton (1958a) suggests that the great cleavage arc in North Wales is shifted sinistrally by the Bala Fault, the amount being as much as 8 miles (13 km). No such evidence has been found during the more recent resurvey. Post-Hercynian movements probably also occurred along the fracture or its branches, and earthquakes have occurred in this century near Bala and Dolgelly.

The nature of the Hercynian movements along the two important faults near the Menai Straits is also debatable. The Berw Fault in south-east Anglesey separates Carboniferous, on the north-west, from Monian and Ordovician. Shackleton allows the possibility of wrench-faulting. The Dinorwic (or Bangor-Caernarvon) Fault separates Carboniferous from Arvonian volcanics, again an apparent north-westerly downthrow. This line, however, appears to have suffered a large *south-easterly* downthrow in pre-Arenig times (Shackleton, 1958). At the other extreme, earthquakes occurred here as recently as 1903 and 1906. In the Plynlimon district, faults such as the E.-W. Ystwyth Fault, with a northerly downthrow of 3000 ft (915 m) may be Hercynian, as they cut mineral lodges which could be of post-Carboniferous age (O. T. Jones, 1954). Wrench-faults in the Harlech Dome could also belong to this orogeny.

The Church Stretton Fault-zone trends from the Wrekin in the north-east at least as far as the Kington district and, according to O. T. Jones (1927) and Kirk (1947), is continuous with the Careg Cennen Fault which continues to at least the head of Carmarthen Bay (see Ch. 7.2). The Church Stretton fracture belt has almost certainly suffered sinistral shifts and vertical movements, though Trotter (1947) believes that the Careg Cennen Fault is a major thrust emerging from beneath the South Wales Coalfield. In the Caer Caradoc district, Cobbold proved a complex zone of at least three parallel faults (two vertical and one, central, fault with a north-west hade of 45°). Only the westernmost fracture now cuts Carboniferous or later rocks.

The post-Hercynian structural history of the Welsh Block is the most difficult portion on which to comment, and would involve much guesswork. Post-Triassic down-faulting has certainly occurred, as in the Vale of Clwyd, and some of the tilting of the Clwyd Range is obviously of this more recent date. Large-scale foundering in Tertiary times is also likely to have occurred in areas like Cardigan Bay and off North Wales. Lastly the stairway of high-level and coastal platforms recognized by Brown (1960) and others points to repeated uplifts of the Welsh Block in Neogene times as these platforms are claimed to be unwarped. This late geological history of the Welsh Block has been discussed by George (1974a, 1974b) and by Owen (1976).

6.2 The Lake District and the Isle of Man

The Lake District proper consists mainly of Lower Palaeozoic rocks forming the heart of a Tertiary dome. These rocks, and particularly the Borrowdale Volcanic Series (Llanvirn/Llandeilo), have been eroded into rugged scenery culminating in Scafell Pikes, 3210 ft (979 m), the highest point in England. Structurally, however, the dome affects a wider area; a description is given in the present section, not only of the district between the Solway Firth and Morecambe Bay, bounded to the east by the Pennine Fault, but also of the Isle of Man. The succession is as follows:

Liassic

Triassic

Permian
(unconformity)

Carboniferous
(unconformity)

?Lower Old Red Sandstone
(unconformity)

Silurian	(Ludlow Series (Wenlock Series (Llandovery Series)
Ordovician	(Ashgill Series (Caradoc Series ((unconformity) (Llandeilo Series (Llanvirn Series (Arenig Series
Cambrian	Manx Slate Series

The Manx Slate Series, making up most of the Isle of Man, is probably all of Cambrian age (Simpson, 1963).

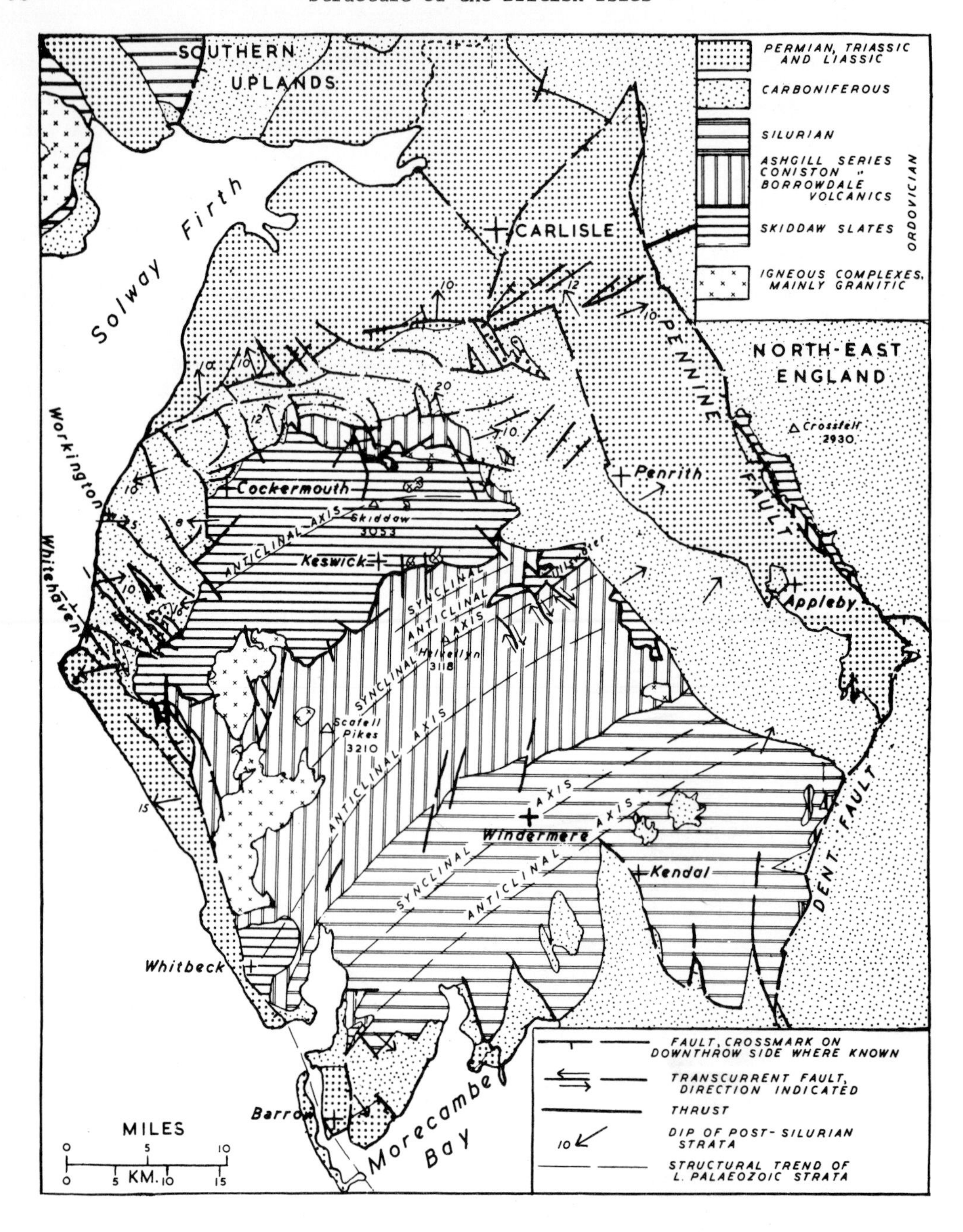

Fig. 6.6. Structural Map of The Lake District.

The Skiddaw Slates, a formation including grits, flagstones and highly cleaved shales or mudstones, form the northern part of the Lake District proper, with detached areas at Black Combe and near Ullswater. They probably belong to the Arenig and Lower Llanvirn, although it has been claimed that the lower part includes Cambrian strata. The main outcrop is followed to the south by the Borrowdale Volcanic Series of andesitic and rhyolitic-lavas, agglomerates and tuffs. These unfossiliferous rocks are held to belong to the Llanvirn and Llandeilo Series. Some of the rhyolites have been claimed to be intrusions. North of the Skiddaw Slates outcrop the Eycott Volcanic Group shows chemical differences from the Borrowdale volcanics. Moreover, micropalaeontological evidence (Downie and Soper, 1972) shows an earliest Llanvirn age making the Eycott group probably earlier on the whole than the Borrowdales. The Caradoc is represented by the Coniston Limestone Series which consists of conglomerates, limestones, calcareous shales, ashes and rhyolite-lava. The Ashgill Series starts with a thin limestone followed by shales. The Silurian consists entirely of sedimentary rocks placed in a number of lithological groups, which palaeontological evidence suggest make up nearly the full succession.

The Lower Palaeozoic rocks are cut by a number of acid to ultra-basic plutonic and hypabyssal intrusions, some of which are of Ordovician date and some of end-Silurian age. Around the complex Carrock Fell intrusion, N. of Skiddaw, there is considerable and varied mineralisation which must have involved later mobilisation as mineral dates are Hercynian or even younger.

Overlying the Lower Palaeozoic and beneath beds of undoubted Lower Carboniferous age there are often conglomerates and grits, the younger of which are probably basal Carboniferous, but the older probably Lower Old Red Sandstone. These are followed by thick limestones and dolomites, overlain by shales and sandstones equivalent in part to the Yoredale Series. The Millstone Grit is poorly developed. There is, however, a fairly full Coal Measure succession in the north-west which includes the Cumberland Coalfield. The highest horizons of the Coal Measures are, however, missing because of pre-Permian denudation.

The Permian consists of red sandstones with "millet-seed" grains - the Penrich Sandstone - of breccias or "brockrams" and of discontinuous beds of Magnesium Limestone. The Triassic contains red shales and sandstones showing much lateral variation; in fact, the boundary between the Permian and Triassic is largely arbitrary. Near Carlisle dark shales and argillaceous limestone of Lower Trias age are present. Lower Carboniferous, Permian and Triassic strata also occur in the Isle of Man.

Structural events were:

Caledonian	(Mid-Ordovician (Late Silurian (? Mid-Devonian
Hercynian	(Pre-Namurian (Sudetic) (Post-Westphalian (Namurian)
Alpine	*(sensu lato)*

<u>Mid Ordovician and late Silurian events</u>

The northerly (Binney-Eycott) succession of the volcanics is probably conformable on the Skiddaw Slates (Moseley, 1975). However, there has been a great deal of debate regarding the relationship of the southerly Borrowdale Volcanics to the Slates. The junction is frequently obscured by drift but in places appears to be

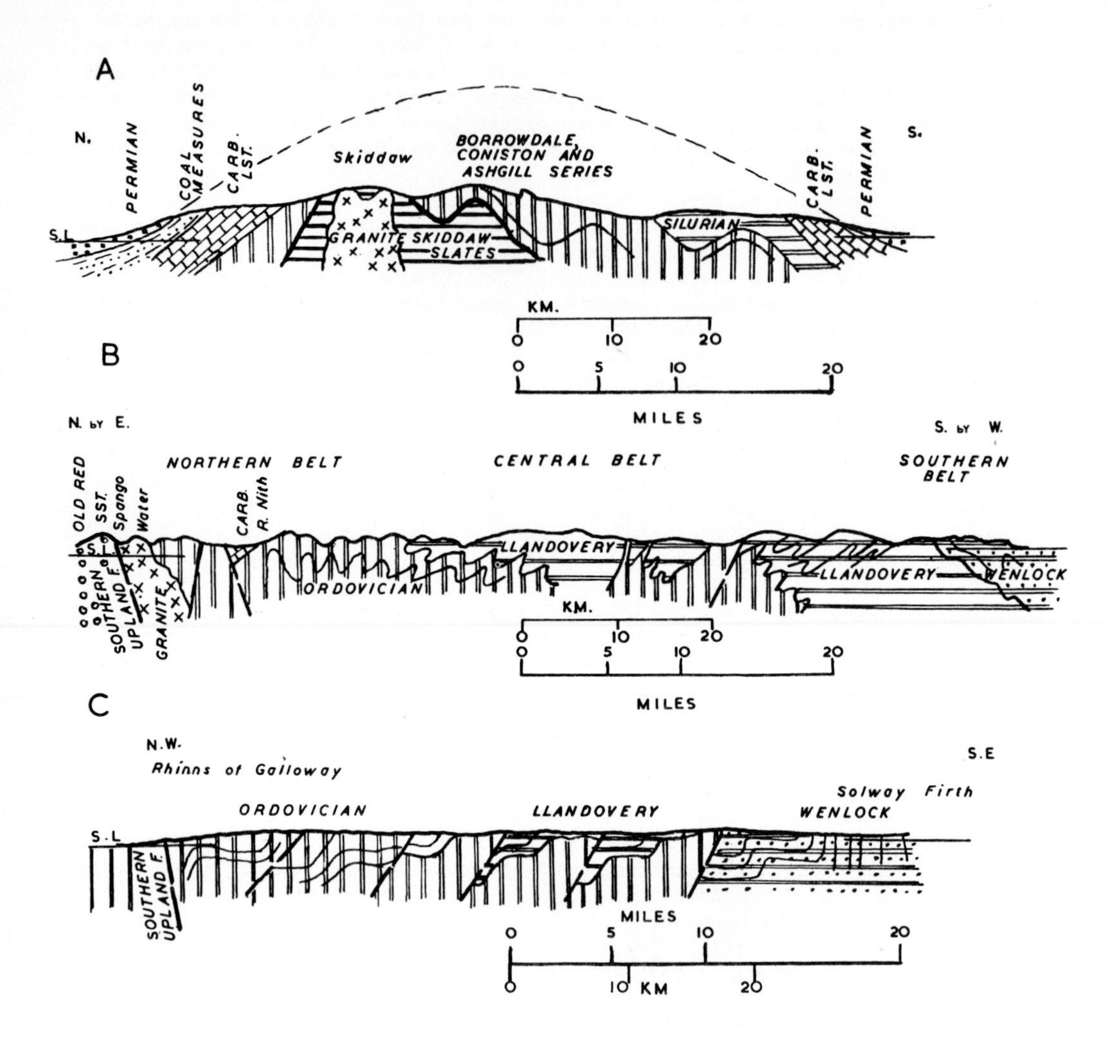

Fig. 6.7. Sections Across the Lake District and the Southern Islands
A. Lake District.
B. Southern Uplands (after Peach and Horne).
C. Southern Uplands (after Walton).

a thrust dipping S.E., e.g. at Ullswater. Moseley (1975) now considers that the junction is an angular but non-orogenic unconformity over the greater part of the S. outcrop. The folding of the two groups is disharmonic, as the tight-folding of the relatively incompetent Skiddaw Slates contrasts with the open folding of the more massive volcanics. Under these circumstances movement at the junction is inevitable. An unconformity at the base of the Coniston Limestone shows that earth movements of pre-Caradoc date took place after the deposition of the Borrowdale Volcanic Series.

The main structures of the Lake District, produced by compression of probably late-Silurian age, trend from N.E. (in the S.W.) to E.N.E.

Soper (1969), while accepting an unconformity at the base of the Borrowdale Volcanic Group concludes that the main deformations, (of which he demonstrates three phases in the Lake District) all post-dated the Borrowdale volcanics and that there was, therefore, no intra-Ordovician orogeny in the region.

In the N. the Skiddaw slates occur in a broad anticlinorium within which the Loweswater Flags (the oldest group, with no exposed base) come up in three or four anticlinal zones with younger groups occupying intervening closely packed synclines. The presence of the Eycott Volcanics in the narrow outcrop on the N. margin of the anticlinorium may be evidence for the oncoming of a Lower Palaeozoic synclinal-zone under the unconformable Carboniferous. The most important structures of the Borrowdale Volcanics are a syncline passing through Scafell and Helvellyn and an anticline to the S.E. through Langdale. Further S.W. near Whitbeck, on the continuation of this upfold, there is an outcrop of Skiddaw Slates forming Black Combe. To the S. of the Langdale Anticline the Silurian occurs in a broad synclinal zone complicated by subsidiary flexures.

The finer-grained Lower Palaeozoic rocks, particularly the shales and tuffs, show strong cleavage. Its relationship to the axes of the main folds is not always simple; as the late-Silurian folding involved three differing stress-phases (Soper, 1969). In the Helvellyn area cleavage is parallel to the axial planes of the folds, and S.E. of Ullswater steep cleavage also strikes parallel to the synclinal fold. In the Dunnerdale district, on the other hand, Mitchell found that the cleavage dips generally N.N.W. agreeing neither with the south-easterly dip of the Coniston Limestone or the north-easterly strike of the Borrowdale rocks.

The Ordovician and Silurian rocks are cut by a number of faults, dominant directions being N.W. and N.N.E; near Ullswater there are also N.-S. and E.-W. fault-trends (Moseley, 1964). These faults may have been initiated during the Caledonian movements, although there are probably also Hercynian and Tertiary components.

In the Isle of Man, Simpson (1963) has shown that the earliest structure is a major north-easterly trending syncline with vertical axial plane; this was followed by a second phase of more open folds, the axial planes of which have the same caledonoid strike but dip gently N.W.; thirdly, there were large open cross-flexures with axial planes dipping steeply E.N.E. After the last fold-phase, a renewal of compression caused sinistral transcurrent movement along N.W. or N.N.W. faults.

?Mid-Devonian events

The evidence for separating Devonian movements from late Silurian movements in the Lake District is tenuous, but the absence of proved Upper Old Red Sandstone strata could point to mid-Devonian disturbance.

Events of pre-Namurian (Sudetic) age

Although palaeontological evidence is incomplete, it seems likely that the Millstone Grit succession is only partially developed in places in north-west England, suggesting that movements occurred in the mid-Carboniferous.

Events of Post-Westphalian (Asturian) age

Unconformity at the base of the Permian shows that important post-Carboniferous movements occurred, but on the whole Hercynian folding in north-west England was comparatively gentle. The Lower Palaeozoic central block was arched into a dome with the Upper Palaeozoic strata dipping off it to the west into the basin of the Cumberland Coalfield. The inclination of the Carboniferous strata does not in general exceed 30° and some of this is due to later Tertiary doming. This coalfield is cut by numerous faults, dominant directions being north-west and north-east. These may have been initiated during the Hercynian movements but as many cut the Triassic their displacement is probably largely Tertiary.

On the east side of the Lower Palaeozoic core the Carboniferous Limestone dips at about 30° into the New Red Sandstone basin of the Vale of Eden. This is bounded to the east by the Pennine Fault, initiated during the Hercynian movements with upthrusting on the west side. Differences in pebble contents between lower and higher Permian breccias have been taken as evidence of further movement of the Pennine Fault during the Permian but it has also been held that this is merely due to local accidents of denudation and sedimentation. Further movements of the Pennine Fault took place in the Tertiary (see below).

Both north-east and north-west faults within the Lower Palaeozoic Block may also be Hercynian or at any rate have important Hercynian components. Moseley (1960 and 1964) has described north-west faults near Ullswater which are thought to have an early phase of dextral or wrench-faulting and a later phase of normal faulting. The normal faulting is associated with mineralization and is probably post-Carboniferous, and it is possible that the wrench faulting may also be Hercynian. Normal faulting of post-Carboniferous age also occurred in the Isle of Man (Simpson, 1963, p.395).

Alpine *(sensu lato)* events

Final uplift of the Lake District dome was almost certainly of Tertiary date. Triassic strata dip away from the Palaeozoic core over almost three-quarters of the periphery and Liassic strata are affected in the Carlisle Basin, where the synclinal amplitude exceeds 5000 ft (1525 m). Moreover, the striking radial drainage, superimposed on the Caledonian trend-lines of the Lower Palaeozoic, can best be explained by Tertiary initiation on a thick Mesozoic cover.

Bott (1974) attributes the uplift to a mass deficiency of a composite granite batholith underlying the Lake District and connecting exposed granites. A negative gravity anomaly supports this view.

Numerous faults, mainly with north-westerly trend, cutting the Cumberland Coalfield and extending into the Triassic, are probably of Tertiary age. Further south a powerful north-north-west fracture brings down Triassic of the coastal strip against the western edge of the Lower Palaeozoic Block. The main movement of the Pennine Fault was also Tertiary, resulting in the downthrow of the New Red Sandstone of the Vale of Eden against the Carboniferous Limestone to the east by not less than 4000 ft (1220 m) in reverse sense to the Hercynian displacement.

Gravity measurements in the Irish Sea (Bott, 1964) suggest that thick Carboniferous, New Red Sandstone and possibly post-Triassic rocks form the major "fill" of a number of deep basins separated by horst-like areas of Lower Palaeozoic of which the Isle of Man forms a part.

6.3 THE SOUTHERN UPLANDS

The Southern Uplands are important in the history of structural geology for they include the unconformities of Upper Old Red Sandstone on Silurian at Siccar Point (Plate 3) and near Jedburgh made famous by Hutton. Moreover, it was in this region that Lapworth discovered the zonal value of graptolites and from the stratigraphy went on to interpret the structure. The Caledonian suture probably runs under the Solway Firth between the Southern Uplands and the Lake District (Fig. 1.1). The wide separation of these two blocks in Ordovician times is supported by palaeomagnetic evidence (Piper 1978).

The Southern Uplands include a number of hills over 2500 ft (762 m) high, mostly with smooth, grass-covered slopes. More rugged scenery and the highest point, the Merrick 2764 ft (840 m), occur in the S.W. where Lower Palaeozoic strata have been altered by granite.

The succession is as follows:

	Triassic
	Permian
	(unconformity)
	(Coal Measures
	(Millstone Grit
Carboniferous	(Carboniferous
	(Limestone Series
	(Calciferous Sandstone Series
	(Upper Old Red Sandstone
	((unconformity)
Devonian	(Lower Old Red Sandstone
	((unconformity)
	(?Ludlow Series
Silurian	(Wenlock Series
	(Llandovery Series
	(Ashgill Series
	(Llandeilo Series
Ordovician	((unconformity)
	(Arenig Series

Since Lapworth demonstrated that 200-300 ft (60-90 m) of pelitic sediments near Moffat were deposited at the same time as 2000-3000 ft (610-915 m) of coarse sediments in the northern part of the region (and in the Girvan area, 7.5) it has been realised that the Lower Palaeozoic strata of the Southern Uplands provide striking examples of facies-variation. A number of modern studies have been directed to this subject and to related problems of provenance and sedimentation. In the present context, however, only a general account of the stratigraphy will be given.

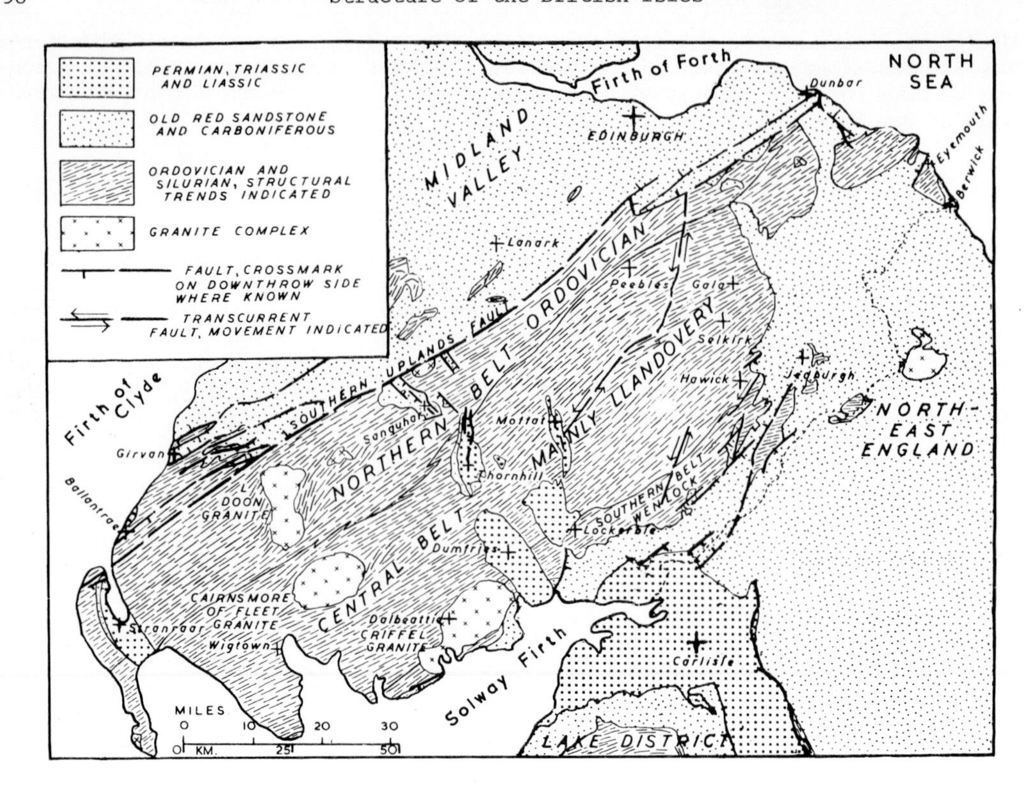

Fig. 6.8. Structural Map of the Southern Uplands.

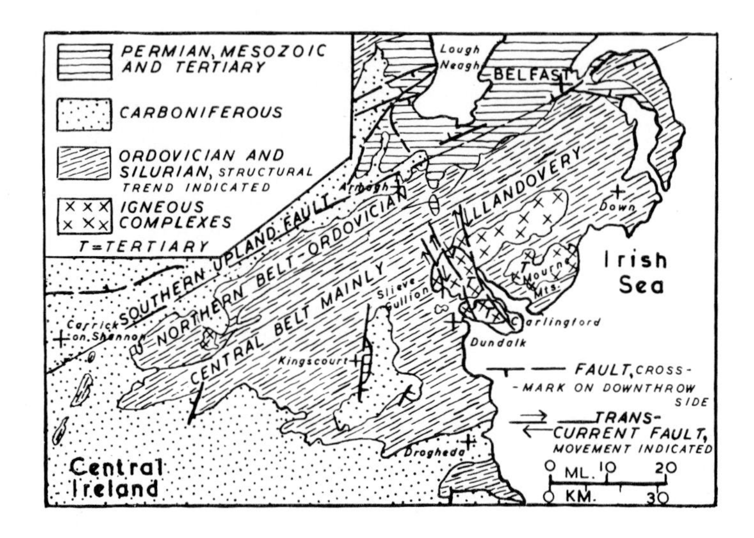

Fig. 6.9. Structural Map of the "Southern Uplands" in Ireland.

The Ordovician rocks are seen in the northern and central belts of the Southern Uplands (Fig. 6.8) and the Silurian in the southern belt (see below). Arenig strata are common in anticlines in the northern belt and consist of mudstones and cherts with lavas (including pillow-lavas), tuffs, and agglomerates. There is a sedimentary break above the Arenig but its extent is disputed. Some authors hold that the Caradoc rests directly on the Arenig (cf. the Girvan district further N.W. 7.5); other authors believe that the Upper Llandeilo is present. It is, however, clear that in the central belt and the southern part of the northern belt the Upper Ordovician consists of a thin succession of dark peilites - the Glenkiln and Hartfell shales - while further N.W. it consists of thick greywackes, shales and conglomerates with occasional volcanics.

In the central belt the Hartfell shales are followed without structural, but with marked palaeontological, break by the Birkhill shales which, together with thick flaggy greywackes and grits, belong to the Llandovery Series. The Wenlock Series occurs in the southern belt and consists of shales, greywackes, grits and conglomerates: it is doubtful if any Ludlow strata occur in the Southern Uplands.

The Lower Old Red Sandstone includes conglomerates, sandstones and volcanic rocks, mainly andesites: the Upper Old Red Sandstone consists of conglomerates and red sandstones in the lower part and paler sandstones with cornstones in the upper part.

Carboniferous strata occur on the fringes of the Uplands and in a few interior basins. The succession is essentially similar to that of the Midland Valley (7.5). East coast sequences also show resemblances to the Lower Carboniferous of North-East England. At Cove thin coals are correlated with those of the Scremerston Coal Group (7.4).

The largest outcrops of Permian rocks in Scotland occur in the Southern Uplands, consisting of breccias, red, dune-bedded sandstones and local basalt lavas. Some of the sequences range down into the Stephanian (Brookfield, 1978). The Triassic is confined to a smaller area north of the Solway, and includes red marls, shales and sandstones.

A few intrusions, mainly of dolerite, are associated with the Arenig volcanics. The most important intrusives of the Southern Uplands are the calc-alkaline plutons which post-date the main Caledonian folding and may be very late Silurian or early Devonian. Associated with some of the plutons are dykes of felsite, porphyrite and lamprophyre. There are also acid and basic hypabyssal intrusions of Carboniferous age, including alkaline types, and north-westerly basic dykes of Tertiary age.

Structural events were:

Caledonian	(Mid-Ordovician (Late-Silurian (Structures due to post-tectonic (Caledonian intrusion (Mid-Devonian
Hercynian	(Pre-Namurian (Sudetic) (Post-Westphalian (Asturian)
	Alpine *(sensu lato)*

The S.W. striking Southern Uplands Fault (Fig. 6.8) forms the N.W. boundary of the region for 130 miles (208 km) from the North Sea coast near Dunbar to the Irish Sea, with an en echelon shift towards the N., S. of Edinburgh. Much remains to be learnt about this generally drift-covered fracture which appears to be reversed, although this is not certain. The main movement, with downthrow towards the N.W., was post-Lower Devonian and probably, like that along the Highland Boundary, mid-Devonian. Later movements (see below) also occurred.

Mid-Ordovician events

Post-Arenig/pre-Caradoc movements took place in the Southern Uplands; these cannot be dated with certainty, as there is doubt whether any Llanvirn and Llandeilo are present (see above). At the end of the Ordovician there was also disturbance; although there is no angular discordance there is palaeontological hiatus between the Hartfell and Birkhill shales. Furthermore, at least part of the Ordovician of the northern part of the Uplands rose about this time to the level of erosion for the Silurian sediments, derived according to current determinations from the N.W., contain material of Ordovician (and of metamorphic, perhaps Dalradian) provenance.

Late-Silurian events

The main folding in the Southern Uplands, resulting in the production of dominantly north-easterly structures, was post-Wenlock and pre-Lower Devonian.

Following Peach and Horne (1899) it has long been accepted that the Lower Palaeozoic strata form an anticlinorium in the north and a synclinorium in the south, made up of numerous isoclines. The structural units comprise a northern belt, extending from the Lammermuir Hills to Loch Ryan and consisting of Ordovician rocks, a broad central belt, from St. Abb's Head to the Mull of Galloway, made up of the Llandovery Series with the Ordovician occurring in long narrow anticlines, and a southern belt from Jedburgh to Kirkcudbright Bay occupied by the Wenlock strata.

The Lower Palaeozoic strata of the Southern Uplands do not, in general, show cleavage comparable with that developed in strata of the same age in Wales. Nevertheless, strong cleavage, amounting almost to schistosity, has been developed in narrow zones, for example in the Leadhills area.

As was shown by Lapworth (1878) and by Peach and Horne (1899), strike-faults are developed in association with the folds (Fig. 6.7). Recent workers, including Craig and Walton (1959) and Walton (1961, 1963), have shown that these play a major role and that the structural pattern is essentially that of north-westerly facing monoclines with strike-faults bringing up older strata to the N.W. One of the most powerful faults within the region is the Ettrick Valley Thrust which brings up the Moffat shales to the N.W. and carries wedges of this formation south-eastwards across the younger strata of the Gala Group (Fife and Weir, 1976).

Three phases of Caledonian folding have been recognised in some areas (Lindstrom, 1958; Kelling, 1961; Weir, 1968). The F1 structures may have formed a decollement on an oceanic basement (Weir, 1979).

Structures due to post-Tectonic Caledonian Intrusion

On both sides of the large dumb-bell-shaped Loch Doon Granite Complex the north-easterly Caledonian structures swing through angles of 30° towards a N.-S. direction. Diversion of strike of 25° also occurs further S. on the north-western side of the Cairnsmore of Fleet mass. These are tectonic effects due to horizontal pressure accompanying magmatic intrusion. Dykes round the Cheviot granite, intruded into Lower Devonian lavas on the Border of England and Scotland, suggest radial tension accompanying intrusion.

Mid-Devonian events

The marked discordance between the Lower and the Upper divisions of the Old Red Sandstone in the Eyemouth district shows that the mid-Devonian folding, so strongly developed in the Midland Valley, affected at least part of the Southern Uplands.

Elsewhere the Upper Old Red Sandstone lies directly on the Lower Palaeozoic, for example at Hutton's classic Siccar Point unconformity (Plate 3).

Pre-Namurian (Sudetic) events

Instability of the Southern Uplands during mid-Carboniferous times is shown by the manner in which the Upper Carboniferous of the Sanquhar, Thornhill, Spango and Canonbie basins rests unconformably on the Lower Carboniferous; the Upper Carboniferous in fact overlaps at several localities onto the Lower Palaeozoic. The angular unconformity between Lower and Upper Carboniferous is, however, slight. A gravity survey of the Sanquhar Coalfield (McLean, 1961) suggest the presence of older Carboniferous strata under known Coal Measures, lying in a pre-Westphalian basin trending north-west and antedating the later Hercynian fold. It also suggests that pre-Upper Carboniferous movements took place along a north-easterly fault near Kirkconnel.

Post-Westphalian (Asturian) events

All the Upper Carboniferous rocks of the Southern Uplands show folding which can be attributed to late-Carboniferous stress. In the Sanquhar Coalfield folding is north-westerly, in the Thornhill Basin northerly. North-easterly folding occurs in the Carboniferous on the north shores of the Solway, and at Cove the Lower Carboniferous strikes N.E. with dips of up to 60°. Such variations suggest that the effects of Hercynian compression were controlled by the different frameworks within which the Carboniferous rocks were deposited.

A number of north-easterly to north-north-easterly and north-westerly to north-north-westerly faults cut the Southern Uplands. In some cases these displace Upper Carboniferous rocks, and they may therefore be shears, i.e. transcurrent faults, developed in the relatively rigid block by N.-S. Hercynian compression.

Alpine *(sensu lato)* events

In the Thornhill, Canonbie and Loch Ryan basins folded Carboniferous rocks are unconformably overlain by Permian. In fact, folding, of likely Tertiary date, on north-westerly axes affects the Permian basins of Dumfries, Lochmaben and Loch Ryan and also, probably, the Sanquhar Coalfield. In the Thornhill and Annan basins folding took place on more northerly axes. The Triassic north of the Solway dips southwards into the Carlisle Basin.

Post-Permo/Triassic faulting has also occurred, in some cases along older fractures. The fault near the E. coast of Loch Ryan (Kelling and Welsh, 1970) has a probable westward downthrow of 5000 ft (1525 m)-it causes an offset of the Southern Uplands Fault. Tertiary tension-fissuring is shown by the presence of several far-flung dykes of the Mull Tertiary swarm which traverse the region.

Lumsden and Davies (1965), from levels of the buried channel of the River Nith, have suggested that there may have been renewal of movement along the Southern Uplands Fault in Pleistocene times, amounting to vertical displacement of nearly 200 ft (61 m).

6.4 The "Southern Uplands" in Ireland

Lower Palaeozoic strata, the continuation of the Southern Uplands of Scotland, form a large triangular outcrop S.W. of Belfast (Fig. 6.9). The nearly straight N.W. side is determined by the Southern Uplands Fault, marginally overlapped by Upper Palaeozoic strata. This fault certainly continues as far as Carrick-on-Shannon (McKerrow, 1959) although Leake (1963), on geophysical grounds, suggests it goes on at least as far as the neighbourhood of Galway. The less regular southern margin, extending from near Carrick-on-Shannon to the Irish Sea near Drogheda, is determined by the unconformable overlap of the Carboniferous Limestone of Central Ireland. The region, as a whole, has lower relief than the Southern Uplands, and parts are heavily drift-covered. Tertiary intrusions, however, form mountainous country near the Irish Sea.

The succession is as follows:-

	Tertiary Lavas
	(unconformity)
	Triassic
	Permian
	(unconformity)
	Coal Measures
	Millstone Grit
	(unconformity)
	Carboniferous Limestone
	(unconformity)
Silurian	(Wenlock Series
	(Llandovery Series
Ordovician	(Ashgill Series
	(Caradoc Series

The Ordovician consists mainly of greywackes and siltstones of turbidite facies which are largely unfossiliferous, although a sparse shelly fauna, including corals and brachiopods, has been found (Wilson, 1972, 19).

The Ordovician forms a strip, up to 6 miles (9.5 km) wide in the N.W., and also appears in a number of anticlinal inliers within the Silurian further south. At a few localities black graptolitic shales similar to the Glenkiln/Hartfell shales, occur. A basic igneous rock with vestiges of pillow structure, of probably Caradoc age, is present at Helen's Bay.

The Silurian, outcropping S.E. of the Ordovician, consists of similar rocks, the separation being based mainly on sporadic occurrences of graptolites. Most of the fossils are Llandovery, although there is some evidence for the presence of the Wenlock. Bedded tuffs have been recorded at two localities.

Lower Carboniferous strata overlap the landward margins of the district and form a considerable area in the Kingscourt Outlier. The succession, which often starts with thick grits and sandstones, generally includes both Tournaisian and Visean strata similar to those of the neighbouring parts of Central Ireland. To the north, however, the Tournaisian is mostly missing.

In the Kingscourt Outlier there is Namurian up to 2000 ft (610 m) thick consisting of pebble-beds, sandstones and shaley marine bands. The same outlier also contains a thin succession of Coal Measure (Ammanian) argillaceous sandstones with subordinate shales and siltstones.

New Red Sandstone (Permian and Triassic) strata occur in the north-east of the area and in the Kingscourt Outlier. In the latter some 550 ft (168 m) of strata, consisting of a basal conglomerate followed by grey shale with traces of plants and marls with two thick gypsum/anhydrite beds, are referred to the Permian. These are followed by about 1500 ft (457 m) of sandstones, conglomerates, breccias and marls classified as Triassic.

A tongue of the main Tertiary basalt outcrop of Antrim overlaps on to the Silurian S. of Lough Neagh, and outliers continue as far S. as Kingscourt.

The Silurian rocks are cut by the Caledonian Newry granite and by the Tertiary complexes of Slieve Gullion, the Mourne Mountains and Carlingford (Ch. 9), and by numerous Tertiary dykes.

Structural events were:

Caledonian	Late Silurian
Hercynian	(Pre-Namurian (Sudetic) (Post-Westphalian (Asturian)
	Alpine *(sensu lato)*

Late Silurian events:

Although the pre-Caradoc folding of the Scottish Southern Uplands probably extended into north-east Ireland, this cannot be proved, as Lower Ordovician rocks have not yet been recognised.

Both the Ordovician and Silurian strata are strongly folded along north-easterly axes; in fact, the main Ordovician outcrop and the Silurian outcrop about as far south-east as a line passing south-westwards through Downpatrick are held to represent the continuation of the Southern Uplands anticlinorium and the rest of the Silurian the prolongation of the synclinorium. Ordovician appears in periclinal inliers within the Silurian. Recent developments in the tectonic interpretation of the South Uplands (5.6) may, however, necessitate reconsideration of the Irish structures.

Three compressional phases have been recognised in the Silurian strata of the County Down coastal sections (Wilson, 1972). During the first phase N.W. maximum stress generated F1 isoclinal folds with axes trending N.E. and extensive strike-faulting. The second and third phases resulted in the formation of cross-folds. The F2 folds trend N.W. and the F3 folds, fewer in number, have N.N.W. axes. The isoclinal folds have a well-developed axial-plane cleavage strong enough to allow the mudstones to be used as roofing slates. In fact, in some areas, the presence in both mudstones and greywackes of sericite, chlorite and muscovite indicates low-grade regional metamorphism; schistose texture is locally developed.

Pre-Namurian events

It is possible that a trough with caledonoid trend developed in the Kingscourt area and that this further evolved in mid-Avonian times. Movements of the latter age are evident from the spread of the Visean sedimentation around the end of the Lower Palaeozoic Block and along its N.W. margin. Further development on the Kingscourt trough took place in post-Avonian times as is shown by the overlap of the Namurian.

Post-Westphalian events

Late Hercynian folding and faulting affected the Carboniferous on the margins of the region, but the only good evidence within the block comes from the Kingscourt Outlier. Here Namurian and Coal Measures, resting discordantly on Lower Carboniferous, are themselves unconformably overlain by Permian.

Alpine *(sensu lato)* events

Tertiary folding and faulting are well authenticated N. of the present region and these movements also affected the Lower Palaeozoic Block, leading to development of the Strangford Lough trough and the further deepening of the Kingscourt Basin. On the W. side of the latter a northerly fault was formed with downthrow to the east of 1200 to 2000 ft (365-610m). Tertiary tectonic effects in the neighbourhood of the igneous complexes are described in Chapter 9.

6.5 South-East Ireland

This is largely an area of Lower Palaeozoic rocks (Fig. 6.10) occupying the counties of Wicklow and Wexford. These older rocks were later invaded by a number of Caledonian intrusions of which the most important is the Leinster Granite, the largest of its kind in the British Isles. The intrusions probably accompanied the later deformation stages of the Caledonian earth movements in this part of Ireland. Tremlett (1959a) and Brindley (1957) have unravelled the detailed phases of this complex structural and intrusive history. Apart from the belt of contact metamorphism around the Leinster Granite and the cleavage imposed on the more argillaceous Lower Palaeozoic strata, the rocks have suffered none of the great alterations so evident in the north-western parts of the Irish area.

In Devonian times, the area may well have behaved in a more positive manner, and the original cover of Old Red Sandstone was either thin or petered off around the margins of this "Leinster Massif". There is, however, the possibility that more vigorous Old Red Sandstone deposition occurred again on the eastern side of the massif, i.e. in the St. George's Channel area. Present studies of the geology of the British Seas are constantly showing that the "blocking-out'" of hard areas was accomplished by late-Palaeozoic (and in some cases even somewhat earlier) times. In Lower Carboniferous times, too, the Leinster Massif behaved firstly as a major headland (possibly resurrected by uplift in later-Devonian times), but became gradually drowned under the advancing Dinantian sea. George (1960a) has suggested that "the heights of the massif, if subsidence was effectively eustatic, was of the order of thousands of feet. Within the area of the Wexford outcrops in a distance of 20 miles or less from Hook Head, the north-eastward overlap of the Old Red Sandstone, about 1000 ft thick, and of the Lower Limestone Shales and the lower part of the Main Limestone, together about 650 ft thick, indicates subsidence of the Leinster coastal slopes of some 1600 ft".

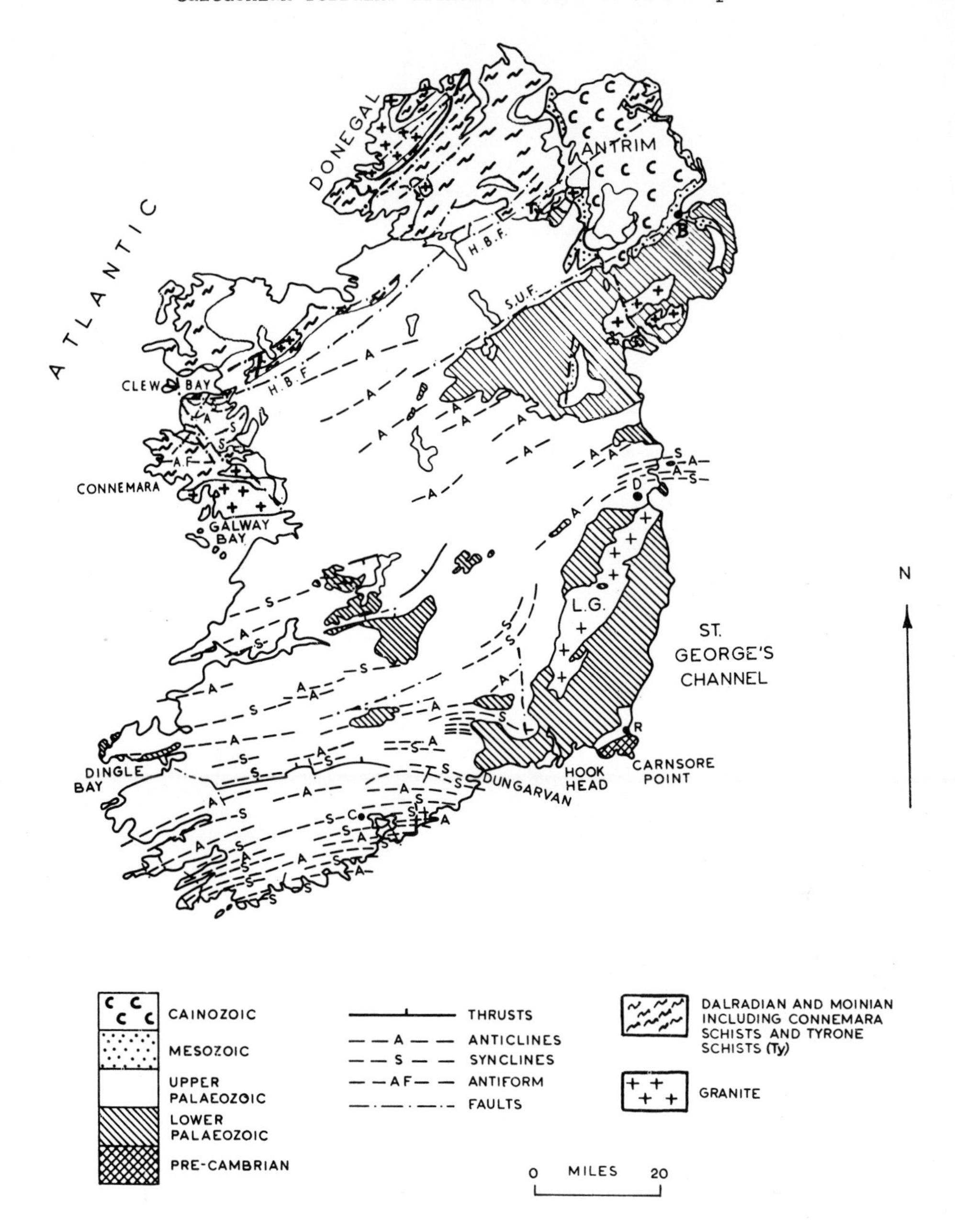

Fig. 6.10. Tectonic Map of Ireland. Key: B, Belfast; C, Cork; D, Dublin; R, Rosslare, L.G. Leinster Granite; L.F. Leannan Fault; H.B.F. Highland Boundary Fracture-zone, S.U.F. Southern Uplands Fault.

Fig. 6.11. Locality Map of Ireland.

The Carboniferous Limestone thus today unconformably wraps around the western and northern edges of the Leinster Massif, transgressing the Old Red Sandstone near Bagenalstown (Fig. 6.11), on the western side. George (1960a, Fig. 3) suggests the possibility that pre-Seminulan rocks may be preserved in St. George's Channel, which could be a persistent relic of a caledonoid downfold initiated in Carboniferous (or earlier) times. The Leinster Massif is generally considered as a salient of the larger competent mass embracing also North and Central Wales during the Carboniferous and post-Carboniferous history of Britain. The water separating Wales from Leinster may, however, hide the evidence in favour of the separation of the two massifs.

The region described in this section is broadly bounded on the western side by the River Barrow and on the north by the great curve of the Liffey. In the extreme south, however, the area extends westwards along the coast to the margin of the Dungarvan Syncline. The region is one of rolling relief, but with a wide coastal plain in Wexford. Generally speaking, the ground rises northwards across the whole area, and reaches its broadest and highest development in the Wicklow Mountains formed out of the Leinster Granite and the tough Bray quartzites. Kippure, Mullaghcleevaun and Lugnacuillia (3039 ft) are high points on the Leinster granite, the last named height revealing the sedimentary roof of the granite batholith. The southern half of the area varies in height between 300 ft (90 m) and 800 ft (244 m) O.D., but Mt. Leinster (2610 ft, 790 m) forms an imposing height along the granite, further west. The inlet at Wexford receives the waters of the Slaney and the harbour of Waterford marks the seaward end of the Barrow, Nore and Suir rivers. Down the steeper eastern slopes of the Wicklow Mountains tumble the rivers Dargle and Avoca.

The rock succession in south-east Ireland is generally as follows:

Carboniferous Limestone
(unconformity)
Old Red Sandstone
(unconformity)
Silurian
(unconformity?)
Ordovician
(unconformity)
Cambrian
(unconformity)
Pre-cambrian

One important point must be stressed about the above generalized succession. Large parts of the sub-Devonian sequence are poorly fossiliferous and precise stratigraphical positioning is often impossible.

The oldest Precambrian rocks in S.E. Ireland are those of the Rosslare Complex. These high grade metamorphic gneisses are probably over 1600 million years old (Max, 1976). This Lewisian basement is overlain by a group of schistose quartzites, greywackes and greenschists (the Cullenstown Group) which have been compared to the Monian of Anglesey (Baker, 1969). The Rosslare Complex is invaded by the Carnsore Granite, dated at 429 m.y. Intense Precambrian movements gave rise to steep-limbed (but often consistently directed isoclinal) E.N.E.-W.S.W. folds, many of which pitch eastwards. Intense minor contortions occur in these more incompetent Knockrath Beds and lenses of recrystallized siltstones are frequently squeezed into the cores of the folds. These Knockrath Beds and the overlying Bray Group (thicker greywackes and quartzites) occupy the south-western part of the Bray Anticlinorium.

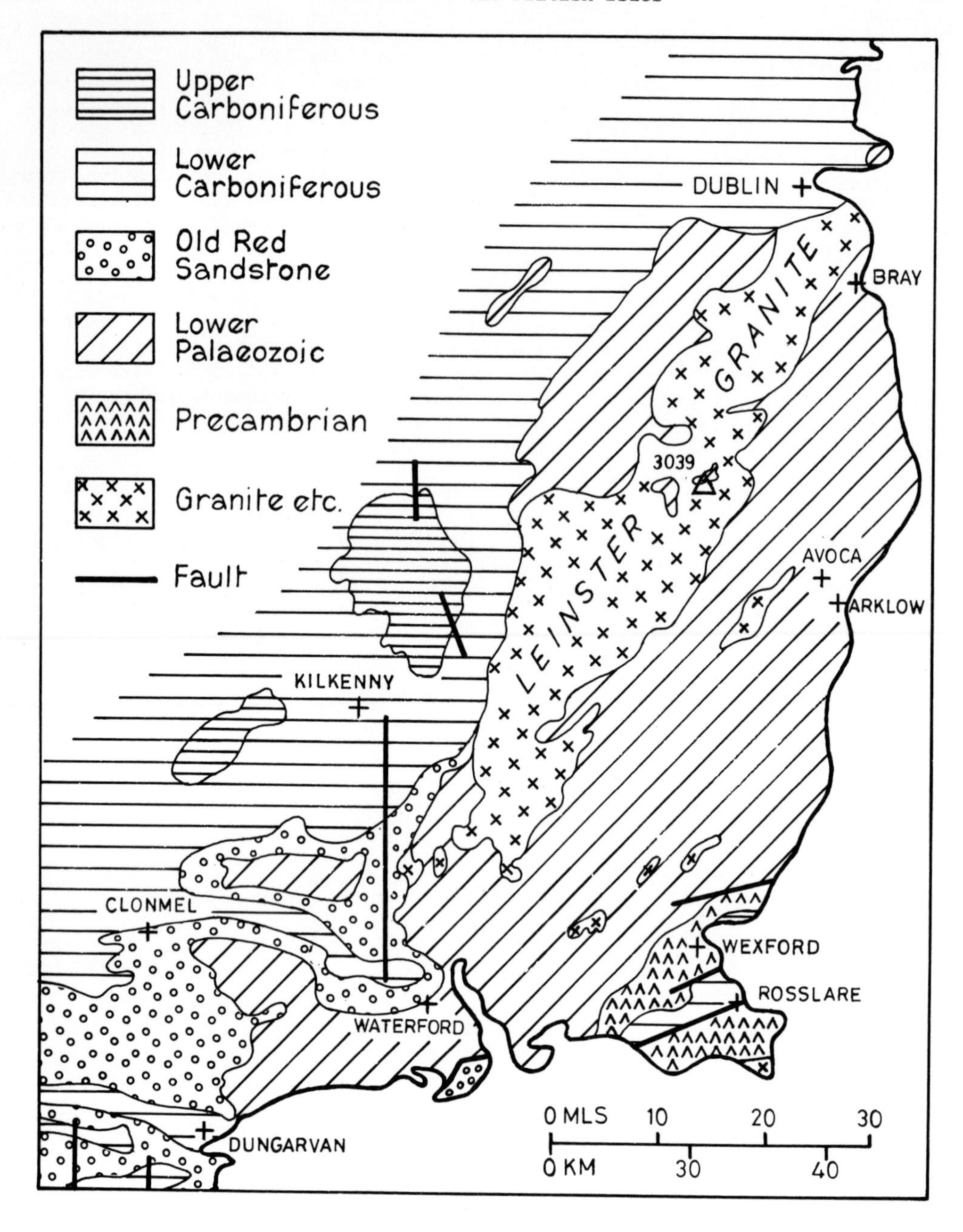

Fig. 6.12. Geology of S.E. Ireland.

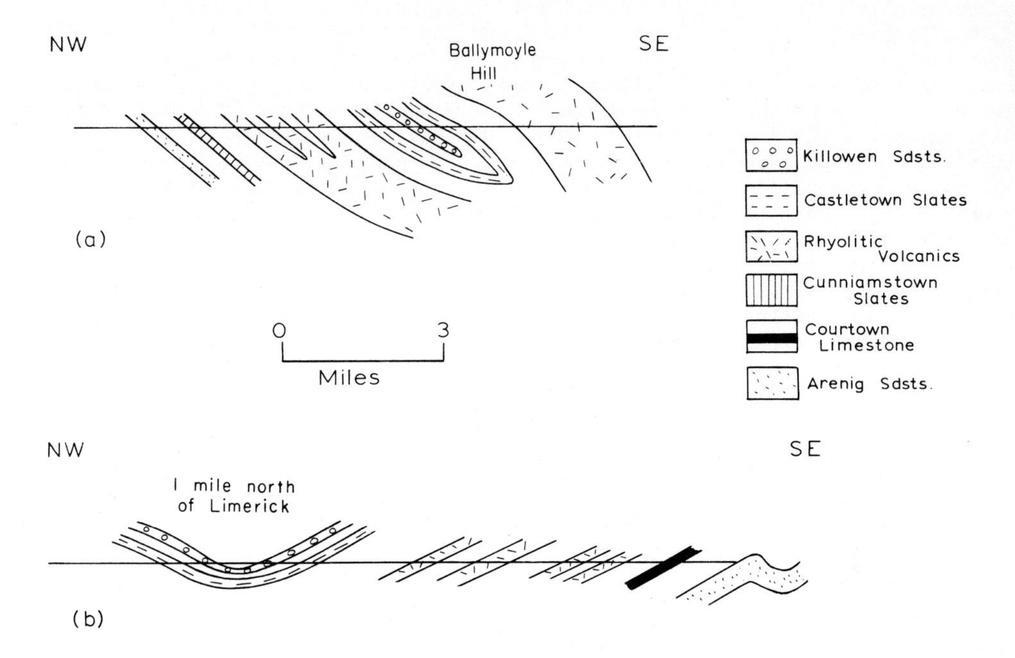

Fig. 6.13. Sections Across The Arklow Syncline (after Tremlett).

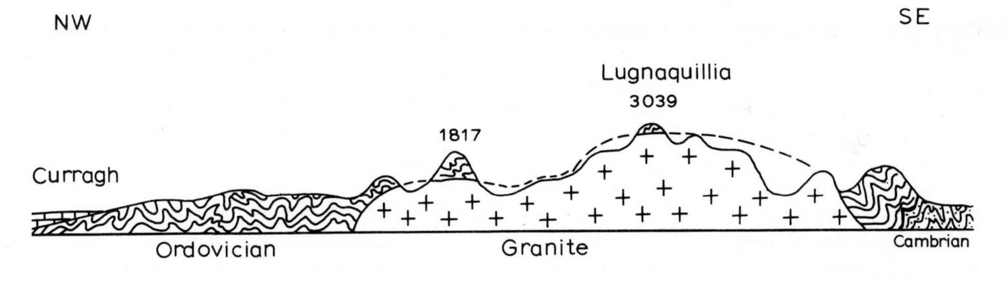

Fig. 6.14. Section Across the Leinster Granite (after Charlesworth).

Unconformably overlying these Precambrian rocks in the Bray Anticlinorium are the dark slates and mudstones of the Clara Group, up to 8500 ft (2593 m) thick and probably of Cambrian age. These appear to pass up into sandstones, siltstones and slates of Arenig age. North-north-eastward trending fractures have in places let down masses of the Clara Beds into the Precambrian areas. The Bray Series is well exposed at Bray Head, showing steady dips to the north-west of 40° to 70°, though this simple dipping is probably deceptive and tight isoclinal folding must repeat the successions. Inland, massive quartzite ribs are responsible for such prominent hills as the Great and Little Sugar-loaf. North of Dublin, the peninsula of Howth (Fig. 6.11) exposes another tract of ancient (possibly Precambrian) quartzites, greywackes and cleaved slates. The rocks are extremely crushed and contorted, and quartzite masses in places break through the slates almost like intrusives. Nevertheless, the dips are predominantly in this case towards the south. Similar crumpled strata emerge in Ireland's Eye, just to the north. As suggested below, the contacts of all these older rocks of Wicklow and Howth with other series may be of thrust origin, the forward movements accompanying later stages of the Caledonian Orogeny. Crimes and Crossley (1968), in their studies of the L.Palaeozoics of County Wexford, show the Cambrian succession to be some 2600 m thick. These sediments were formed in a northeast-southwest-trending trough bounded to the southeast by an Irish Sea landmass. Crimes and Crossley have also demonstrated repeated eastward upfaulting of regional blocks from Cambrian to Permian times along this Wexford coastal area of Ireland.

Ordovician strata, possibly up to 20,000 ft (6100 m) thick in places, occupy large expanses of south-east Ireland occurring on both the north-western and south-eastern sides of the Leinster Granite and forming its foothills. The rocks are in the main black, dark-grey and green slates interbedded with thin, fine-grained green and dark grey grits and cherts, with Caradocian limestone beds at some localities, more particularly at Tramore on the south coast and north of Courtown on the east coast. An important break probably interrupts the Ordovician sequences and, as in many British areas, Llandeilian strata are probably absent, as also may be rocks of Llanvirn age (in places at least). The Arenigian Tramore slates of the south coastal area and Ribband Group of Wicklow are, therefore, probably overlain unconformably by a thick succession of Caradocian sediments and volcanics. The latter are confined to the Ordovician belt south-east of the Leinster Granite and form two great lines, the easternmost extending for 80 miles (128 km) from Wicklow Head to Great Newton Head and probably reaching 8000 ft (2440 m) thick north of the Avoca River. The extrusives are probably all of Caradocian age and include rhyolitic lavas, pillow-lavas, tuffs and ashes, though many of the latter have subsequently been shown to be intrusive igneous rocks. The best exposures occur on the Tramore coast, the *Nemograptus gracilis* calcareous beds being followed by a great thickness of felsitic lavas, keratophyres, pillow-lavas and basic tuffs. Many massive dolerite sills also occur.

Stillman *et al* (1974) have shown that Caradocian volcanics north and west of Dublin were mainly tholeiitic andesites and basaltic andesites whereas to the south-east, in the Wicklow-Waterford area, the Caradocian extrusives were calc-alkaline. In a plate tectonic context this places the ocean trench to the north-west of Dublin.

Ashgillian rocks are unproved in the region and the relationship of Ordovician to any Silurian strata in the area is probably an unconformable one. It is likely also that only the Llandovery Series is represented, these rocks being probably restricted to slates in the extreme south-west of the region (east of the Comeraghs) and, further east, to sheared conglomerates above the Bala slates to the north of Courtown.

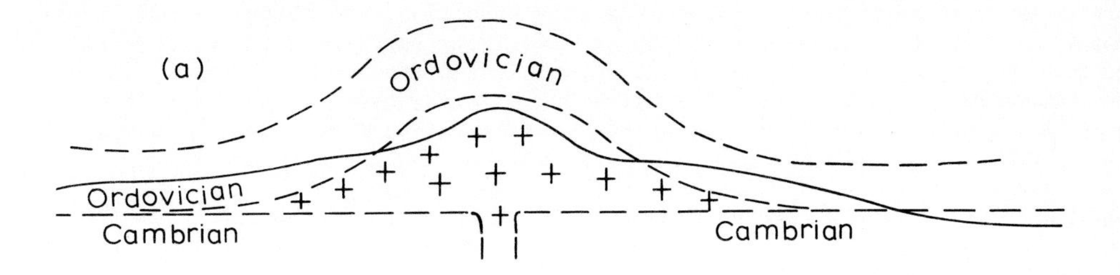

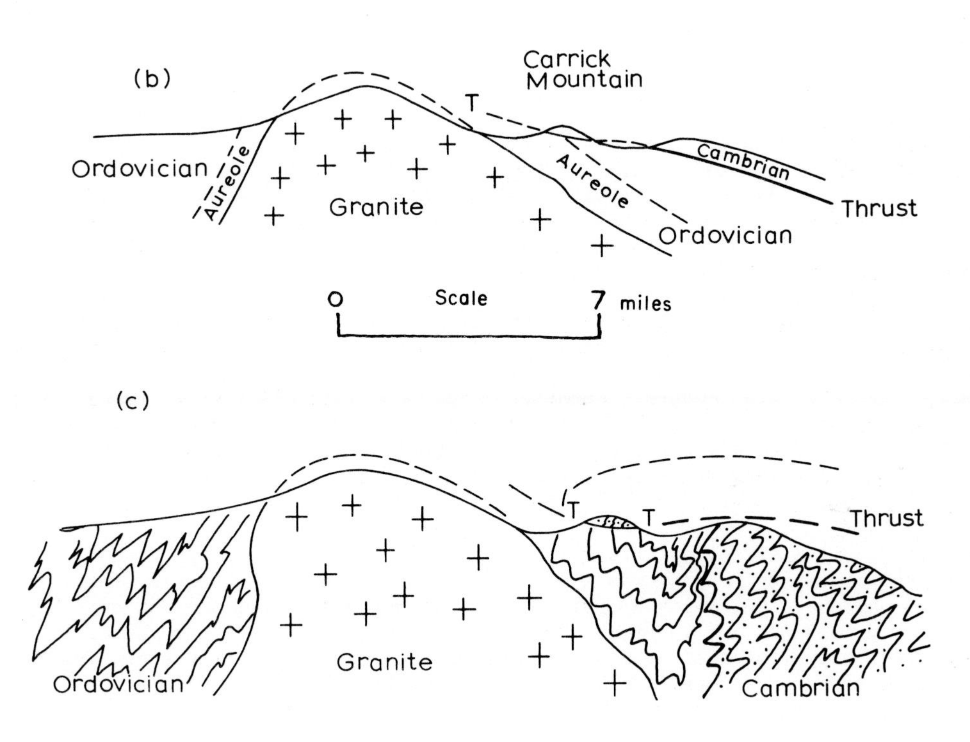

Fig. 6.15. Possible Explanations of the Relationship of the Bray Series to the Leinster Granite (see text).

There is, however, no doubt about the presence of Old Red Sandstone rocks as they form massive ring-like walls in the Comeragh Mountains, and their very discordant relationship to the more steeply dipping underlying Lower Palaeozoics is well demonstrated along the Suir Valley at Waterford and again on the northern limb of the Dungarvan Syncline, near Ballyvoyle Head. As, however, these rocks form only the outer margin of the area considered at present, their description is delayed until Chapter 7 (sec. 7.6).

Carboniferous Limestone follows conformably on Upper Old Red Sandstone at the peninsula of Hook, east of the Waterford inlet, and comprises some 1300 ft (396 m) of sandstones, shales, limestones and dolomites, the sequence extending upwards to low Visean horizons. Further east, Dinantian rocks occur again at Wexford (George, 1960a), reaching a thickness of just over 1000 ft (305 m). Mid-Tournaisian basal conglomerates here rest on the Bray Series, and are followed by limestones, shales and dolomites, and then by Lower Visean calcite mudstones, algal limestones and crinoidal dolomites. These Carboniferous strata have thus overlapped and overstepped the Old Red Sandstone north-eastwards from Hook against the slopes of the Leinster Massif.

Three major orogenies have affected South-east Ireland, namely late-Precambrian, Caledonian and Hercynian. Of these, the second was the one to leave its greatest mark, being also the time of great igneous intrusion. Late-Precambrian movements involved intense WNW.-ESE. compression with some metamorphism and even migmatization in South-east Wexford. The style and intensity of the folding is strongly reminiscent of that in the heart of the Longmynd Plateau of Shropshire and may therefore belong to the same Precambrian phase.

The main Caledonian climax here in South-east Ireland may, as elsewhere, have been preceded by strong compressive (and even intrusive) phases, one result being the absence of Llandeilian and perhaps Ashgillian strata. The major Caledonian movements in Leinster are, however, considered to be of late-Silurian to Lower or mid-Devonian age and the Leinster Batholith was intruded during this climax. An absolute date of 386±6 million years has been obtained for the Leinster Granite (Kulp *et al.*, 1960). Fragments of the intrusion occur in the Lower Carboniferous of the Dublin district.

The first obvious effect of this Caledonian Storm was to up-end the Ordovician slates and volcanics to dips which are seldom less than 50°. In such Ordovician areas as Kildare, the Arklow district and the lower reaches of the Barrow and Suir rivers, dip readings are generally 50°, 60°, 70° or 80°, with minor contorted dips being frequently reported, as for example near Duncannon and Arthurstown, on the banks of Waterford Harbour. Whilst the detailed fold structure of many of these areas is as yet still unknown, it is obvious that widespread intense (often isoclinal and overturned) folding is involved. The similar style of folding in the Lower Palaeozoic inliers of Central Ireland (see Ch. 7), here involving strata of Wenlock and even Ludlow age, supports the belief that the Caledonian orogenic climax occurred no earlier than late-Silurian times.

The broad compressive effects are more or less determinable from a geological map of South-east Ireland, with major anticlinorial uplifts along the lines of the Bray and Wexford outcrops of Precambrian and Cambrian strata and an intervening synclinorial downwarp along the Arklow region. A broader, and perhaps later, upwarp occurred also along (and perhaps due to) the great Leinster intrusion.

A more detailed account of the Caledonian structural evolution of one region - the Arklow district - on the south-eastern side of the Leinster Granite has, however, been given by Tremlett (1959). The initial compression resulted in a synclinal fold trending NE.-SW. This buckle was closed and overturned north-westwards

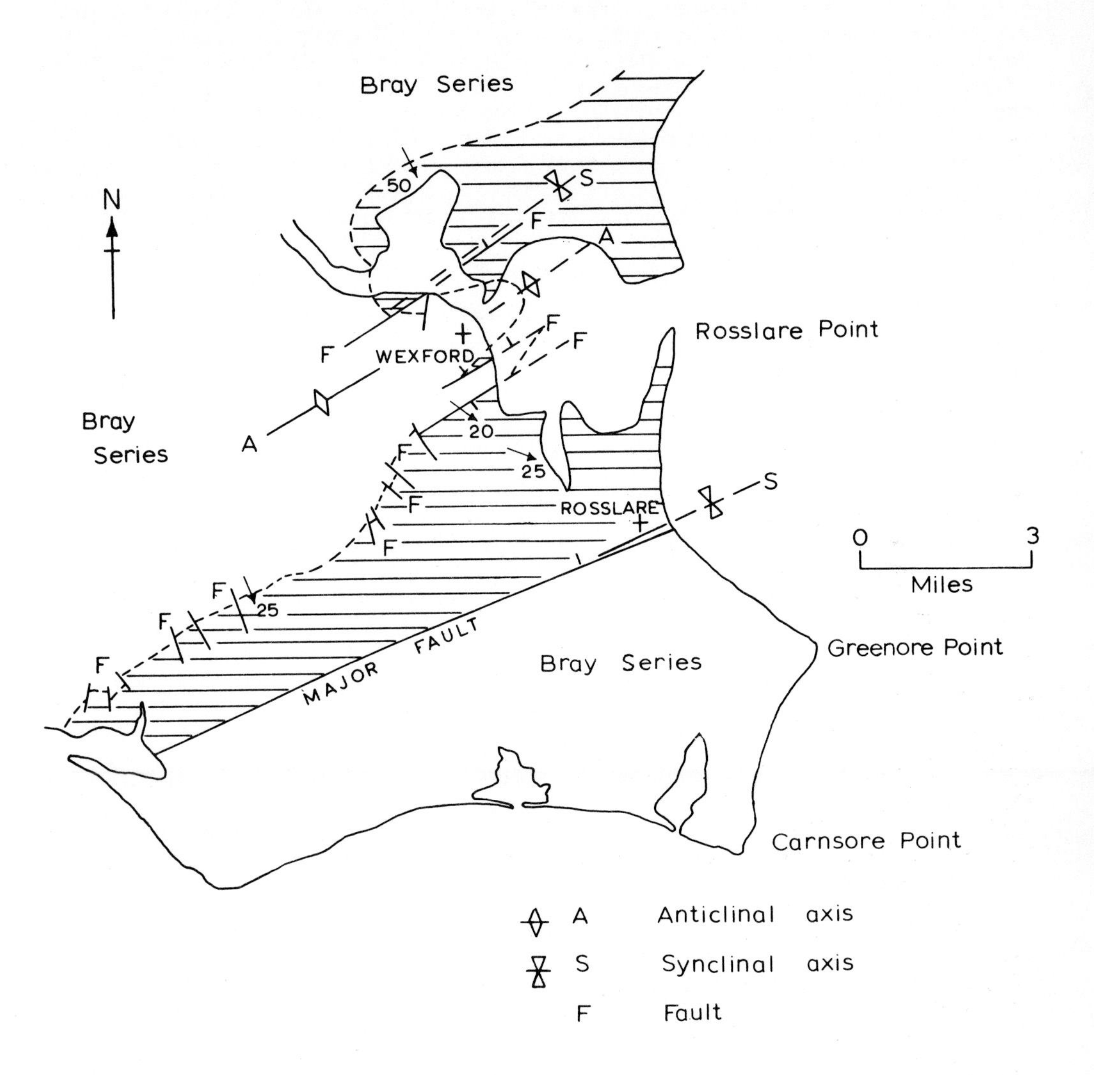

Fig. 6.16. The Structure of the Wexford Area (after T.N. George).

(Fig. 6.13a) in the north-east of the area (east of Rathdrum) but became a more symmetrical synclinorium (but still with many small steep folds) further to the south-west (Fig. 6.13b). The strong cleavage generally coincides with the strike of the beds, is normally greater than 60° and often vertical, and was formed at the time of initial compression. Basic dykes and stocks were then intruded parallel to the north-easterly cleavage, consolidating as dolerite or alkaline gabbro. This intrusion was succeeded by further strong compression causing thrusting, first to the south-east and then north westwards. Then followed faulting, including the great reverse-type Arklow Fault and numerous "north-westers" of oblique-slip type. This fracture phase mainly preceded (but in places outlasted) a new basic intrusive phase, the magma differentiating to diorite and even syenite.

In the north-west of the area, the slates were converted into sericite and chlorite-schists, whilst sphalerite, galena and copper sulphide mineralization was widespread over the whole region, particularly in the Avoca district. North-south dextral tear-faults which affect these mineral veins could be late-phase Caledonian fractures. In the discussion of Tremlett's paper, the possibility of some of the lead-zinc veins being Hercynian was pointed out (Moorbath, 1962, has obtained an absolute age of 280 million years). Gill, in the same discussion, stressed also the presence of important caledonoid dextral wrench-faults in areas to the north and south of the region mapped by Tremlett. These fractures were accompanied by folds pitching at angles of up to 70°.

The rise of the Leinster Granite was probably contemporaneous with the late phases outlined by Tremlett. Its absolute age, as determined by Kulp and others, appears to post-date that obtained for some, at least, of the copper sulphide ores of the Avoca district (420 million years). The granite is 70 miles (112 km) long and up to 12 miles (19 km) wide, and may even continue at depth to the south-west, e.g. a muscovite-biotite granite associated with a fault has also been discovered in the Slievenaman Inlier. As shown in Chapter 7, sect. 7.6, a rigid underlying mass, like granite, may underly (and be responsible for) the structural style of the Old Red Sandstone in the Comeragh Mountains.

The Leinster Granite is probably a batholith but is asymmetrical in outward slope about its longer axis, being much steeper on its western side. As a result the contact metamorphic zone is much wider on the granite's eastern flank. Small pendants of the sedimentary roof are preserved as, for example, on Percy's Table, Lugnaquillia. The northern margin of the granite is probably faulted against the Carboniferous Limestone of the Dublin district (Cole, 1921; Turner, 1950). The pluton initially burst through the already folded sediments in a direction slightly oblique to the regional Caledonian trend, and later deformed dyke and vein systems on the pluton's eastern flanks. On the eastern side of the granite, low-angled thrusting was directed from the east towards the pluton's northern portion. A klippe of Bray slates and quartzite was carried on to the metamorphic aureole of the granite (Cole, 1921; Smithson, 1928; Brindley, 1957). This thrust block lies 700 yards (640 m) away from the plutonic boundary. The thrust boundary appears to dip north-eastwards at about 20°. This klippe raises the possibility, of course, of the whole of the Bray Precambrian (and/or Cambrian) mass being a thrust outlier. One has to explain the presence of Ordovician strata and not Cambrian or Precambrian against the flanks of the pluton. If the granite were intruded into an upfold of Lower Palaeozoics then one might expect the Bray-type rocks to occur on the immediate contact flanks of the intrusion. Three possibilities are worth considering, as illustrated in Fig. 6.15. The first is that suggested by Sollas (1891) and by Cole (1921). They believed the granite to be intruded as a laccolithic mass between a floor of Bray Series and a roof of Ordovician (Fig. 6.15a).

The second possibility is, as also suggested by Cole (and later by Smithson), that the whole of the pre-Ordovician mass of Bray Head has been carried westwards by powerful regional low-angled thrusting (Fig. 6.15b). The last remaining possibility (and, bearing all the evidence in mind, probably the correct one) is that the granite was pushed up into a complex fan-folded area of Ordovician rocks, the place of intrusion being just west of an anticlinorium in the core of which occurred the Bray Series (Fig. 6.15c). Renewed lateral compression just after the emplacement of the pluton caused a splinter of the Bray rocks to be carried westwards to the Carrick Mountain, this thrusting being the compressive action to the resistant core afforded by the central granite mass.

The effects of the Hercynian Orogeny in South-east Ireland can be discerned only from those of earlier earthstorms when Devonian or Carboniferous strata are present and this limits one to the extreme south of the region. That the Leinster Massif (with its central "stiffening" of granite) acted as an effective buffer to the northward-advancing Hercynian ripples is generally agreed, and the result is seen in the sharp tectonic contrast occurring at about the latitude of Dungarvan. Nevertheless, some slight renewal of movement along Caledonoid lines is suggested in the southern part of the Leinster Massif, with a gentle arching occurring more or less along the line of the longer of the two Ordovician volcanic belts, i.e. NE.-SW. through the Tramore district. To the north-west of this line, the Old Red Sandstone and Lower Carboniferous rocks immediately north-west of Waterford town dip north-westwards, whilst those of the Killae-Rathmoylan and Hook peninsulas dip generally south-eastwards. The same south-eastern flank of this caledonoid Armorican flexture is represented in the vicinity of Wexford (Fig. 6.16) where, as shown by George (1960a), the main synclinal axis occurs along the southern boundary fault which brings up the Precambrian. This caledonoid upwarp could, of course, have been determined by the interior resistance of the tough Ordovician and Caledonian igneous belt. The north-eastward trend of the minor folds in the Wexford area are, however, perhaps also a pointer to the presence of a deep embayment of Upper Palaeozoic rocks in St. George's Channel, a possibility discussed at the commencement of this chapter. The meeting-place of the main Hercynian front and the Leinster mass is probably today seen as the E.-W. lines of shattering in the Ordovician coastal rocks between Dunabrattin Head and Ballyvoyle Head and in the fault-bounded blocks of Old Red Sandstone, often tiled up to near vertical positions, e.g. in Ballydovane Bay.

CHAPTER 7

Hercynian Terrains

7.1 South-West England

The area to be described extends as far north as the head of the Severn estuary and eastwards to the base of the Middle Jurassic, roughly through Bath, Frome and Sherborne to Bridport (Fig. 7.1). A thick column of rock formations is involved, ranging from Lower Palaeozoic (the Lizard and Start complexes could even be Precambrian) through Devonian, Carboniferous, Permian, Triassic and Jurassic to Cretaceous. Small areas of Oligocene (Bovey Tracey and Petrockstow) and of Pliocene? (St. Erth) occur in Devon and Cornwall, respectively. The Devonian and Carboniferous systems together form a considerable thickness, amounting possibly to over 30,000 ft (9150 m), and also form the largest portion of the surface. A broad distinction may be made at a N.-S. line running from Watchet through Exeter to Paignton. To the west of this line the post-Carboniferous cover has been almost completely removed and the Hercynian folds are exposed. To the east, however, those structures lie largely buried under Permian and Mesozoic formations, except in the Quantock Hills, the Mendips and in numerous inliers over the Bristol area.

Topographically, the area is one of diverse relief with the resistant granites, and the Culm-Devonian grits forming the highest ground. Yes Tor on the north-western flank of the Dartmoor Granite reaches 2039 ft (620 m) O.D. and Dunkery Beacon on Exmoor rises to 1705 ft (520 m) O.D. The lowest lying regions occur on the Triassic-Liassic areas of Somerset and the Bristol district. A number of well-developed surfaces occur in South-west England, at heights of 1000, 750 and 430 ft (305, 230 and 130 m) O.D. The bulk of the Bodmin granite surface represents the higher base level, whilst the lowest of the three surfaces is extensively developed over Devon and Cornwall. Emergence following this planation was rapid and many of the rivers now follow courses in deep ravines.

Devon and Cornwall are both noted for the rugged beauty of their coastline, particularly impressive where the rocks are resistant, as on the Hangman in North Devon, at Pentire Point and the Lizard. Lundy's granite rises impressively out of the sea.

Omitting problematical areas like Dodman Point and the Lizard-Start complex, the oldest (proved) rocks in the area occur in the extreme north, to the east of Berkeley. The relationship of these Tremadocian shales to the Silurian of Tites Point (on the Severn) and Tortworth is not clear, though they may be overlain unconformably by the Upper Llandovery. As pointed out by Kellaway and Welch (1948), these inliers lie on the line of the Lower Severn Axis (Fig. 7.2) and late "Taconic" unrest operated here, as in the Malverns further north. In the Tortworth Inlier, a Llandovery to Lower Ludlow succession is exposed, the beds being synclinally disposed and following round the northern rim of the Bristol Coalfield. The overstep of the Upper Old Red Sandstone across Ludlow and Wenlock points to slight earth movements here during (probably) mid-Devonian times. Other small exposures of Silurian occur west of Wickwar and in the core of Beacon Hill, the easternmost pericline of the Mendips, where 400 ft (122 m) of pyroxene-andesite overlie Llandovery lavas and tuffs. The Upper Old Red Sandstone here rests on Wenlock, the intensity of the break possibly pointing to early uplift along the Bath Axis.

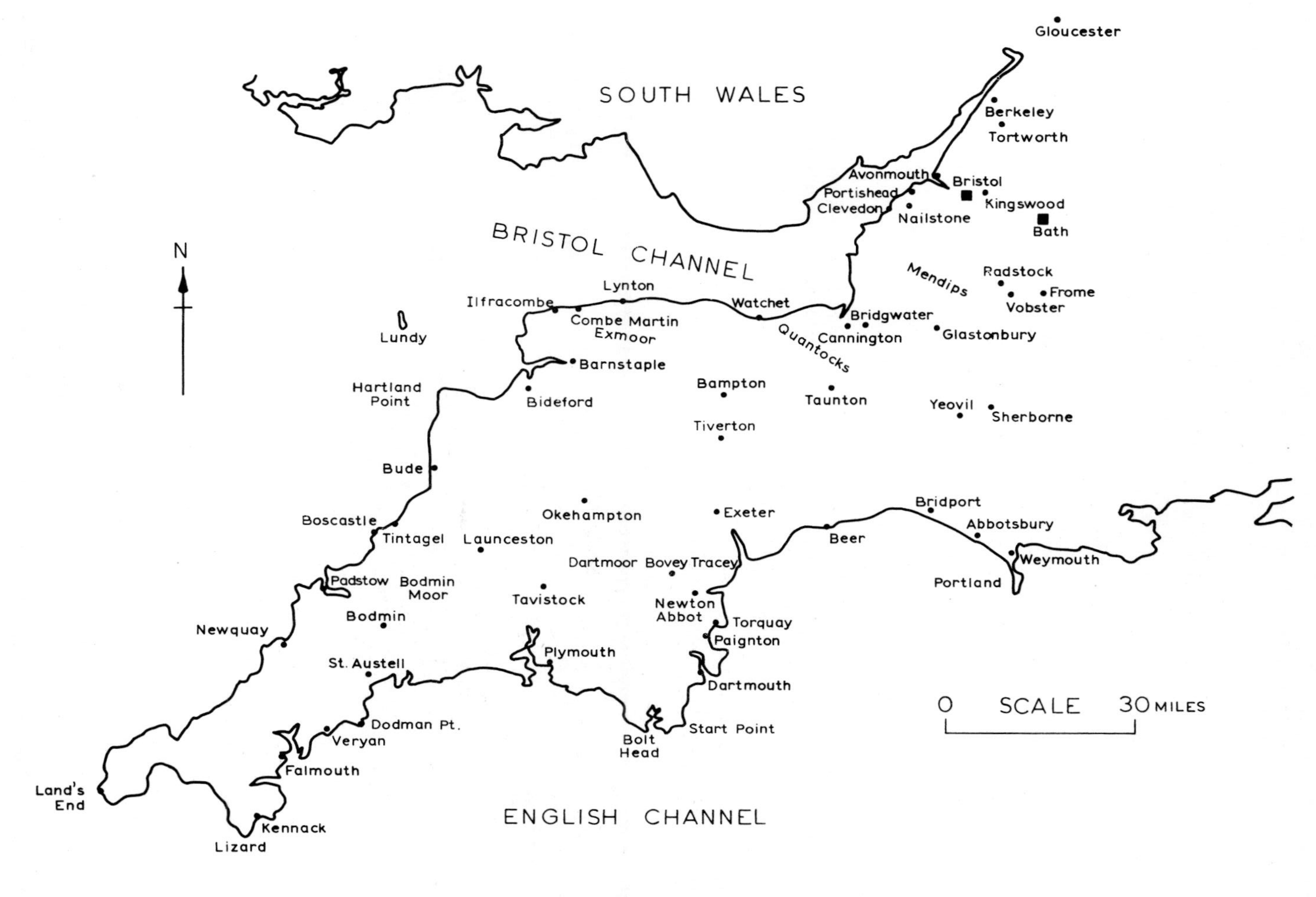

Fig. 7.1 Locality Map for South-West England.

Abundant exotic blocks of quartzite and limestone of Lower Palaeozoic age occur however in South Cornwall, in the "Veryan Series" on the north side of the Lizard Complex and through Roseland to Gorran. The immense size of the blocks would suggest that they have not been transported for too great a distance. The blocks are fossiliferous, the quartzites yielding Ordovician brachiopods and trilobites with the limestones containing a Silurian fauna. The blocks are associated with conglomerates, breccias, mafic igneous rocks and turbidite sandstones of Devonian age. Greenstone phacoids and fragments in the breccias appear to have Lizard associations. Recent work by Barnes *et al* (1979) on these South Cornish melanges and olistostromes show that the structural trend of the melange units is ENE. Their observations are consistent with the development of olistostromes in an unstable trough adjacent to an advancing Lizard ophiolite. The deposits were subsequently tectonised in a shear zone beneath the overriding ophiolite.

The Lizard Complex forms the southernmost headland of Cornwall. The complex, which might well be, at least in part, of Precambrian age, is faulted on its northern side (Mullion to Porthallow) against Devonian rocks. The oldest rocks are now the Old Lizard Head Series (mica-schists, hornblende schists and quartz-granulites, all once part of a series of sediments and volcanics). They were locally intruded by the (now) Man of War Gneiss. Other oldest areas include the Llandewednack and Traboe hornblende-schists, probably all part of the same (Precambrian?) depositional area. The geological events from these early times onwards have been listed by Edmonds *et al* (1975). Precambrian to Lower Palaeozoic orogenesis and regional metamorphism followed, with the intrusion of peridodite, now altered to the famous Lizard serpentines, and then intrusion of gabbro. Radiometric dates of 492 and 442 m.y. indicate this Lower Palaeozoic igneous and metamorphic history. A cluster of younger dates (371 - 350 m.y.) point to further movements, however, (probably of Mid-to-end-Devonian times) during which intrusion of basic dykes was closely followed by acid intrusion (Kennack Gneiss). Still further deformation with appreciable faulting of the complex against Devonian sediments occurred during the Hercynian Orogeny, to be followed by post-orogenic vertical adjustments.

Another area of metamorphics is seen (above water) on Eddystone, a small exposure of garnetiferous gneiss some 14 miles (22 km) south of Plymouth. This narrow (underwater) belt also includes mica schists.

Structural studies (showing only one major phase of deformation) suggest that the green schists and mica-schists of Start Point (S. Devon) are younger than the metamorphic areas described above and that they are probably altered Devonian. There is no evidence that the Start Schists are thrust against the (Devonian) Meadfoot Beds to the north. The contact is now known to be a steep northward-dipping fracture. Sadler's (1974) work on the Lizard's northern margin demonstrates similar steep-dipping faulted contacts, with in one locality the presence of Menaver (Roseland-type) conglomerate resting directly (unconformably?) on a Lizard base.

Devonian rocks occur in two main areas, (a) south of a line from Boscastle to Tavistock and Newton Abbot, and (b) Exmoor. These two large areas represent essentially upfolded areas (but on a complicated scale) on either flank of the Central Devon Synclinorium which has preserved the Carboniferous Culm Measures. Smaller detached inliers peep through the Mesozoic blanket in the Quantocks, at Cannington, in the cores of the Mendip periclines and in the Bristol-Portishead districts.

The Devonian sediments were laid down in a complex depositional area which embraced Devon and Cornwall, Brittany, Belgium and the Rhineland. The northern fringe of this trough lay near to Exmoor but was subject to northward or southward migration. Southward movements led to the influx of Old Red Sandstone-type deposits, represented, for example, by the Hangman Grits. The trough was often subdivided into a

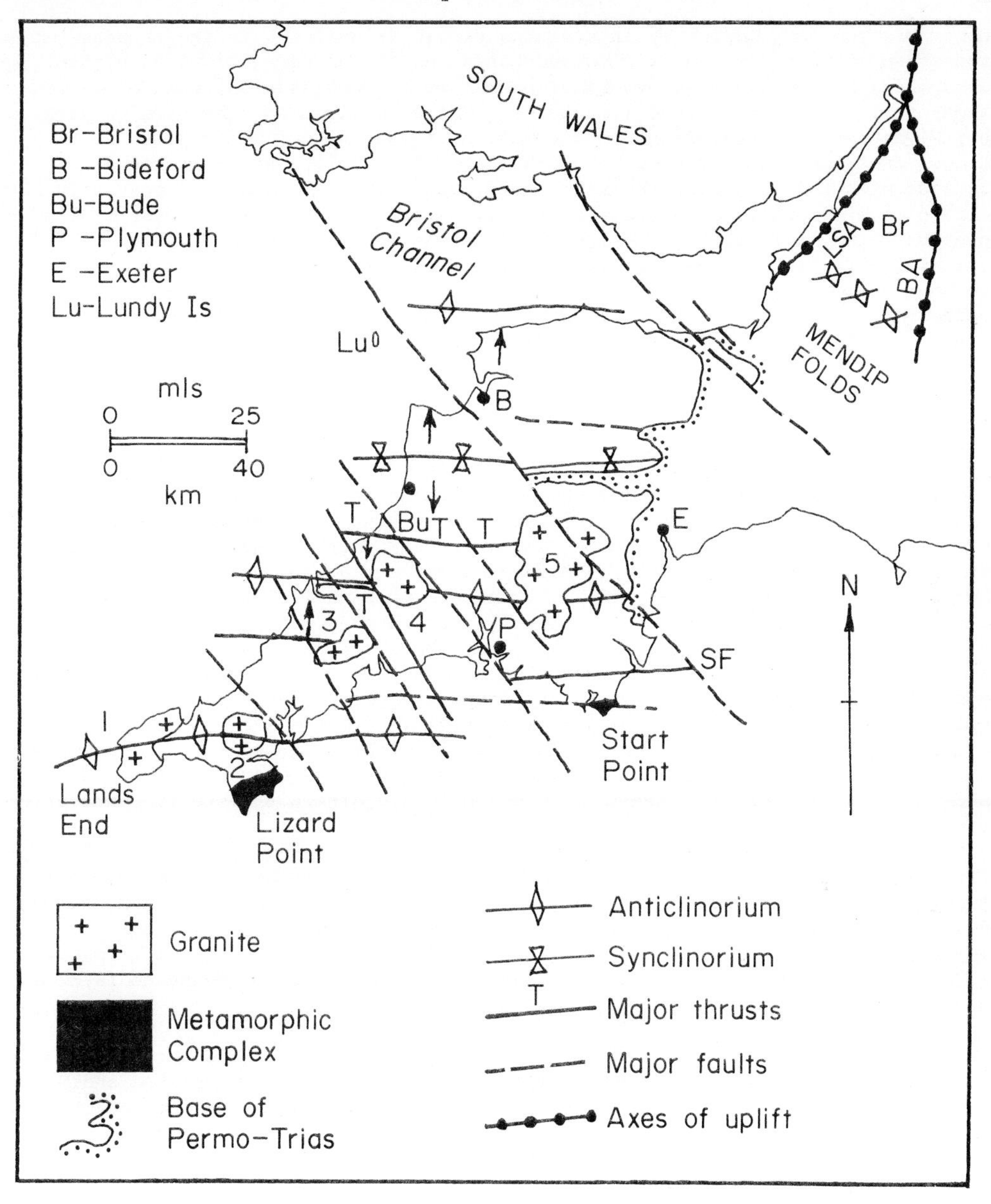

Fig. 7.2 Map Showing some of the Major Structural Elements in S.W. England (in part after Freshney and Taylor).

Key: B. Bideford; Br. Bristol; Bu. Bude; E. Exeter; Lu. Lundy; L.S.A. Lower Severn Axis; B.A. Bath Axis; Key to granites: 1. Lands End; 2. Falmouth; 3. St. Austell; 4. Bodmin Moor; 5. Dartmoor.

number of secondary basins by intervening ridges or swells. In the troughs, monotonous successions of grey marine mudstones and shales were laid down without interruption, but over the geanticlinal swells were deposited red, purple or green sandstones, conglomerates and limestones. One such ridge was the Staddon area, near Plymouth. In general the growing ridges tended to become progressively younger from south to north in Cornubia (Hendriks, 1959). Volcanic rocks were associated with the flanks of the ridges and were probably contemporaneous with movements of the trough's corrugated floor. Over large areas of South-west England the Upper Devonian at least and the lowest Dinantian was marked by a "bathyal lull" (Goldring, 1962), but these bathyal conditions tended to move northwards with time and the possibility remains that as Devonian times progressed more and more of the southernmost areas of the depositional floor in Cornubia were becoming raised into firstly geanticlinals and later growing nappes.

A good deal of confusion has arisen concerning the identification and correlation of various Devonian horizons in South-west England, but the work of Hendriks, House, Sadler, Dineley and others has done much to clarify the picture. It is now appreciated that great lateral changes occur, e.g. that the Staddon Grits and Meadfoot Beds are in part lateral variants of each other, as also are the Meadfoot Beds and the Dartmouth Slates. It is also known that the Gramscatho Beds of South Cornwall include what used to be the Falmouth, Portscatho, Veryan, Mylor (etc.) series and that they are mainly of Middle Devonian age, and, at least in part, of flysch origin (Hendriks, 1937, 1959). The work of Hendriks has also shown that previous views of the structure of the Plymouth Sound-Bolt Tail coast need to be seriously revised, the folding probably being of a severe recumbent and thrust type rather than frilled "fans". The Devonian facies in the Padstow area is closely comparable with rocks and faunas in Brittany, supporting the view that exotic thrust regions of more southerly deposited sediments exist along this Tintagel-Pentire coast. Moreover, considerable areas previously mapped as Upper Devonian are known to belong to the Middle Devonian division (House, 1956, 1960). Such big changes in stratigraphical identification obviously affect the precise unravelling of the geological structure.

Similar problems, though perhaps not as great, exist in connection with Exmoor stratigraphy and structure. A fairly agreed succession has been reached though faulted contacts still cause some worry. Precise thickness are difficult to achieve, and it is already known (Holwill, 1961) that Evans' original thickness (1922) of about 20,000 ft for the exposed North Devon Devonian sequence is excessive.

The precise ages of the supposed Devonian rocks of South Cornwall, north of the Lizard Complex, have also presented problems. These Veryan, Gramscatho and Mylor Beds occupy parts of the Cornish peninsula as far north as a line from about Perranporth to Pentewan. The Veryan Beds are mainly made up of the melanges of Meneage and Roseland and have already been described above. They are probably of Lower Devonian age. The Gramscatho Beds (slates and greywackes of flysch origin with northward transported debris) are now thought to be mainly of Middle Devonian age (Hendriks *et al*, 1971 and Sadler, 1973). The Mylor Beds (slates with volcanics), occupying a wide area from Penzance to near Truro, have been given ages ranging from Lower Palaeozoic (early workers) to Lower Devonian (Wilson and Taylor, 1976) to Upper Devonian-Lower Carboniferous (Simpson, 1969). Recent palynological finds (Turner *et al*, 1979) now support a Famennian (Upper Devonian) age for at least the Mylor Beds of the Mount Wellington area (S.W. of Truro).

In South Devon, the oldest Devonian beds are the mainly non-marine Dartmouth Slates (of Dittonian or Siegenian age at the oldest). These beds occupy much of the axial zone of the main Watergate Anticlinorium, a major E.-W. structure running from Newquay to Dartmouth. Above these fluvio-deltaic green and purple sandstones and siltstones, with pyroclastics, come the marine Meadfoot Beds and their lateral equivalents, the Staddon Grits and Newquay Slates. The Middle Devonian of South Devon is a complex of slates with intercalated lenticles (often of considerable size) of limestones and volcanics. The larger carbonate masses occur at Brixham, Plymouth and Torquay. To the west, in Cornwall, much of the Middle-Upper Devonian sequence is made up of the Trevose Slates (and the underlying pillow lavas of the Padstow area), the pillow lavas at Pentire and the slates and volcanics of the Tintagel area. The Upper Devonian of South Devon is made up largely of ostracod-bearing slates with thinner cephalopod-bearing limestones and shales taking their place along the northern edge of the Devonian tract (for example, at Chudleigh).

In N. Devon and W. Somerset, the oldest Devonian exposed occurs at Lynmouth. These marine Lynton Beds are followed by the thick (1300 m), mainly Eifelian, Hangman Grits. This advance of continental conditions from the north, was separated from another, later, O.R.S. incursion (the Pickwell Down Sandstone) by the marine Ilfracombe Beds and the cleaved Morte Slates. The highest Devonian sediments (the Pilton Beds) pass up, without a break, into the Lower Carboniferous.

The Lower Carboniferous portion of these Pilton Beds (shales with limestones and calcareous sandstone) range up to just above the Tournaisian-Visean boundary to be succeeded in the Bampton-Westleigh areas by cherts, shales and limestones, and westwards by the Codden Hill Cherts. Further south, along the south rim of the Culm outcrop from Boscastle in the west to Newton Abbot in the east, the Lower Carboniferous sequences have acquired a complex nomenclature of local formations. In the west, slates and thin limestones pass up into the Tintagel tuffs and lavas and thence into cherts and slates. In the Launceston-Okehampton districts the various Meldon formations include slates, quartzites, tuffs and agglomerates, and, in the higher Visean, the Meldon Chert Formation. In the eastern (Exeter-Newton Abbott) areas purple and green slates pass up generally into cherts and shales with limestones and tuffs.

The Namurian portion of the "Culm" occupies a fairly large area of the southern portion of the Central Devon Synclinorium with a corresponding but much narrower outcrop on the north side of the main downfold. Over the southern area, the lower portion of the Namurian succession is shaley and contains chert. By H1 (Chokierian) times however, turbidites begin to appear, the debris probably being derived from rising areas to the south. The direction of turbidity flow however was often westwards or eastwards along the axis of the depositional trough. On the northern side of the synclinorium, Namurian muddy deposition lasted longer (till R2 times) but then greywackes flooded into these areas also.

Only the lower portions of the Westphalian sequence are now preserved in Devon and in NW. Cornwall and it is possible that higher portions were never deposited as Hercynian deformations spread further and further northwards into Cornubia. The basinal Bude Formation, comprising thick-bedded, sandstones, with shales, includes a fauna comparable to the Lower Coal Measures of South Wales. The Bideford Formation of that region is a more deltaic-paralic sequence of sandstones, siltstones and shales. They embrace the Abbotsham and Northam Beds.

The Palaeozoic rocks of Devon and Cornwall were intruded by a major granite batholith, cupolas of which outcrop as the separate plutons of Dartmoor, Bodmin, St. Austell, Carnmenellis, Lands End and Scillies.

These six great granite masses are connected at depth (Bott, Day and Masson-Smith, 1958). The whole "batholith" trends ENE.WSW. and is believed to have an inverted L-shaped vertical cross-section (Exley, 1961). The model had an envelope of cooler contaminated magma and a core of hotter, more sodic and volatile-rich magma which broke out at the St. Austell hinge of the "L". Gravity surveys (Bott *et al.*, 1958) indicate a possible northward limiting base for the Dartmoor granite at a depth of 10 km. The age of this granite is believed to be 280±10 million years, i.e. the intrusion occurred at the end of the Carboniferous Period. The granite intrusion was accompanied by granitic "elvan" dykes, tourmaline and aplite veins, and the mineralization of fissures, the majority of these lines tending ENE.-WSW. or NE.-SW. (see Hosking, 1962, fig. 9/11). The ultimate controlling boundaries of the granite injection may have been the Lizard-Start fracture-zone in the south and the Tintagel-Trevose-Scilly line (a thrust?) in the north-west. Hill and Vine (1965) believed that both upthrust of the metamorphic basement and the batholith intrusion are accommodations of a change in structural trend between the Western Approaches and South-west England.

South-west England lies on the northern edge of the Armorican Arc, and Hercynian effects in this Cornubian area were quite intense. Briefly, "a large structural depression was formed in which the rocks were intricately folded and faulted. At depth, intense folding and refolding was accompanied by mild metamorphic recrystallisation which has given Devonian rocks a slaty cleavage. Nearer the surface, the uppermost Carboniferous rocks were folded into flat-lying to upright, concertina folds. Two successive north-south compressions occurred, the initial north-facing folds meeting the later south-facing folds in a 'confrontation' zone. As the compression waned, the Cornubian granite batholith rose into the rocks, producing a rich copper-tin mineralisation" (Dunning *et al*, 1978).

Two important studies have summed up the main structural features of South-west England. They are summarised in figs. 7.3 and 7.4. Firstly, Dodson and Rex (1971) have shown the different groupings of K-Ar datings across Cornubia (fig. 7.3). South of the facing-confrontation line (at Polzeath on the West Cornish coast) ages are mostly in the range 340-320 m.y. North of this line, they are in the range 300-280 m.y. Moreover the age variations in Cornubia detect three major deformations:

(i) Bretonic (late Devonian-early Dinantian),
(ii) Sudetic (end-Dinantian) and
(iii) late Carboniferous.

These deformations seemed to have moved progressively northwards across South-West England. The spread of Pennant molasse (containing Cornubian-type debris) into South Wales in upper Westphalian times seems to suggest that the third deformation referred to above could at least have begun by that time.

Secondly, Sanderson and Dearman (1973) have subdivided the Hercynides of Cornubia into 12 structural zones (fig. 7.4). North-facing folds pass southwards across N. Devon into upright and then south-verging folds from Zones 1-3, ultimately to become the well-known recumbent chevron style in Zone 4. All these zones form a superstructure resting (on a major Rusey Thrust) above an intensely deformed fifth zone characterised by recumbent isoclinal folds and intense cleavage and the southward-facing recumbent folds in Devonian slates of a sixth zone. Here, however, a major "line of confrontation" occurs (from Polzeath to the southern margin of Dartmoor). South of this line the folds of Zone 7 now face *northwards*, an attitude maintained southward to the Lizard boundary. The axis of the Zone 8 Dartmouth-Watergate Anticline is repeatedly offset by Tertiary wrench-faults. Zones 9-11 have a NE.-SW. Cadomian trend, the eleventh zone being bounded on the southeast by the Lizard Block.

South of the confrontation, the north-facing folds "have a slaty cleavage that dies out northwards. The beds north of Polzeath were not affected by this early phase of deformation. These unfolded beds were later deformed into south-facing recumbent folds, the local F_1 phase, with a well-developed axial-planar slaty cleavage.

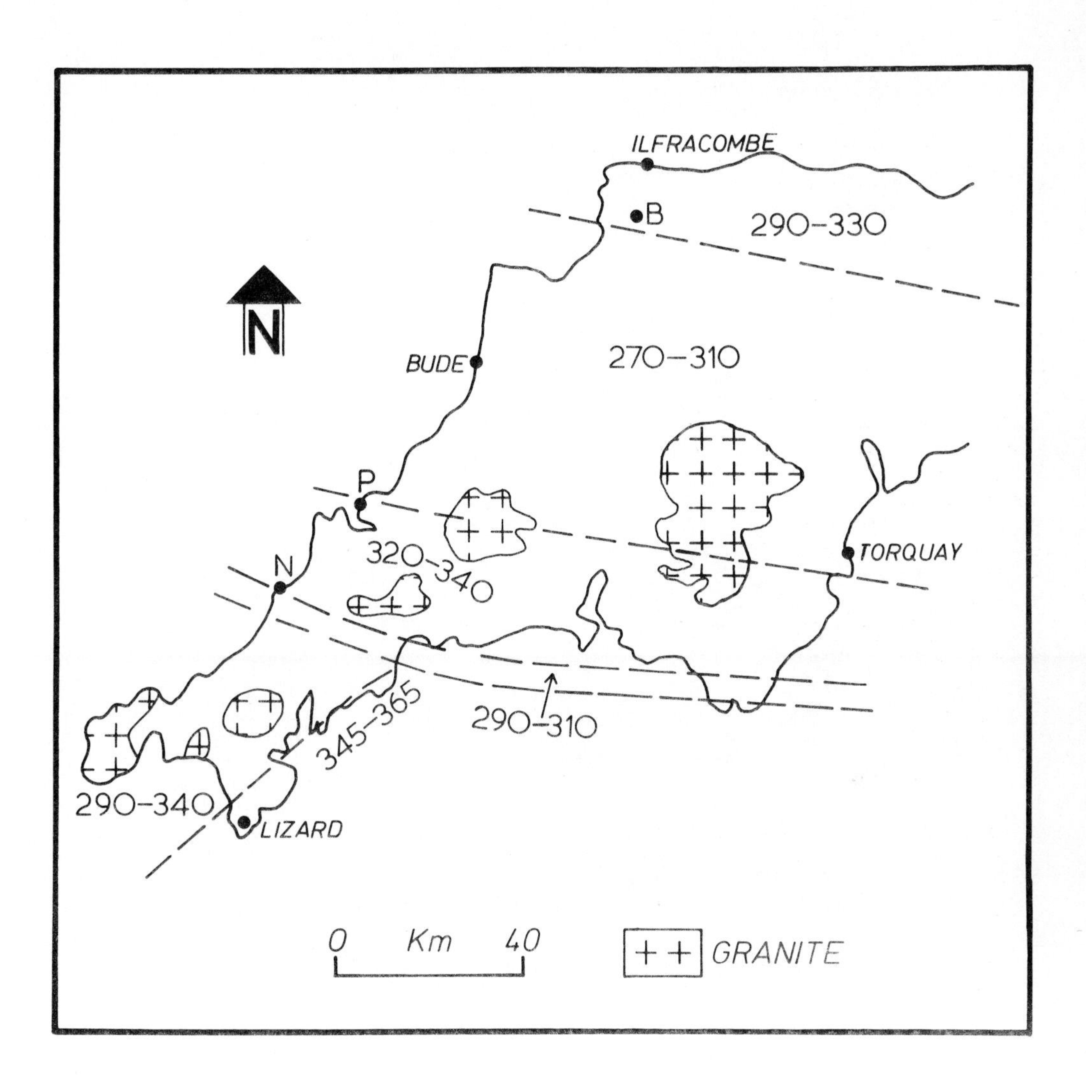

Fig. 7.3 Significant Groupings of K-Ar Ages (after Dodson and Rex) (dates in millions of years).

Key: B. Barnstaple; N. Newquay; P. Polzeath.

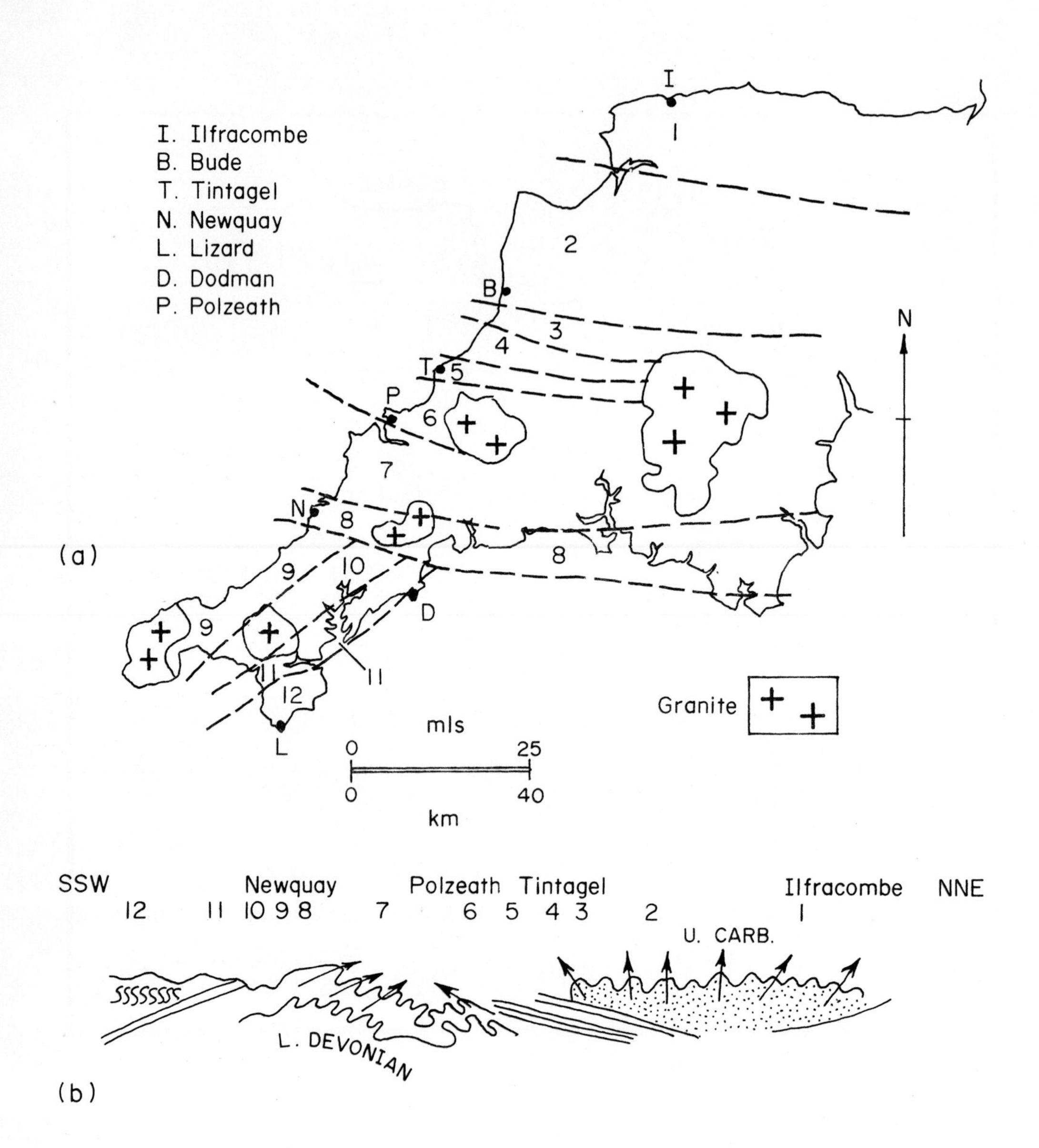

Fig. 7.4 Structural Zones Defined in the Hercynian Fold Belt of South-West England (after Sanderson and Dearman, 1973).

This later deformation refolded the earlier N.-facing recumbent folds, where the northern slaty cleavage is developed as a crenulation cleavage" (Anderton *et al*, 1979, p.165).

Finally, before leaving this main portion of Cornubia, and in particular the more southern regions, one should recall the prolonged debate concerning the plate tectonic context of the area and especially the proposal that the Lizard Complex is a thrust (obducted?) ophiolite mass. Part of the debate has been outlined by Ager (1975) and Owen (1976). More recently, the problem has been discussed by Anderton *et al* (1979, pp. 170-174) and they have provided their own model in order to interpret the Early Devonian-late Carboniferous events across the region. Fig. 7.5 summarises their fig. 12.7 and shows how the Lizard Complex could be a piece of thrust, partly oceanised, Precambrian crust that formed in a marginal basin area.

Exmoor

The major fold axis of this Devonian area is an anticlinorial axis trending WNW.-ESE. and lying close to the North Devon coast but trending inland towards the northern end of the Quantock Inlier. Over Exmoor the major overall regional dip is a southward one averaging 30° to 50°, but reversals of dip direction occur fairly frequently, as, for example, near Exton and Pixton Park and especially in the more argillaceous members. A strong cleavage dipping between 50° and 90° is developed in these argillites. Drag-folds are common in thin calcareous bands as, for example, just above the David's Stone Limestone in the Combe Martin area (Holwill, 1962), and rolled-up cylinders were once reputed to be "fossil trees". The axial planes of the minor folds dip southwards, parallel to the cleavage, and many of them pitch westwards at angles of 15° to 30° (Evans, 1922). The most prevalent fault direction is between west and north. These fractures mostly hade south-westwards and are probably dextral wrench-faults displaying slickensides which dip up to 30°. These faults are of relatively late age. Of earlier origin are NE.-SW. fractures, mostly downthrowing south-eastwards. The Middle Devonian formations are frequently traversed by metalliferous-filled fractures also downthrowing southwards. The important Cothelstone Fault, a wrench-fault with a 3-miles dextral shift (Webby, 1965), defines the north-west alignment of the Quantock Hills. This fracture is blanketed by New Red Sandstone.

Gravimetric work by Bott, Day and Masson-Smith (1958) suggested that the Devonian rocks of Exmoor might be *underlain* by Coal Measures, the contact being a thrust-plane, a situation resembling the Faille du Midi on the Continent. This thrust could lie just off the north coast of Exmoor and then trend inland to Cannington Park, near Bridgewater. The flattening of the cleavage dips at the North Devon coast could then be due to the thrusting. Another thrust could underlie the Glastonbury area (Bullerwell, 1964) or just north of Cannington Park (Webby, 1965).

Bott *et al* (1958) did however point out that the negative gravity anomaly over the Exmoor area could also be alternatively explained by a very thick mass of (L. Devonian?) sandstones beneath the exposed Devonian sequence of that region. Matthews (1974) believes that the Exmoor Thrust does not exist. He thinks that Devonian sequences thickening northwards from the line of a deep fracture situated south of Exmoor (such as the Brushford Fault) could suffice to explain the observed effects. Even more recently, Brooks, Bayerly and Llewellyn (1977) have doubted the existence of a major thrust under Exmoor. They substitute instead a simple geological model in which a thick sequence of relatively low density Lower Palaeozoic or late Precambrian rocks occupies the core of a structural culmination under the southern part of the Bristol Channel. A major fault with a large northerly downthrow has however been detected by Whittaker (1975) just SW. of Cannington Park.

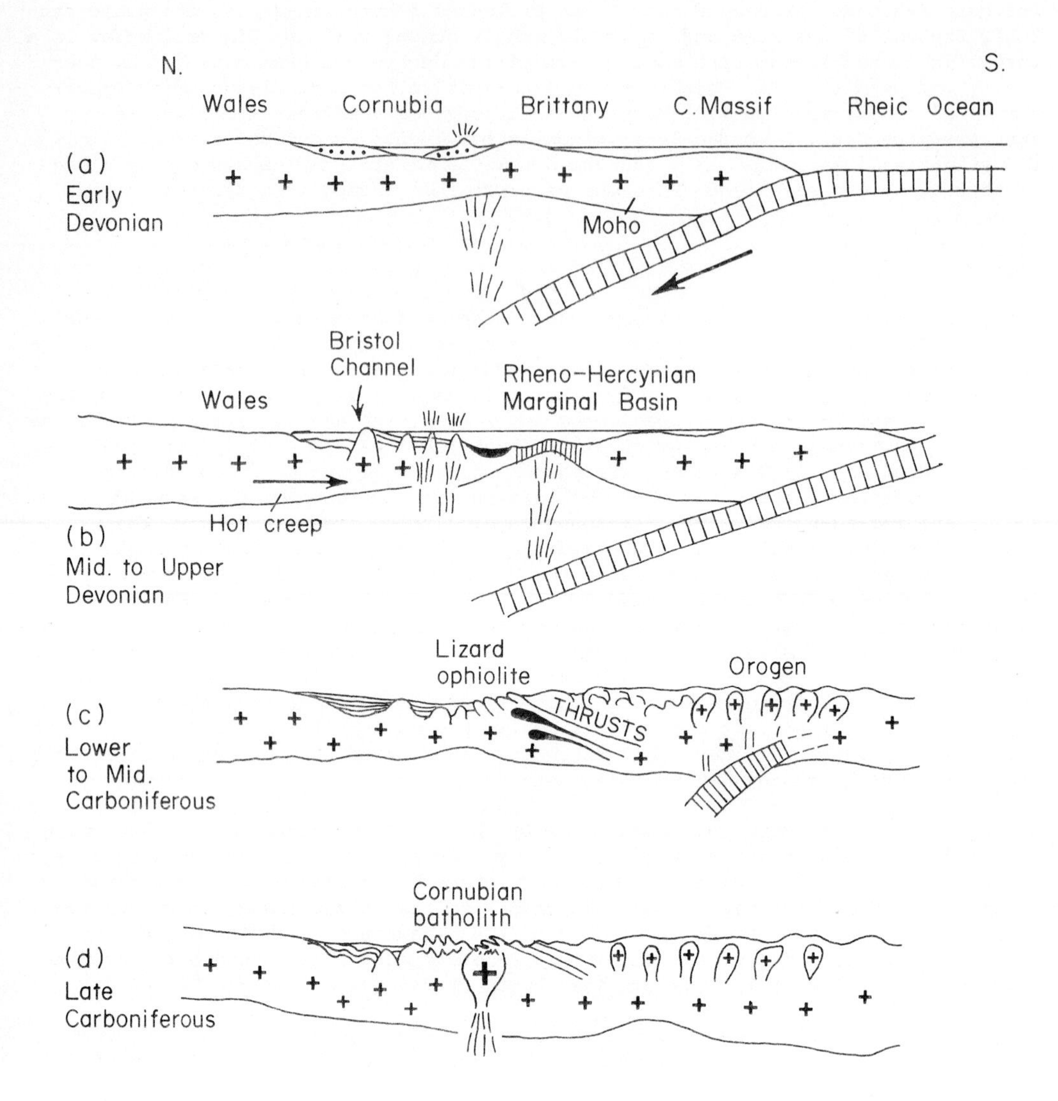

Fig. 7.5 Possible Plate-Tectonic Evolution of South Britain from Early Devonian to Late Carboniferous Times (based on Anderton *et al*, fig. 12.7).

The Mendips

Four great E.-W. periclines form the Mendip Hills and the upfolds still form the topographical "highs". From west to east, they are the Blackdown North Hill, Pen Hill and Beacon Hill upfolds and they are arranged *en echelon*, being offset further north in a westerly direction (Fig. 7.7). The two closest folds are the central pair, the intervening syncline being virtually cut out by the Emborough Thrust. All four folds are assymetrical (see, for example, Fig. 7.6), the northern limbs being steep, with 70° to 80° normal dips along the two outer folds and overturning in the case of the central structures. The southern limbs have dips of 25° to 40°. Much thrusting aligned NW.-SE. occurs in the Westbury area, south-west of the North Hill fold, as, for example, the Ebbor Thrust. The Blackdown upfold is relatively free of cross-faulting but such fractures increase in frequency eastwards across the Mendip belt. Two important cross-faults cut across the North Hill Pericline.The western (Stock Hill) fracture is clearly pre-Triassic whilst the eastern (Biddle) fault has moved again in post-Rhaetic times. Strike thrusts, especially on the southern limbs of the upfolds, have frequently originated at the lubricant horizon of the Lower Limestone Shales (Welch, 1933).

The echelon pattern of the Mendip flextures is obviously the result of a northward push against the N.-S. aligned bulwark of the Bristol-Radstock downfold and its parallel eastern upwarp, the Bath Axis. This resistance held back the eastern end of the Mendip ripple allowing the Blackdown Pericline to form freely to the west. With increasing northward push, cross-shearing accompanied by forward thrusting took place. The most severe final results of the increasing resistance in the east were (a) the thrusting of the Pen Hill fold tightly against the North Hill Pericline and (b) the heaving forward of klippen of Carboniferous Limestone and Millstone Grit on to vertical, overturned and highly contorted Coal Measures on the north side of the Beacon Hill fold (Fig. 7.6). It is indeed fortunate that such klippen as those of Vobster have been preserved to today. A further slight lowering of the topography and perhaps their existence would never have been believed!

The Bristol District

The Coal Measures of the Bristol area are only partly exposed being covered in places by Triassic and Jurassic beds. The geological structure is complex and now the folds begin to "box the compass" (Fig. 7.7). This is because the Hercynian compressive forces were beginning to be modified and twisted by older foundational structures such as the Lower Severn and Bath axes. Their influence had been felt earlier, e.g. in the Portishead-Clapton district, Pennant Measures rest on the Old Red Sandstone. Caledonoid folding occurred on a fairly severe scale in this western part of the Bristol area. A sharp anticlinal axis lies near to the Portishead-Clevedon coast, followed south-eastwards by a tight syncline south-east of Portishead Down. The Naish House and Tickenham thrusts then intervene before the Nailsea basin in Coal Measures is reached. North-eastwards the Clapton Basin gives way to the steep-sided Westbury-on-Trym Pericline, whose north-western limb virtually stands up on end as, for example, on Kingsweston Hill. The classic Avonian section of Arthur Vaughan forms the south-eastern flank of this upfold, which is succeeded north-westwards by the shallow Avonmouth Coalfield.

Between the Westbury and Blackdown upfolds lies the Broadfield Down Dome (Fig. 7.7) a relatively symmetrical fold exposing Tournaisian rocks in its core. Coal Measures intervene between this structure and the Mendip range.

The eastern side of the Bristol district is dominated by the partly buried Gloucester and Somerset Coalfield, essentially a N.-S. elongated basin but complicated by E.-W. compression belts such as the Kingswood Anticline and the Farmborough Fault-zone. The latter involves three major thrusts having a total throw

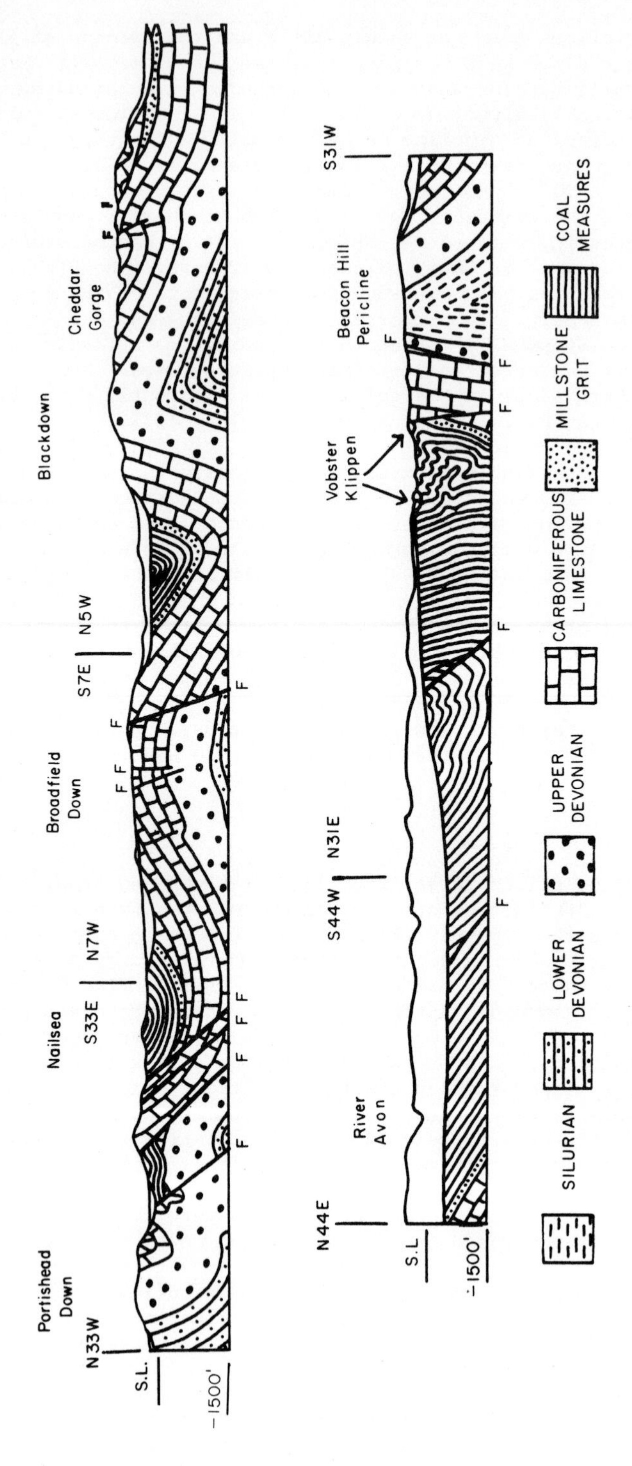

Fig. 7.6 Section Across the Mendip Folds (based on Geological Survey). (Unshaded areas represent Mesozoic blanket).

of 900 ft (275 m) accompanied by a shortening of the beds by nearly ¾ mile (Kellaway and Welch, 1948). Much incompetent folding and thrusting took place in the core of the Kingswood fold. The moderately steep eastern flank of the major coalfield is fairly well exposed in the north near Wickwar but then disappears southwards beneath the Mesozoic blanket, making one brief final appearance at Wick. The main coalfield is extensively faulted, the main fracture trends being N.-S. (e.g. the Iron Acton Fault), E.-W. and ENE.-WSW. Some fractures are buried under the Mesozoic cover, others cut Jurassic strata as high as the Great Oolite.

Post-Hercynian history

The Hercynian structures of South-west England, of course, dominate the geological picture of the region but nevertheless the area has had an interesting post-Carboniferous history though this must here be treated in a very brief fashion.

The present outline of the coast of South-west England appears to have been blocked out by erosion following the late-Carboniferous Orogeny (Hill and King, 1952), "valleys" of New Red Sandstone occurring close to the Bolt-Plymouth-St. Austell coast. The patches of New Red Sandstone near Bideford and off the Cornish coast are further proof that the Devon-Cornish Peninsula is an exhumed Permian upland. How much (if any) of the Triassic and Jurassic systems once covered Cornubia is difficult to say and the area may, through repeated uplifts, have kept "its head above water" until later Chalk times. The eastern regions of the area described in this section were, of course, extensively buried under Permian and/or Mesozoic strata, though certain areas, particularly the Mendips, stood out as "islands" for a time. Intraformational movements along minor "axes" or "blocks" account for thickness and formational changes in certain areas. A NNE.-SSW. axis of intra-Cretaceous uplift has been detected in the Beer district (Smith, 1957).

The full effects of the mid-Tertiary earth movements in South-west England are, as one might expect in areas of Palaeozoic rocks, not easy to detect but there are nevertheless some interesting clues. The NW.-SE and NNW.-SSE. faults, that affect many parts of Cornubia and especially the edges and margins of the granite masses, though initially perhaps of Hercynian origin, appear to have suffered dextral displacements in Tertiary times. One powerful zone, the Sticklepath-Lustleigh fractures, trends into the faulted Bovey Tracey Outlier of Oligocene strata (Blyth, 1962). Tremors on this fracture belt in 1954 and 1955 show it to be still active. Dearman (1963) has calculated the cumulative dextral-displacement along those charnoid faults to be at least 21 miles. "Putting back" the lateral shift places the southern margin of the Dartmoor Granite at the same latitude as that margin of Bodmin and dispenses with the awkward bend of both the Lizard-Start line and the Land's End-Bodmin-Dartmoor granite "batholith". It is also significant that sharp aeromagnetic anomalies trend north-westwards from the Tertiary Lundy Granite. Downfaulting at some time during the Tertiary (perhaps even as late as the Pliocene) has been postulated along the Exmoor coast (Simpson, 1961). The change in the coastal morphology at Combe Martin probably reflects the point at which the Devonian-Mesozoic boundary swings out to sea westwards (Donovan, 1961).

7.2 South Wales

Like South-west England, this region has been largely affected by the Hercynian earth movements and, as in the south-western area, these folds were intensely eroded and covered by a blanket of Mesozoic strata. Large areas of this cover still remain in the Vale of Glamorgan and some traces also occur in Gower and South Pembrokeshire. One important difference between South Wales and Devon and Cornwall, however, is the lack of cleavage in the deformed rocks of South Wales,

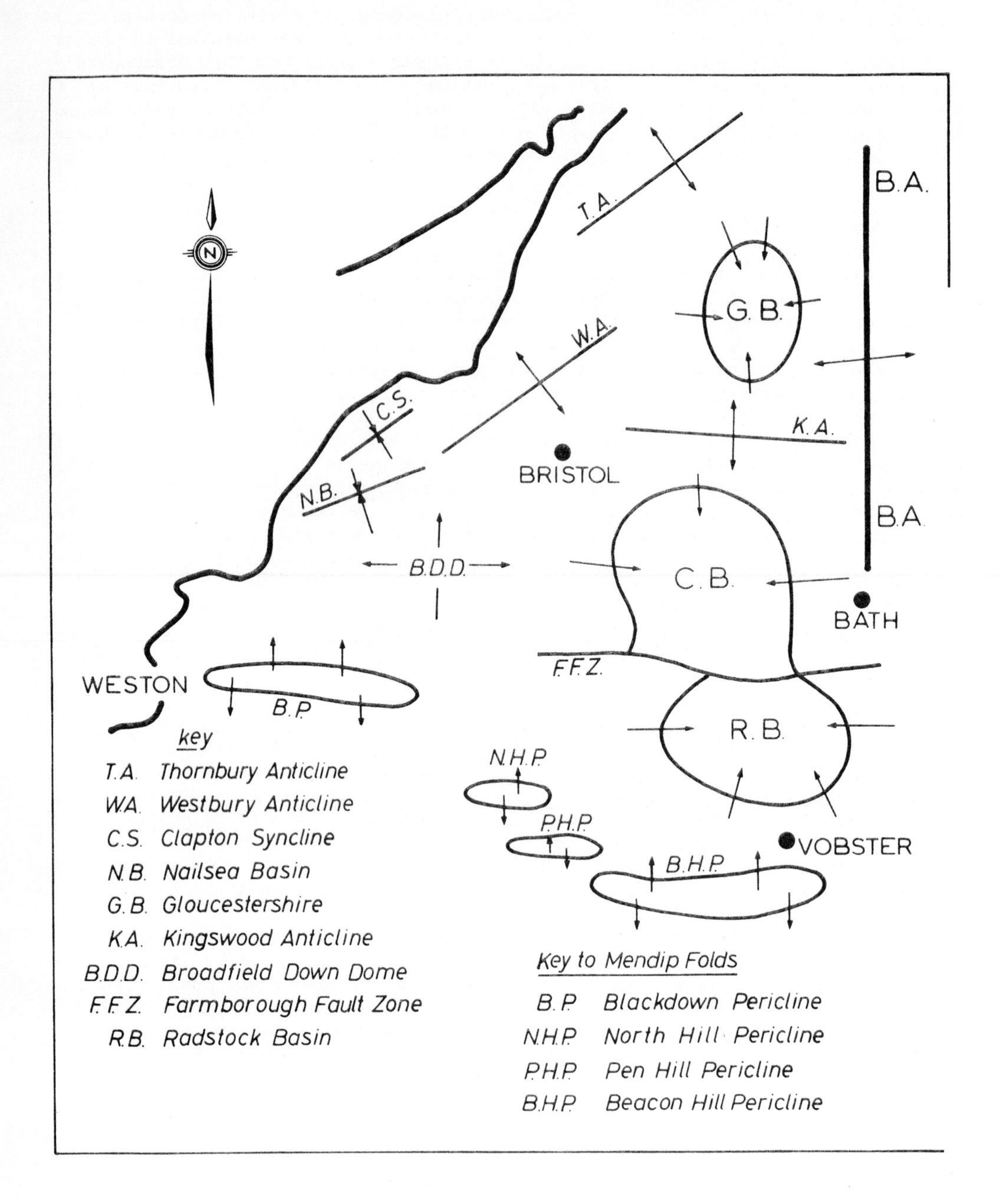

Fig. 7.7 Pre-Mesozoic Structure of the Bristol District (after Moore, 1951). B.A. Bath Axis; C.B. Central Basin.

apart from a partly developed fracture cleavage in some of the more affected rocks e.g. Devonian Red Marls of South Pembrokeshire.

It is convenient to take the northern boundary of the area dealt with in this chapter as being the base of the Upper Palaeozoic from St. Bride's Bay (through Narberth to Mynydd Eppynt and Kington). The eastern boundary is taken as a line from Kington through Hereford and Ross-on-Wye to Newnham, on the banks of the River Severn (fig. 7.8). The Forest of Dean Coalfield is thus included, whereas the Silurian inliers of Woolhope and Mayhill, together with the Malvern-Abberley axis, are best considered in the section on Central England (7.3).

The area is one of diverse relief, but rises generally from west to east. The highest areas are the Coalfield Plateau, reaching almost 2000 ft (510 m) O.D. at the Head of the Rhondda Valleys (and presenting marked outward scarps on its north-eastern, eastern and south-eastern sides) and the high Old Red Sandstone rim which forms the outer ramparts of the Coalfield through the Carmarthenshire and Breconshire Fans, the Brecon Beacons (highest point 2906 ft O.D.) and the Monmouthshire Black Mountains. Several points on this high rim reach 2000 ft (610 m) O.D. or over.

Further west, especially beyond the River Loughor, the Coalfield surface falls appreciably to form dissected relics of coastal plateaux at 183, 122 and (especially in the Pembrokeshire Coalfield) 60 m above present sea-level. Large areas in South Pembrokeshire, Gower and the Vale of Glamorgan are part of the marine-bevelled (200 ft, 60 m) surface. The bevel is more striking here because of the steeper dips of the Hercynian fold-limbs.

The coast of the area is a very indented one, the main inroads by the sea being (from west to east) St. Bride's Bay (carved into softer Coal Measures), Milford Haven (the line of the Ritec Thrust), Carmarthen Bay and Swansea Bay (both these being carved again essentially out of Coal Measure and Millstone Grit shales). A large number of rivers drain the Coalfield, including the Gwendraeth, Loughor, Tawe, Neath, Avan, Rhondda, Taff and Ebbw. The River Usk rises on the flanks of the Towy Anticline and then encircles the eastern fringe of the Coalfield, cutting through, in the process, the Usk upfold (exposing Silurian rocks) which separates the South Wales and Dean Forest Coalfields. The River Wye meanders a most irregular course from Hereford to Ross and around the western confines of the Forest of Dean to Chepstow. The deeply incised nature of these Wye meanders affords striking proof of the fairly recent uplift (or eustatic fall) of the region.

The greatest sea inroad is, of course, the Bristol Channel, which represents a downfolded and faulted belt of Triassic and (especially) Jurassic strata, unconformably overlying folded and eroded Upper Palaeozoic strata. A more detailed discussion of these Channel structures follows later(Ch. 10).

The succession in the area is as follows:

Lower Lias
Triassic (Keuper and Rhaetic)
(great unconformity)
Coal Measures
Millstone Grit
Carboniferous Limestone
Old Red Sandstone
(unconformity in places)
Silurian
(unconformity)

Ordovician
(unconformity)
Cambrian
(unconformity)
Precambrian

In Central Pembrokeshire, Precambrian igneous rocks are represented by an extrusive Benton Series (felsites and rhyolites) and an intrusive Johnston Complex (quartz diorites, quartz dolerites and hybrid granites). The Johnston Complex has recently been dated at 643 m.y. (Patchett and Jocelyn, 1979). Recently Precambrian volcanics and sediments (the latter bearing an Ediocaran fauna) have been identified near Carmarthen by Dr. J.C.W. Cope and colleagues.

Some Upper Cambrian strata have also now been identified near Carmarthen. Ordovician rocks occur in two small anticlinal inliers in South Pembrokeshire (at Freshwater East and Freshwater West). In these fold cores, only Llanvirn Beds are exposed and appear to be overlain unconformably by Wenlock strata (Walmsley and Bassett, 1976). The Llanvirn shales are dark grey or blue-black, and occasionally cleaved, the cleavage ranging W.15°N. with a steep southerly dip.

Only the upper portion of the Llandovery Stage is represented in the area, and these Lower Silurian rocks occur only to the north of Milford Haven. Near Rosemarket, on the southern side of the Pembrokeshire Coalfield, the Upper Llandovery arenites rest unconformably on the Benton Series. At Marloes Bay the Upper Llandoverian is represented by conglomerates, sandstones and mudstones. At Marloes and Wooltack these high Llandoverian strata rest on the Skomer Volcanic Series, a complex alkali-basalt volcanic pile best seen on the island of Skomer. These extrusives were once thought to be of Arenig age but are now known to be of late Ordovician to Llandovery age (Ziegler *et al*, 1969).

The overlying Wenlock and Ludlow are well exposed at Marloes Bay but much thicker sequences occur in the Usk Inlier, where the Wenlock comprises at least 800 ft (244 m) of calcareous mudstone, micaceous sandstone and massive nodular limestone, whist the Ludlow Series includes up to 2500 ft (762 m) of siltstones with bands of limestones. At Marloes, the highly calcareous Wenlock mudstones are overlain by greenish-grey Wenlock sandstones and mudstones passing up gradually into red beds of probably Ludlow age. The importance of the Ritec line through Milford Haven is shown by the higher stratigraphical base of the Silurian to the south of the estuary than to the north of it (Dixon, 1921) and an important fundamental basement fracture probably exists in this vicinity. Another - the Benton Fault - lies, as was shown in Chapter 6.1, just to the north. The effect of the Ritec line is greatest at Freshwater West where only 36 ft of Wenlock Beds separate the Lower Llanvirn and the Downtonian.

Early Caledonian movements appear then to have occurred in South Pembrokeshire in mid- and/or late Ordovician, early Llandovery, pre-Wenlock and pre-Ludlow times. Late Silurian-earliest Devonian movements appear to have concentrated on the area south of the Ritec line where the Old Red Sandstone is unconformable on the Salopian and the south side of the Towy Anticline (especially its western continuations to the south of Carmarthen and Whitland) where the Downtonian oversteps the Silurian to rest on horizons low in the Arenig. In other areas, e.g. Marloes and Usk, there is virtual conformity of Silurian and Devonian. The intensity of the break at Carmarthen, however, is not as severe as it first seems, because much of the Ordovician and Lower Silurian could have been removed by late-Ordovician and early to mid-Silurian unrest along the Towy Anticline.

The Devonian System occupies a large part of the surface of South Wales, more especially north-east and east of the Coalfield. The total thickness is about

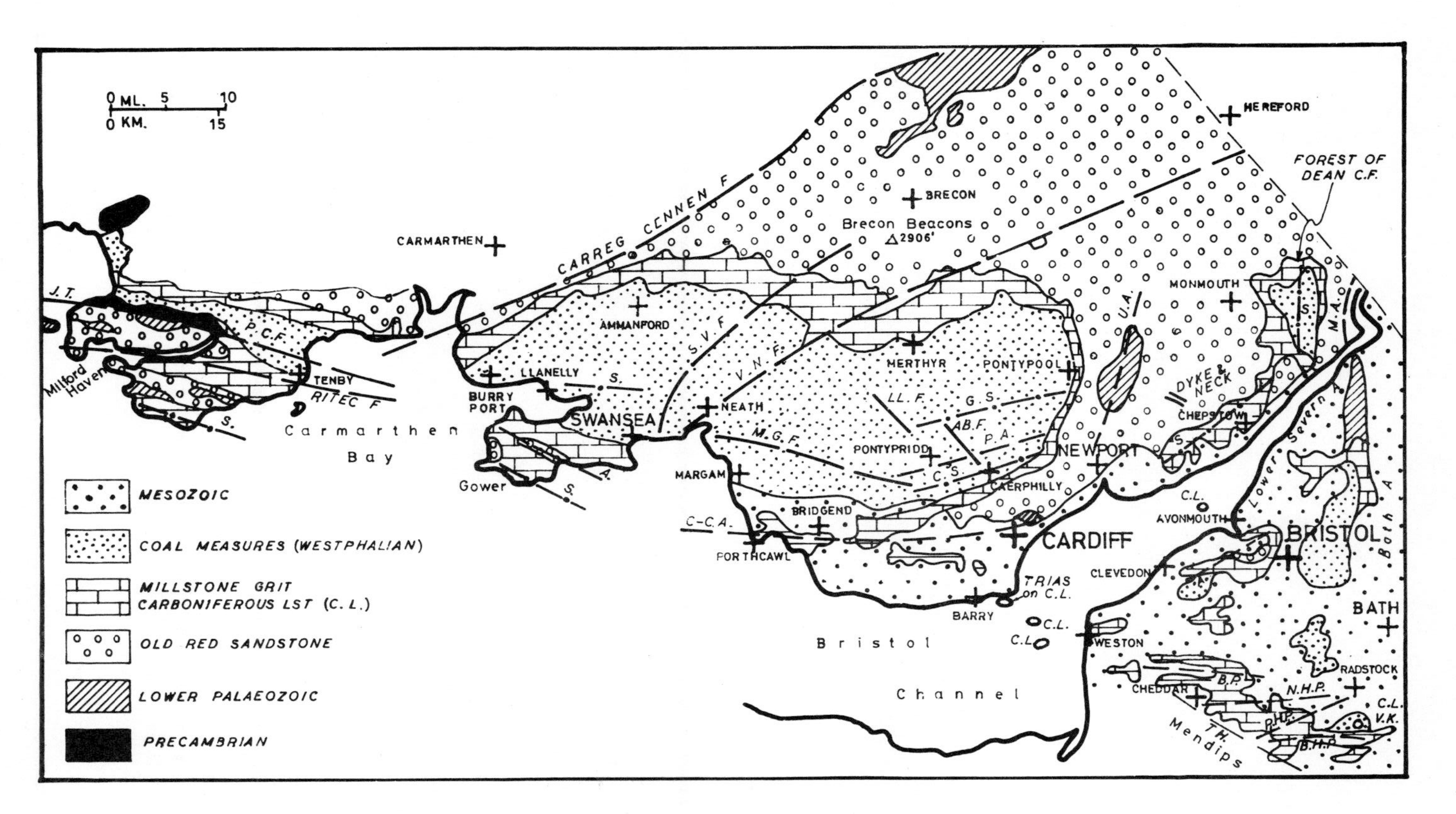

Fig. 7.8 Structural Map of South Wales, Pembrokeshire, Forest of Dean and Bristol Coalfields.
AB.F. = Aber Fault; B.H.P. = Beacon Hill Pericline; B.P. = Blackdown Pericline; C.-C.A. = Cardiff - Cowbridge Anticline; C.L. = Carboniferous Limestone; C.S. = Caerphilly Syncline; G.S. = Gelligaer Sycline; J.T. = Johnston Thrust; LL.F. = Llanwonno Fault; M.A. = Malvern Axis; M.G.F. = Moel Gilau Fault; N.H.P. = North Hill Pericline; P.A. = Pontypridd Anticline; P.C.F. = Pembrokeshire Coalfield; P.H.P. = Pen Hill Pericline; S.V. = Swansea Valley Fault; TH. = thrust; V.K. = Vobster Klippe; V.N.F. = Vale of Neath Fault.

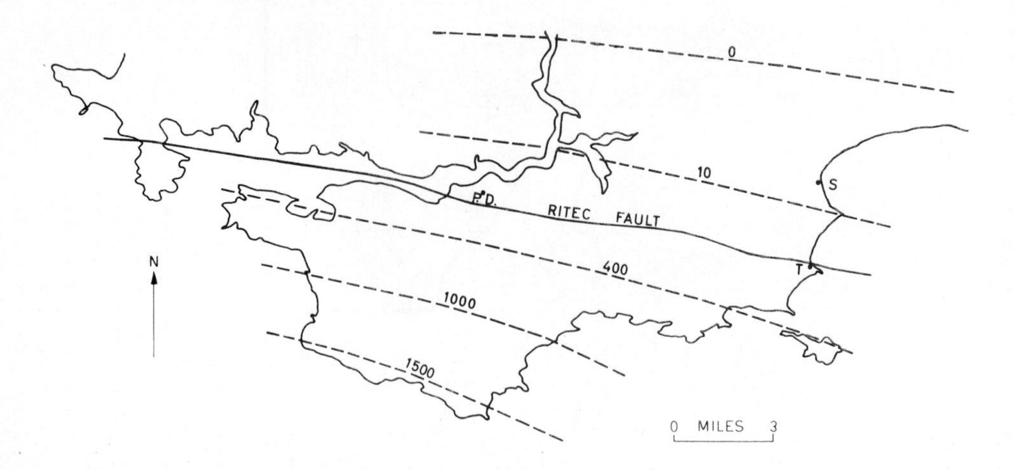

Fig. 7.9 Variations in Thickness of the Upper Carinia Zone (C_2S_1) of the Dinantian in South Pembrokeshire (based on Sullivan). Thickness in feet.
P.D. Pembroke Dock; S, Saundersfoot; T, Tenby.

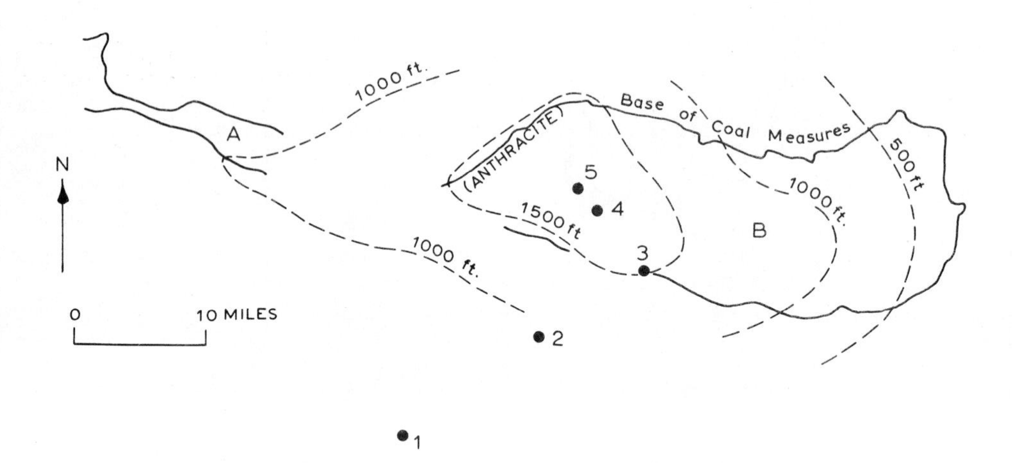

Fig. 7.10 Diagram Showing the Varying Position of Maximum Thickness During Carboniferous Times in South Wales (after Owen).
1. Dinantian; 2. Namurian; 3. Basal Westphalian; 4. Middle Coal Measures; 5. Upper Coal Measures.
Isopachytes Refer to Middle Coal Measures Only.

5000 ft (1525 m) in Carmarthenshire and Breconshire, but was reputed to be thicker in the Neyland-Cosheston area just north of the Milford Haven. Here again the influence of the Ritec Fault is seen, the Lower (and Middle?) Devonian sequence being different on either side of the estuary. The southern derivation for the Ridgeway Conglomerate (Williams, in Owen *et al*, 1971) points to appreciable mid-Devonian uplift and erosion of an area somewhere to the south of Pembrokeshire (see also George, 1962a, p.26).

The general absence of Middle Devonian strata in South Wales suggests widespread uplift, and on the dip slopes of the Fans at the head of the Tawe Valley the slight discordance in dip between the Upper Devonian Plateau Beds and the Lower Devonian Brownstones can be clearly discerned. More violent uplift may well have been located along the Towy Axis and the very steep dips of the Red Marls and Senni Beds along the Careg Cennen Disturbance may be, at least in part, of pre-Upper Devonian origin.

The Lower Carboniferous rocks comprise thick successions of calcareous rocks - oolites, crinoidal limestones, dolomites, calcarenites - which reach maximum thicknesses (5000 ft, 1525 m) in south-west Gower and southernmost Pembrokeshire. The only hints of unrest in these southern areas are the lagoonal character of the basal Visean sediments and the slight scouring of the underlying Caninia Oolite, as for example, in Caswell Bay. The best signpost to the greater unrest along the northern and eastern rims of the Coalfield is perhaps again the Ritec line, shown by Sullivan (1965) to have separated very thin Upper Caninian deposits to the north from suddenly thicker sedimentation to the south (Fig. 7.9). Pulsatory uplifts along the Towy Axis and the Neath Disturbance (George, 1954, 1963a) in mid-Dinantian times were forerunners of movements which in mid-Visean to early-Namurian times were to raise, for the first time, an Usk upfold (George, 1956; Owen and Jones, 1961) causing flooding of the calcareous flats with sand and mud on the north-eastern rim of the Coalfield, gaps in the Visean-Namurian sequence in areas such as Llanharry and Rudry, and fairly intense folding in the Lydney-Tiddenham area.

Namurian uplift of St. George's Land supplied further sand to the northern sections of the Coalfield, whilst the possibility of an evolving barrier between South Wales and North-east Devon has been noted (Owen, 1964). Very attenuated and mainly arenaceous sequences on the north-eastern and eastern rims of the Coalfield point to continued uplift along either an Usk or a broader Malvern upfold. Slumping along the line of the Neath Disturbance (see Owen *et al.*, 1965) was probably a seismic effect, and similar effects occurred at the same time in Pembrokeshire. Namurian overstep on to St. George's Land was very rapid in Pembrokeshire (George, 1962a), the Dinantian disappearing beneath a transgressive Reticuloceras Stage which eventually rests on Lower Palaeozoic strata in the St. Bride's coastline.

By Westphalian times, then, a number of eventual Hercynian elements were already being anticipated including the Towy Axis, the Neath Disturbance, the Usk Anticline and the Ritec Fault. What of the Coalfield itself? Sedimentological and stratigraphic studies suggest that the South Wales Coalfield became, in terms of differential subsidence and *periodic* uplifts of surrounding areas (even to the south), a basin in its own right as Westphalian times progressed (see Bluck and Kelling, 1963; Kelling, 1964; Owen, 1971). Moreover, as Coal Measure times progressed (Fig. 7.10) the centre of greatest subsidence appears to have moved progressively into the present Coalfield and towards the north-western portion, i.e. towards the anthracite field (see Owen, 1971). The possibility of very thick uppermost Westphalian and Stephanian sediments being deposited over the anthracite belt from Ystradgynlais to Kidwelly cannot be ignored (see also O. T. Jones, 1956; Wellman, 1950). The load hypothesis could still account at least in part for the unique three-dimensional occurrence of the low volatile coals in South Wales.

Today the Coal Measures are thickest in the Swansea district (9000 ft, 2745 m) and tend to thin off eastwards and north-eastwards.

The South Wales Coal Measures are subdivided as follows:

	(Upper Pennant Measures
(iii) Upper Coal Measures	(
	(Lower Pennant Measures
(ii) Middle Coal Measures	
(i) Lower Coal Measures	

The Lower and Middle Coal Measures are predominantly argillaceous with occasional sandstones, more particularly towards the top. Coals and fireclays are numerous and a number of marine horizons form useful markers. The Upper Coal Measures are predominantly arenaceous, particularly the Lower Pennant Measures, which comprise the thick felspathic Pennant sandstones, responsible for much of the plateau in the centre of the Coalfield and shown by Kelling to be derived largely from southerly or easterly sources. On the south-eastern rim, Moore (1947) has claimed a sharp unconformity at Risca, with the Pennant overstepping almost on to Namurian. More recently, Squirrell (1965) has demonstrated that the Lower and Middle Coal Measures are present but in a very condensed sequence. In other words the Usk Anticline still exerted an influence, but acting in a more continuous way with prolonged slow positive uplift. Short-lived, more intense movements are perhaps represented by the occasional influx of coarser conglomerates into the East Crop Pennant succession.

In the Forest of Dean Coalfield, the Upper Carboniferous succession is limited to the Morganian (i.e. Upper Coal Measures) though Sullivan (1964) has shown that a thin Ammanian remnant is present near Mitcheldean. Whether the Namurian and the bulk of the Ammanian was ever deposited over the Dean Forest area is not known. The probability is that such deposition was very slow because subsidence was slow or periodically halted, so that any deposits were themselves removed or "handed on" to adjacent depositional areas. Sullivan's remnant represents measures that were fortunately preserved from such periodic removal.

Towards the close of Carboniferous times, South Wales was affected by the main Hercynian climax. Because the impulse was from the south or south-south-west, the greatest effects were felt in the three southern peninsulas of Pembrokeshire, Gower and the Vale of Glamorgan. Because of the south-westward trend of the Welsh Block, and because of the relatively sheltered behaviour of areas like east Breconshire and north Monmouthshire, the Hercynian folds were driven much tighter against the Caledonian structures of Pembrokeshire, this tight effect then decreasing eastwards so that in the Vale of Glamorgan, (a) the folds are much gentler (dips of over 30° are in fact exceptional here) and (b) the folds are beginning to abandon their WNW.-ESE. trend and turn more into a north-easterly attitude - the influence of the prolongation of the Usk line. Traced northwards, too, the intensity of the folding decreases into the Coalfield so that the northern flanks are relatively uniformly dipping except where crossed by such Caledonoid belts as the Neath, Tawe, Ammanford and Careg Cennen disturbances (Fig. 7.8).

The major structural elements produced by the Hercynian Orogeny in South Wales were:

The intense folds of South Pembrokeshire
The folds of Gower and the Vale of Glamorgan
The main South Wales Coalfield
The Pembrokeshire Coalfield
The Neath and Swansea Valley Disturbances
The Careg Cennen Disturbance
The Usk Anticline
The Forest of Dean Coalfield.

South Pembrokeshire

Here the folds are closely packed and vertical limbs are common. The folds trend about 10°N. of W. and include the Sageston, Ridgeway, Freshwater East and Castlemartin Corse anticlines, and the Pembroke, Angle and Bullslaughter Bay synclines. Most of the folds pitch eastwards, the exceptions being the Angle Syncline (a boat-shaped trough), the Bullslaughter Bay downfold and the Freshwater East Anticline. The more northern folds have axial planes inclined southwards whilst those of the southernmost folds are inclined northwards. Buckling of folds along the axis, or in a limb, is common (Dixon, 1921). Axial buckling is well marked along the middle of the Freshwater East-Castlemartin Corse anticlinal complex (often referred to as the "Orielton Anticline").

The majority of the cross-faults are wrench-faults. Those that trend NNW.-SSE. are dextral, whilst the majority of the NNE.-SSW. are sinistral. The NNW. fractures are by far the more abundant. The dips of these faults range between 70° and 90° and horizontal slickensiding is common, as also are fault breccias (Anderson, 1951). The horizontal movement appears for the most part to be post-folding, as there is an obvious match of flexures on either side of the faults, e.g. the Popton Fault. In a few cases the wrench-faults pass into thrusts, e.g. the north-north-west trending Manorbier Fault turns west-north-westwards into the Manorbier-Newton Thrust. The Ritec Fault, and its branches, constitute one of the most important thrust-belts in the area. It trends from the Tenby district through the Milford Haven and crosses the narrow isthmus north of the Dale Peninsula. The northerly downthrow of the chief Ritec Fault increases westward, being 1500 ft (458 m) near Flemington and over 3000 ft (916 m) near Llanreath. It is difficult to estimate its throw at Dale because of its position within the Red Marls. Shear-planes in Seminula Oolite, close to the fault on St. Mary's Hill, dip southwards at angles between 55° and 90°. Shear-planes on the south side of Castle Hill, however, dip at less than 40°.

Hancock (1973) has delineated a number of Variscan structural zones. His sub-zone Ia, e.g. lies south of the Ritec Fault and is characterised by large asymmetric folds, some periclinal, trending WNW. There are well defined belts of both south and north verging folds. Parasitic folds are very common.

More recently, Hancock has added a fourth sub-zone (Id) to his 1973 work. This sub-zone Id, 1 km wide, is characterised by outward-facing conjugate folds and defines the Variscan Front (personal communication).

Gower and the Vale of Glamorgan

The majority of the folds in the Gower Peninsula trend WNW.-ESE., but tend to turn into a more E.-W. direction when traced eastwards to Caswell Bay and Mumbles. The major anticlines are those of Cefn Bryn, Llanmadoc Hill, Harding's Down (all exposing Old Red Sandstone), Oxwich Point, Coltshill and Langland. The main synclines, Oxwich, Port Eynon and Oystermouth, preserve the Namurian. The exact form of the structure of Rhosilli Down, another Devonian inlier, is not clear, though this is probably due to the cutting off of the western side of the structure by the Broughton Fault. The limbs of the folds are frequently steep and overturning occurs on the northern limb of the Langland Anticline.

Thrusts trend parallel to the fold axes and include the Cefn Bryn, Port Eynon and Bishopston fractures. They dip southwards, a notable exception being the Caswell Thrust which dips northwards at 50° to 60°. A large number of cross-faults trend NNE.-SSW., N.-S. and NNW.-SSE. At first glance it seems that horizontal movement along these dip faults post-dated completion of the folds but in detail it can be seen that the fold intensity usually differs across the fracture, e.g. the Oxwich Point Anticline (see George, 1940, p.175). George has clearly demonstrated that the folding, thrusting and cross-faulting were more or less contemporaneous and marked contrasts in structure occur across many of the cross-faults. Rotation has also occurred in blocks between the wrench-faults, e.g. in Mewslade Bay (George, 1940, fig. 13).

Roberts (1979) has described the jointing and minor tectonics of the South Gower coastal areas and has shown that a systematic pattern of up to 6 sets of joints exists. The systematic joints are thought to have developed during the folding under the influence of a N.-S. compression.

In the Vale of Glamorgan the Carboniferous Limestone and Old Red Sandstone can be seen to have been folded before the (now partially removed) blanket of Mesozoic rocks was laid down. The main fold is the Cardiff-Cowbridge Anticline, exposing Old Red Sandstone and, in the Cardiff area, the Silurian. Further west, the Candleston Anticline runs parallel with the main Cowbridge fold but lies some miles to the north-west. Within the major Cowbridge flexure, minor elements are arranged in similar echelon form. The majority of the faults in the area are clearly of post-Liassic age and will, therefore, be discussed later.

The South Wales Coalfield (fig. 7.8) is a major downfold extending for 60 miles (96 km) from the Burry Estuary in the west to Pontypool in the east, and preserving a width of about 20 miles (32 km) over most of its extent. This major basin structure includes a number of minor folds, the Pontypridd and Maesteg anticlines, the Gelligaer, Caerphilly-Llantwit and Llanelli-Gowerton synclines. As noted earlier, a marked north-eastward turning in axial trend takes place eastwards across the Coalfield. On a small scale, sharp deflections sometimes occur locally, as, for example, the Llantwit Syncline just north of the Miskin Fault. East-west faults, for example, the Moel Gilau and Meiros faults, sometimes replace the fold axes. East of the Llanwonno Fault, the Pontypridd Anticline is comparatively gentle and symmetrical, but further west the southern limb steepens to 50° and the fold becomes monoclinal. At the horizon of the Yard Seam, intense compressional disruption was encountered in the core of the anticline at the Maritime Colliery. The Maesteg Anticline is a continuation *en echelon* of the Pontypridd upfold.

The Coalfield is extensively faulted, the main fault direction being NNW.-SSE. (backing to a more north-westerly directions in Monmouthshire), NE.-SW. and E.-W. (becoming WNW.-ESE., in Gower). A good classification has recently been given by Woodland and Evans (1964) with respect to the faults in the Pontypridd-Maesteg districts. They have recognised the following groups:

(i) strike-thrusts, confined to the South Crop;
(ii) normal strike-faults;
(iii) dip or cross-faults;
(iv) incompetence faults.

The first group include the Margam and Llanharan thrust-belts. The Margam zone involves four major thrusts (Fig. 7.11) and at least two lag faults. The Newlands Fault is the most important thrust member with two smash belts some 80 yards apart. The stratigraphical throw is up to 1000 ft (305 m) in places and the fracture planes dip northwards at 50° (the regional stratal dip being 20°). The Margam thrusts seem to be associated with synclinal drag-folds developed beneath the major planes of movement. The Llanharan Thrust brings Llynfi Beds against the lighter beds of the Lower Coal Measures near Llanharan Colliery, where again a major synclinal drag-fold has been proved to occur on the underside of the thrust. These large drag-folds beneath the fracture planes suggest to Woodland and Evans that the faults developed by the under limbs of the folds being pressed northwards beneath the upper limbs, i.e. they are underthrusts. Their passage into either tough Pennant rocks or Millstone Grit shows that they are not incompetence structures. They pre-date group (iii) as cross-faults cut the Margam and Llanharan belts.

At Cribbwr Fawr Colliery, two sets of shear-planes are developed, one parallel to the Newlands Thrust, the other (subordinate in incidence) dipping at only 6° to the north. The bisectrix of the included angle therefore dips north at 28° which is close to the regional dip of the strata. This important evidence shows that the thrusts were developed (one plane direction becoming, as is so often the case, the dominant one) when the strata were horizontal and had been later tilted during the folding.

The second group were formed during the growth of the folds, e.g. the Meiros Fault (co-extensive with the Meiros Anticline) and the Cilely group of fractures occurring on the northern limb of the Pontypridd Anticline.

Examples of groups (i) and (ii) occur in other parts of the South Wales Coalfield. Group (i) structures were described by Robertson (1933) in the Rhigos district (p. 196) and these structures were recently well exposed by opencast workings. The anthracite belt of the Gwendraeth Valley and Cynheidre region is also notorious for these thrust and lag structures, as also are the Cross-Hands and Ammanford areas (Trotter, 1947).

Woodland and Evans have thrown a good deal of light on the nature and origin of the cross-faults (group (iii) in the Pontypridd sector of the Coalfield. These dip faults are the most commonly occurring fractures in South Wales and have aroused considerable controversy, with origins ranging from tensional to wrench-faulting. Woodland and Evans show that in the Pontypridd sector their throws are not large, only some four faults having throws greater than 500 ft (or 152 m), and maximum throws are frequently developed along the axes of synclines (which the fractures so often favour). There is an obvious relationship between the linear extent of the cross-faults and the limits of the major synclines. Differential folding on either side of the fractures is implied, and this is shown by the behaviour of the Pontypridd Anticline across the Aber Fault. Not only is the fold shifted dextrally on the west side of the fracture but has at the same time become a much gentler flexure. The fault, therefore, developed at the same time as the folding. Where structures can be matched, eastward-throwing faults are sinistral, and westward-throwing ones dextral.

The development of fault movement contemporaneously with the formation of folds must now, therefore, be the important character of the South Wales cross-faults, as this feature has been stressed by George (1940) for the Gower cross-faults, by Trotter (1947) for those of the anthracite belt, and by Owen (1954) for the dip faults that occur on either side of the Neath Disturbance. Woodland and Evans make one further, very important point, viz. that the Pontypridd cross-faults were incipient tear faults (*arrested* at an *early* stage in their development, whereas those of Gower and especially Pembrokeshire represent a more advanced stage). Both

Trotter (1947) and Owen (1954) have detected later tensional movements along pre-existent, wrench-type, cross-faults with the formation of faulted troughs.

Throughout the Coalfield numerous minor thrusts and lag faults affect the more productive, but also more incompetent, portion of the Coal Measures. Throws range from 1 to 300 ft (92 m) and, as Woodland and Evans point out, their size and frequency give an unenviable reputation to the South Wales Coalfield. Spectacular examples of these incompetence disturbances occur in the Tondu Brick-pit (between Maesteg and Bridgend) at about the horizon of the Amman Marine Band.

A major structure which appears to pre-date the main folding in the Pontypridd sector is the so-called Jubilee Slide, extending for 12 miles (19 km) from the Gilfach Goch area towards Pontrhydyfen. This belt is made up of some sixteen arcuate normal faults trending WNW.ESE. but arranged *en echelon*, and with throws up to 300 ft (92 m). They are branches of a flat-lying continuous dislocation in the midst of the main group of productive seams. This dislocation is not a lag but rather a rotational "landslip" (Woodland and Evans, 1964). Another echelon fault system, also trending mainly WNW.-ESE. (but turning west-south-westwards towards Swansea Bay) is the Ty'n-y-Nant and Moel Gilau Fault System which has the main effect of repeating the Pennant escarpment near Baglan and Aberavan. The appreciable southerly downthrow increases to 3000 ft (915 m) west of Maesteg and reaches 3750 ft (1144 m) near Cwmavon. The normal dip of the fault probably varies from 20° to 30° near Baglan (in the west) to as much as 70° further east. The western, low-dipping portion is claimed by Woodland and Evans to represent a low tectonic level. No cross-faults shift the line of the Moel Gilau Fault and there appears to be a 400-yard (366 m) net sinistral shift of matchable displaced cross-faults. Like the Jubilee Slide, the belt was possibly initiated before folding and later appreciably reactivated.

The Pembrokeshire Coalfield is a much narrower belt with a relatively gentle northern flank but with a much contorted and thrust central and especially southern margin. The southernmost boundary is, in fact, the great Johnston Thrust towards St. Bride's Bay, this fracture throwing Precambrian or Silurian rocks against the Coal Measures. The hade of the thrust is not known. Its highly irregular surface outcrop could indicate a fairly low dip but, on the other hand, the fracture could have been itself involved in late folding. The coastal sections along St. Bride's Bay from Settling Nose to Falling Cliff (including the famous Sleekstone Anticline) and from Saundersfoot to Monkstone Bay display the spectacular folding and thrusting of the Ammanian rocks. (These structures fall within Hancock's Zone Ic whereas the Coal Measures north of Druidston Haven fall into his Zone II). These sections have been described in detail in the Milford and Haverfordwest *Survey Memoirs* and more recently by Jenkins (1961). The majority of the anticlines have steeper (even overturned) northern limbs and most of the thrusts dip southwards at 20° to 50°. Many thrusts occur along coals and fireclays and some thrusts are themselves folded.

The Neath and Swansea Valley disturbances trend NE.-SW. across the central portion of the Coalfield and cut the northern Limestone and Millstone Grit rims. They then cut across the Old Red Sandstone belt of Breconshire-Monmouthshire into Herefordshire, the Neath Disturbance extending as far as the Malvern-Abberley line at the point where the River Teme breaks through. The Swansea Valley Fault probably extends through Glasbury and is probably represented on the south-eastern flank of the Titterstone Clee Outlier (see Weaver, 197). Proto-Hercynian movements are known to have occurred along both these disturbances (see Owen, 1971). Weaver (1976) has shown that contemporaneous movements occurred along the Swansea Valley belt in Coal Measure times.

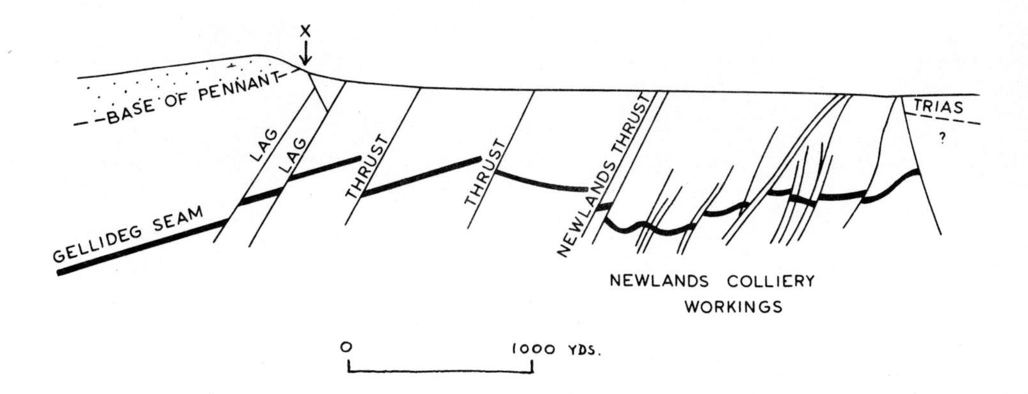

Fig. 7.11 The Margam Fault-Zone, South Wales (after Woodland and Evans).

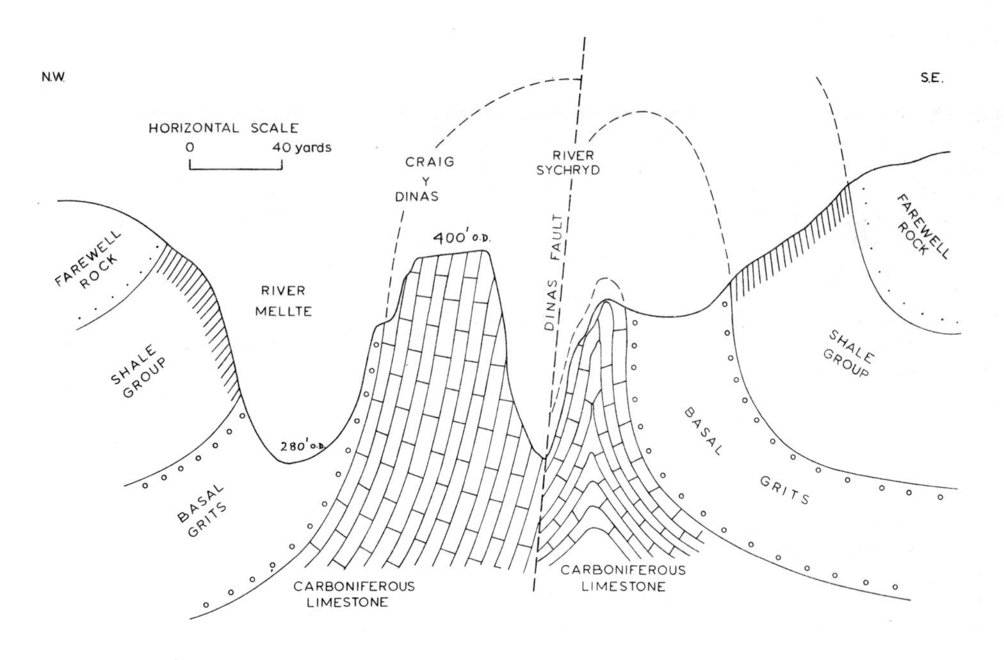

Fig. 7.12 Section Across the Neath Disturbance, South Wales (after Owen).

S.S.E.

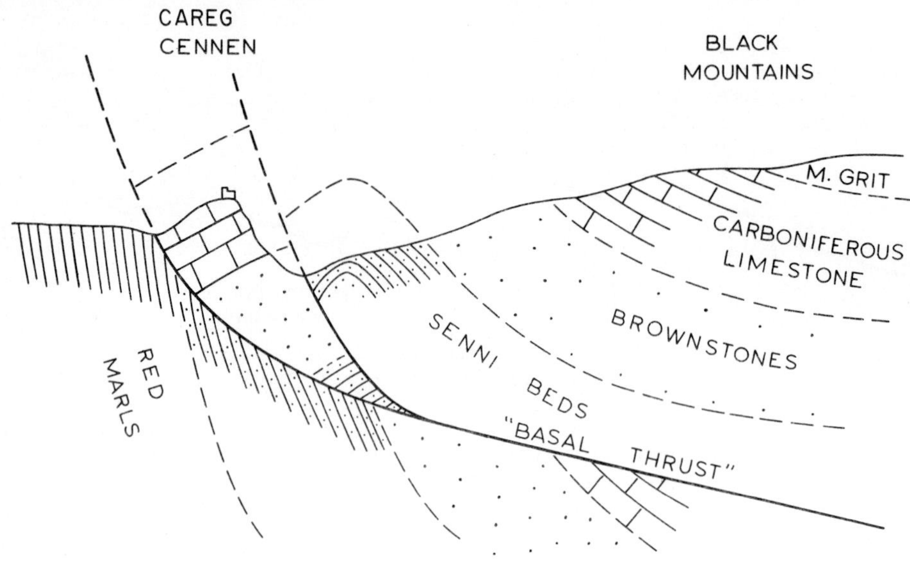

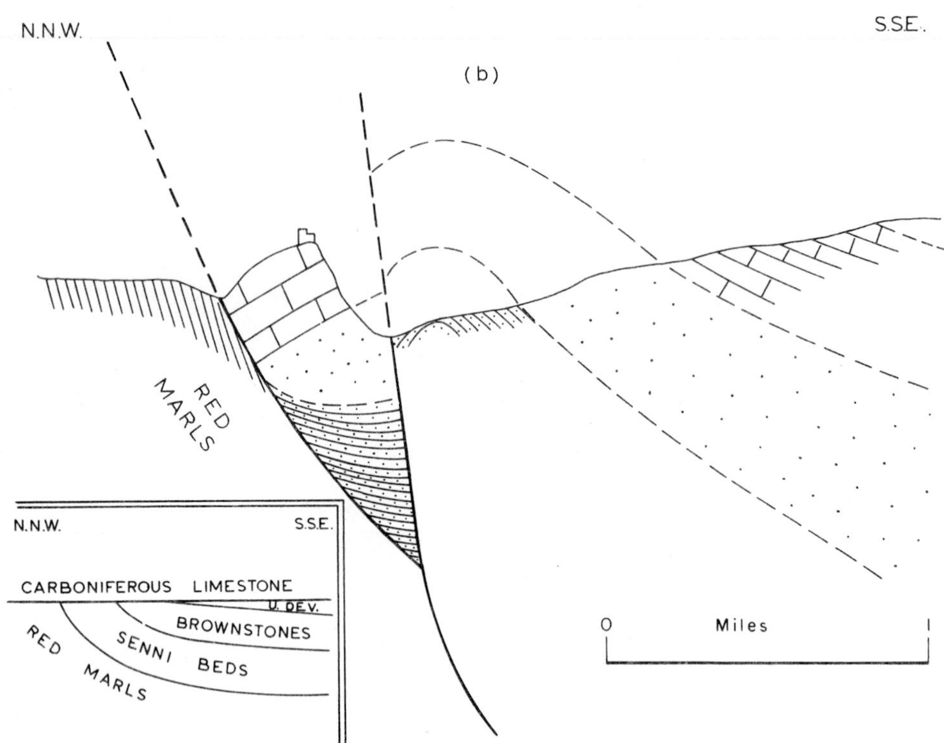

Fig. 7.13 Sections Across the Careg Cennen Disturbance, South Wales. (a) After Troter, 1948. (b) Alternative explanation? Inset: Possible Northward Oversteps Beneath Upper Devonian and Carboniferous Limestone.

These two disturbances involve very narrow belts of Caledonoid folding and faulting. The folds are somewhat impersistent (see Owen, 1954) but often locally sharp (Fig. 7.12). The main fractures in both disturbances are wrench-faults with sinistral displacements of about three-quarters of a mile. The Duffryn Trough is in turn shifted by this amount at Pontardawe and Neath by the two wrench-faults. The folding, and somewhat later vertical fault adjustment, are responsible for the preservation of small Carboniferous outliers to the north of the main rim of the Coalfield, as, for example, on Fan Fraith, in the case of the Swansea Disturbance, and Dolygaer and Bryniau Gleision along the Neath Disturbance. Owen (1954) has suggested that the anomalous trend of the latter is due to basement control by a deep-seated fracture belt which resolved the NNE.-SSW. Hercynian compression along its length. A similar conclusion (and comparable sinistral shift) is reached by Woodland and Evans for that portion of the Neath Disturbance which falls within the Pontypridd (1-inch) Geological Survey sheet.

The Careg Cennen Disturbance is another lengthy belt of folding and faulting. Its north-eastern extension as the Dolyhir-Church Stretton line has been claimed by O. T. Jones and Kirk (see Ch. 6.1). This fault, too, probably continues into Pembrokeshire (through Amroth into the Johnston Thrust). A number of limestone outliers occur along the Careg Cennen Disturbance - at Careg Cennen (Fig. 7.13), Bryn-yr-Odyn and Llandyfaelog. Trotter (1948) claims that the main fracture is a great basal thrust which underlies the Coalfield and emerges on the surface at Careg Cennen (Fig. 7.13a). The vertical attitude of the Red Marls north of the fault at this locality seems to support Trotter's view. On the other hand it must be remembered that mid-Devonian folding could have sharply tilted up the Lower Devonian rocks on the inner flanks of the Towy Anticline prior to the deposition of the higher Devonian strata (Fig. 7.13b inset) and the major (wrench?) fault of the disturbance at Careg Cennen could be the southernmost one, with the northern one a later vertical adjustment (compare Dolygaer).

Potter (1965) has demonstrated the existence of a number of rotational strike-slip faults, of probable Armorican age, in the Llandeilo region, to the north-west of the Careg Cennen belt.

The Usk Anticline has been described by Walmsley (1959). The main structural element is a periclinal axis trending NNE.-SSW., but complicated by sub-parallel minor folds on its south-eastern limb. Several faults displace the axes of these minor folds and there is some suggestion of contemporaneity of folding and cross-faulting. According to Walmsley, the main movements were Hercynian with Sudetic and Malvernian preliminary phases. The main local compression was ESE.-WNW. and deep-seated basement control probably locally deflected the compressive forces in the Usk area. The Usk fold is believed to turn south-westwards as the Cardiff-Cowbridge Anticline (George, 1966).

The Forest of Dean Coalfield has suffered folding in pre-Coal Measures and post-Coal Measures times. Nearly all the folds, however, trend N.-S. and the main structure is a N.-S. aligned basin, sharply monoclinal near its eastern rim, the strata here being downfolded a maximum vertical distance of 700 ft (214 m). To the west of the main axis occurs the Cannop Fault-belt, a zone involving some twenty-five faults with predominant westerly downthrow (Moore and Trueman, 1954). These are almost N.-S. fractures. The other main faults trend WNW.-ESE. as, for example, the Woorgreen Trough.

Following the Hercynian Orogeny, the Upper Palaeozoic rocks were appreciably uplifted and deeply eroded, the greatest erosion occurring in the present-day southern peninsulas (Plate IV). Thus at Cardiff the Triassic rests on Silurian, involving the previous removal of probably some 20,000 ft (6100 m) of Devonian and Carboniferous strata. Keuper, Rhaetic and Lower Liassic sediments were then deposited over the Vale of Glamorgan and probably also over Gower and Pembrokeshire. Other Mesozoic deposits may even have eventually extended (at least temporarily) over the whole of South Wales. Apart, however, from a small pocket of Tertiary (Oligocene?) pipeclay at Flimston, Pembrokeshire, no depositional evidence remains for the great Lias-Pleistocene gap. It is therefore difficult to date the faulting and slight flexuring of the remaining Mesozoic cover of the Glamorgan Vale. The structures have always been referred to an Alpine Orogeny, and this may well be true. On the other hand, they could be the results of early to mid-Cretaceous movements, known to have been very effective in Southern Britain and in the surrounding waterways. In Chapter 10 it is suggested that the main Bristol Channel downfold is the result of Cretaceous movements and the flexures (and perhaps even the fractures) in the Triassic and Liassic cover of the Vale may be contemporaneous.

These folds in the Mesozoic are broad and shallow, the arching sometimes coinciding fairly closely with older (Hercynian) lines, e.g. the Cowbridge Anticline. The majority of the post-Liassic faults of the Vale of Glamorgan trend WNW.-ESE. or E.-W. and many of them are normal (gravity) fractures. Some are well displayed in the cliffs at Southerndown and Penarth, though reverse faults are also seen along the Triassic-Liassic cliffs of the Vale, as, for example, at St. Mary's Well Bay and in the eastern corner of Seamouth (Southerndown). This latter fracture is a complex zone with associated spectacular drag-folds.

7.3 Central England (Fig. 7.14)

This complete, kite-shaped area extends from the Craven Lowlands in the north, to the head of the Severn estuary in the south (the tail of the "kite"), and from the western edge of the Cheshire Plain and the line of the Malvern Range in the west, to the long sinuous line marking the base of the Jurassic System in the east. This latter boundary trends through Stratford-on-Avon, Leicester, and Market Weighton. The region is complex for two reasons: (a) the great time-range of the rocks (Precambrian to Lias) and (b) the presence of at least three structural trends (caledonoid, malvernoid and charnoid). Although late-Taconic and Caledonian movements have played some part in the moulding of the regional structure, the main effects were produced during the Hercynian and Alpine orogenies. In fact the greatest movements along several important fractures occurred during the last (Tertiary) movements.

The region is an important one because of: (i) the saline deposits within the Permian and Triassic formations, the Cheshire field being perhaps the best-known example; (ii) the reservoirs of oil and natural gas in the Chesterfield-Eakring-Gainsborough area; (iii) the important coalfield basins, including the largest British field - the East Pennine Coalfield (Fig. 7.15), some 60 miles from north to south in even its exposed western portion.

From what has been said above one can expect the region to be one of diverse relief, though there is a general fall in height southwards. The high regions form the backbone of the Pennine upfold reaching to 2000 ft (610 m) as on Kinderscout. Large areas over 1500 ft (457 m) O.D. extend between the North Staffordshire and Lancashire Coalfields on the west and the East Pennine field on the east. Southwards lies a broad crescent of lower ground, the Midland Plain, surfaced largely by "the New Red Sandstone". This region embraces some six counties and lies largely below 400 ft (122 m) O.D., but is broken by occasional higher regions, formed either of older rocks (including the tough Precambrian masses of Charnwood) or of tougher Permo-Triassic varieties (as, for example in the Clent Hills, 1036 ft., 315 m O.D. The area is drained by such important rivers as the Ouse, Trent, Avon, Severn and Mersey.

A. Folded and Thrust Coal Measures. Little Haven, Dyfed (Pembrokeshire), South Wales. (Geological Survey photograph).

B. Triassic Unconformable on Carbonferous Limestone. South-east Corner of Sully Island, South Wales. (Length of section in photograph about 20 yds). (J.G.C.A.).

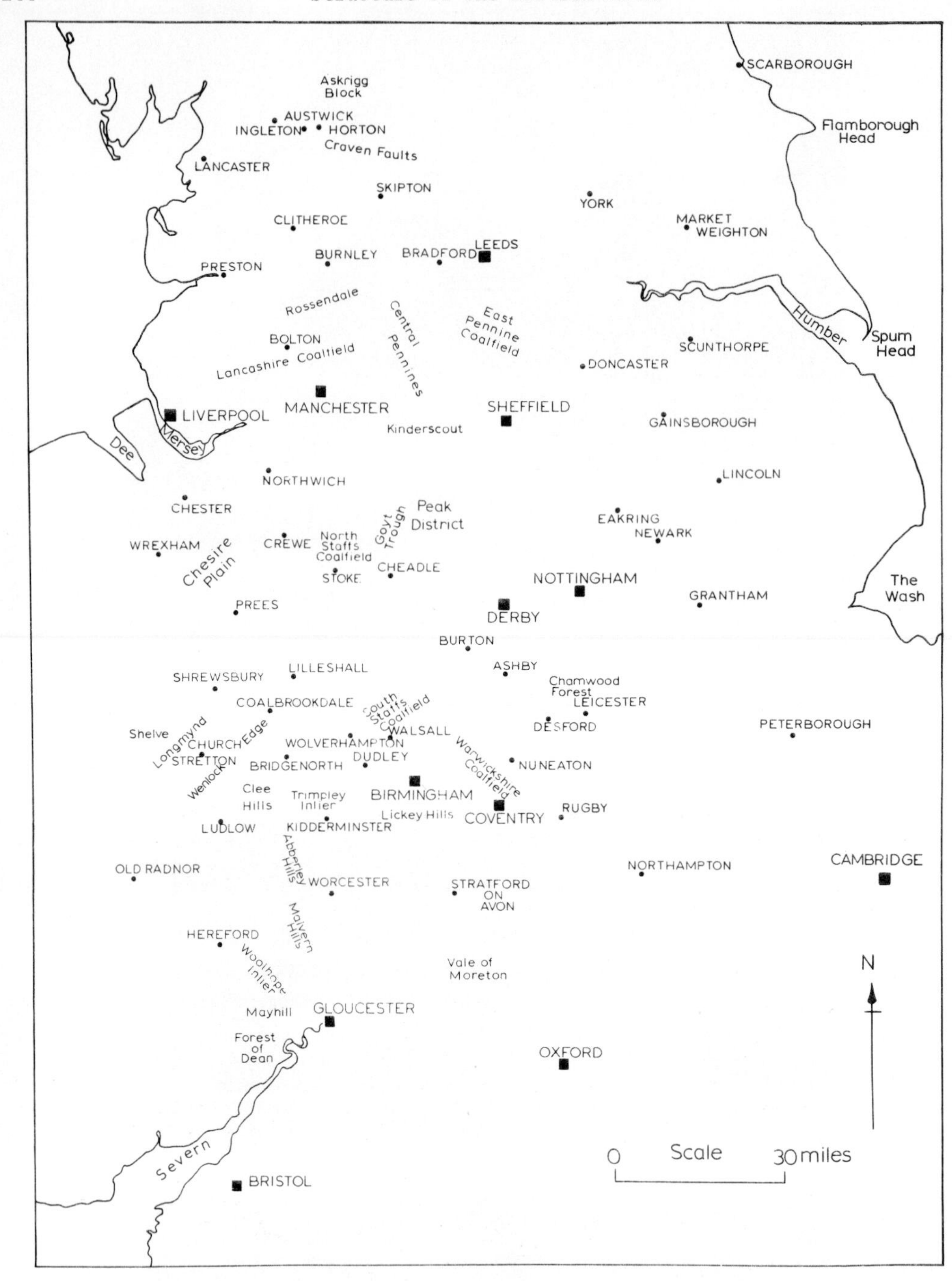

Fig. 7.14 Locality Map for Central England.

The rock succession in the area is as follows:

Lias
Triassic
Permian
(unconformity)
Coal Measures
Millstone Grit
Carboniferous Limestone
(great unconformity in places)
Old Red Sandstone
Silurian
(unconformity)
Cambrian
(unconformity)
Precambrian

Exposed Precambrian areas are small, the largest being Charnwood Forest. Others include the Malvern Hills and small areas in the Lickey Hills and near Nuneaton and at Lilleshall.

The Malvernian gneisses and schists with veins of syenite and granite pegmatites represent the base in the area but the true antiquity of this floor is not known. Radiometric dates denote an important metamorphic event at almost 600 million years ago (Lambert and Rex, 1966). The relation of the associated Warren House volcanics (rhyolites, trachytes and tuffs) is also vague.

In the Midlands the Charnian volcanics and sediments reach their greatest exposed thickness (over 2500 m) in Charnwood Forest, Leicestershire. Lower sequences are composed largely of tuffs with hornstones and agglomerates whilst upper portions comprise coloured slates, grits and conglomerates. It may be that the volcanics correlate with the Shropshire Uriconian whereas the higher sediments are Longmyndian (?). Many intrusions occur, including dacitic plugs and granophyric diorites. The plugs are definitely Precambrian whilst the diorites range from Cambrian to Upper Palaeozoic. The Charnian rocks were folded at the close of Precambrian times into a great NW.-SE. trending pericline, its northern half being blanketed by Trias. The inlier is extensively fractured, mostly parallel to the main fold axis, long overthrusts accompanying the folding.

The Charnian igneous activity may possibly be linked in a plate tectonic context with Cadomian closure and subduction (see Anderton *et al.* 1979, fig. 6.3, in part reproduced as fig. 6.4 in this chapter).

These late-Precambrian movements were widespread and the overlying Cambrian quartzites rest unconformably above. The finer grain of the Middle and Upper Cambrian sediments (Stockingford Shales) of Nuneaton and Dosthill point to either lower land relief or more extensive marine deposition, but instability at the close of the Middle Cambrian is suggested by the calcareous conglomerate at the top of the Abbey shale member. This break correlates with a similar one in North and South Wales and was even more prolonged in the Comley area of Shropshire.

Virtually only the Tremadoc Series of the Ordovician System is represented in the Midlands, and there is therefore a great gap between the Cambrian and the Silurian. The orogenies responsible are probably of pre-Caradoc and pre-Llandovery age, but little is known of their Midland effects mainly because of lack of exposure. Violent local lateral pressures appear to have affected the pre-Rubery Sandstone (Upper Llandovery) rocks in the Lickey Hills, but at Nuneaton the equal dip of the Cambrian and the overlying Coal Measures suggests that the main pre-Llandovery

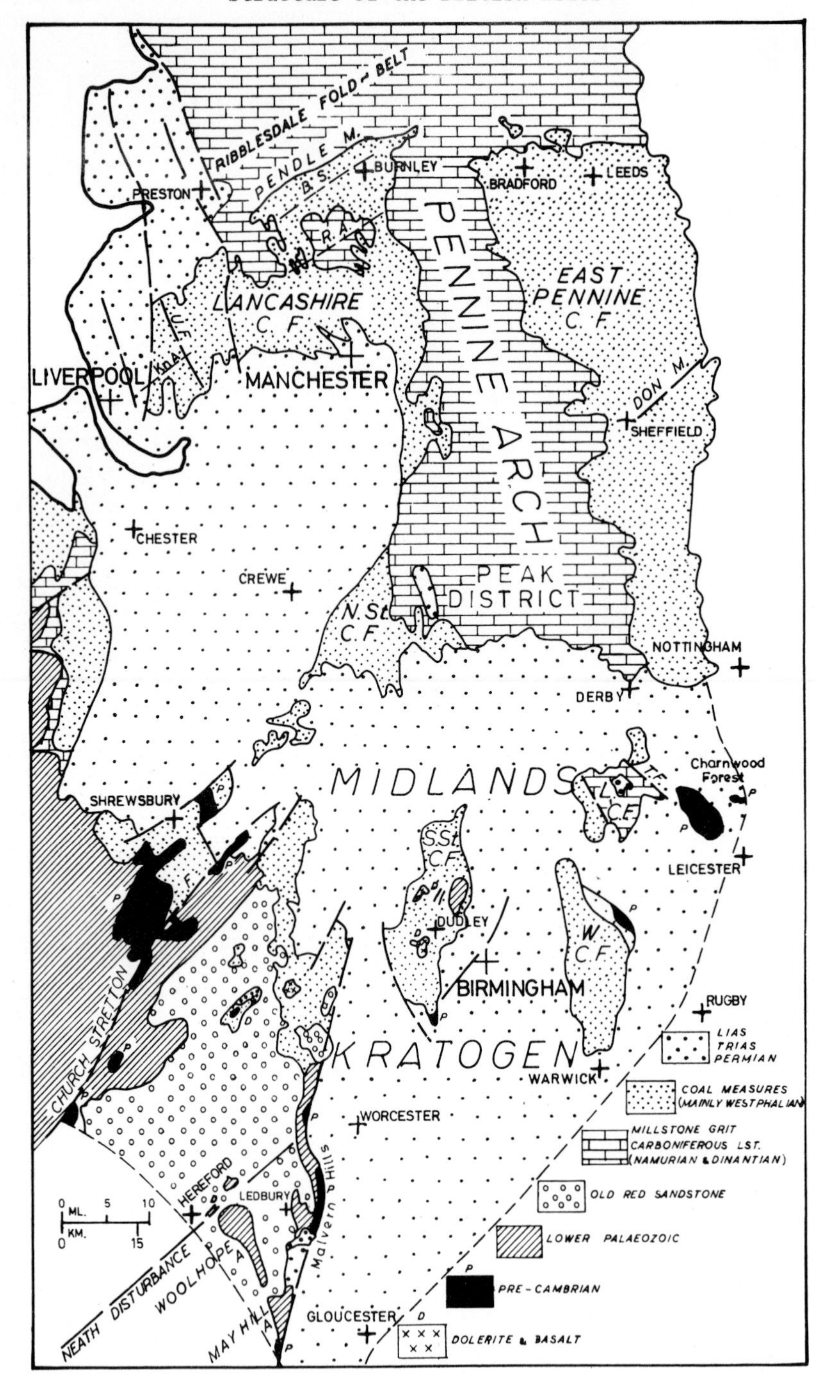

Fig. 7.15 Structural Map of Central England.

movements were of a vertical, probably block-faulted, type. The passage from Upper Llandovery sandstones and shales through Salopian limestones and calcareous shales into the Lower Devonian in all the Shropshire and Worcestershire exposures points to a much delayed Caledonian effect in the Midlands. In fact that effect was ultimately not great in terms of compression, though there was some NE.-SW. folding in the Clee Hills area before the deposition of the Upper Devonian. The Caledonian movements in the Central England area were pre-Upper Llandovery (mainly) and probably mid-Devonian, though during both times the Midlands was essentially a stable block suffering mainly uplift and ultimate erosion. A considerable thickness of Old Red Sandstone may have been removed in places (Dudley for example), though some may be still preserved at depth as postulated for the Lancashire Coalfield (George, 1963a). Again the red conglomerates and sandstones encountered at 5000-7000 ft (1525-2135 m) depth in the Eakring borings (Lees and Taitt, 1946) could represent residual Devonian tracts beneath the Carboniferous.

Carboniferous rocks cover an extensive portion of the surface over the northern half of the region as far south as the Peak district, but become more isolated beneath the Permo-Triassic blanket in the southern half. In the north the Carboniferous succession reaches a thickness of over 17,000 ft (5185 m) in places and includes Dinantian strata (down to at least basal Visean levels), a complete Namurian sequence and a virtually complete Westphalian Succession. Lateral variations within the basin facies of the Upper Dinantian and Lower Namurian resulted in spreads of sand or of Pendleside-type shales. Further north, the effect of contemporaneous movements along the southern edge of the Askrigg Block resulted in only a thinner high Visean sequence, oversteps in the lowest Namurian and perhaps a localization of Dinantian reefs.

To the south, also, lateral variations set in (as for example, in the Widmerpool "Gulf"), the main kind being the result of overlap or overstep of various higher Dinantian, Namurian and even Westphalian horizons against the irregular pattern and behaviour of a Mercian or St. George's landmass across the southern part of the area. The basal Namurian becomes more or less absent into the borders of the Peak district, whilst much further south near Wolverhampton and near Desford (Leicestershire) the Dinantian is reduced to a few feet in thickness. The effect of the Dinantian overstep on to older rocks is perhaps best seen on the margins of the Little Wenlock Coalfield in Shropshire, the Carboniferous Limestone eventually coming to rest on the Tremadoc shales near the Wrekin. The thin Dinantian appears in places to extend southwards even beyond the feather edge of the Millstone Grit and eventually by the latitude of the Clee Hills, and the south Staffordshire-Warwickshire Coalfields, the transgressive base of the Carboniferous is formed by the Coal Measures.

Even within the Coal Measures, the horizon of the (local) base varies from place to places over the southern half of the area, and obviously points to contemporaneous movements, locally intense, of the Mercian barrier. They make the precise timing of the major effects of the Hercynian Orogeny extremely difficult. For example, in the Abberley district, fairly late Coal Measure movements were responsible for the infolding of the Hillside Measures (Mykura, 1951) whereas in the Trimpley Inlier further north, small patches of Middle Coal Measures appear to lie across folded Dittonian rocks. In the latter case it might be argued that later-Devonian folding was here responsible but even so the Coal Measures are folded over the Trimpley Inlier about a broad NNE.-SSW. pericline.

The intra-Westphalian and Stephanian folding and block-faulting in the Midlands has been unravelled in detail by Wills (1948 and 1956). Two periods of considerable intensity occurred, one in earliest Morganian times (the end of Wills' palstage P2) and the second in highest Stephanian times, i.e. prior to the deposition of the Clent-Hatfield breccias (the end of Wills' palstage 4b). The famous Symon "Fault" (a locally intense unconformity in the Coalbrookdale Coalfield) exemplifies the earlier phase, whilst the more major effects of the second phase are shown in Fig. 7.16 adapted from Wills' Fig. 13 (1956).

Some of the structures produced in late-Stephanian times were, however, already being anticipated during the First Malvernian movements (early-Morganian). These elements include the Coalbrookdale downfold, with early movements along its north-western border (the Wrekin-Lilleshall Fault), the Abberley and Longmynd axes, the Cheshire downfold, and the Nuneaton, Charnwood and Eakring axes (see Wills, 1956, Fig. 6). In fact much of the erosion recorded by the sub-Triassic unconformity along the Chartley continuation of the Nuneaton Axis may have resulted from uplift during the earlier movements (Wills, 1956, p.61). This axis is continued beyond Chartley as the anticlinal between the North Staffordshire and Goyt synclines and, still further north, as the N.-S. folds defining the sharp eastern edge of the Lancashire Coalfield. Again during the deposition of the Enville Beds, just prior to the Second Malvernian climax, block-fault movements in the surrounding Mercian uplands were giving rise to influxes of coarse conglomerates and occasional breccias. The importance and span of these movements is shown by the considerable thickness of the group (3000 ft, 915 m) in Warwickshire, with five different conglomerates involved (Shotton, 1927, 1929).

The highly variable and locally thick formations of the Permian and Triassic systems in the Midlands bear evidence of continued unrest throughout "New Red Sandstone" times, and in a sense one could say that the Hercynian structural elements were not completely moulded until at least the deposition of the Dune (Lower Mottled) Sandstones of the Midlands and the Collyhurst Sandstone of Lancashire and again the invasion of the east Pennine area by the Zechstein (Magnesian Limestone) Sea (Wills' Palstage 4d). During the deposition of both the Clent (Haffield, Nechells, etc.) breccias and the succeeding Bridgenorth and Collyhurst-Dune sandstones (with local breccias at Hopwas, Moira, Barr Beacon, etc.), block-faulting was widespread and appreciable in the Midlands, Cheshire, North Wales and Lancashire. Horsts included the Longmynd, Worcester-Kidderminster, West Warwickshire, Nuneaton and Charnwood areas in Clent times; and Wenlock, south Staffordshire and the Berwyns in Dune Sandstone times. Even changes in the vertical movement can be inferred during this long span. Thus the Worcester-Kidderminster area was one of uplift at first, but became a graben later. The lines along which the vertical movement occurred were to some extent already initiated during earlier Hercynian phases but were to become the major linesof faulting in Central and North-west England in the later (Alpine) earth movements. Continued unrest along faults like the Birmingham Fault can be made out by detailed studies of thickness and local extent of the Permo-Trias Beds in that area (Wills, 1948, p.78).

Thus, by about the deposition of the (Triassic) Bunter Pebble Beds, the full force of the Hercynian Storm was virtually spent and the area eventually began to enter a quieter period with a widespread lowering of surrounding and intervening hill areas and the deposition of firstly non-marine Keuper sands and muds and, later, marine Rhaetic and Lower Jurassic clays and carbonates. Reversals to areas of non-deposition probably occurred, at least in the north of the area, during Middle Jurassic times and more generally over the region in uppermost Jurassic and lowest Cretaceous times, prior to the widespread Upper Cretaceous marine transgressions culminating in the laying down of the Chalk.

Widespread regional uplift of much of northern, central and western Britain at the end of Cretaceous times began a new continental phase which was to culminate in the Miocene (Alpine) Orogeny. These earth movements were more intense in the southern areas of England but some broad folding and, more especially considerable fracturing occurred in Central England. The sites of the Alpine folds were again influenced by the older axes and by the horst-graben pattern of New Red Sandstone times. Thus upfolding occurred over the Lickey (see Wills, 1948, Fig. 20) and Nuneaton axes, with broad downfolding over the Cheshire Syncline. Renewed movement (often reversed in direction) occurred along earlier, Hercynian, fractures, and many new faults were also formed. Horst-graben effects accentuated the variability in surface geology over the Midlands and isolated the Coalfields of Central England from the intervening New Red Sandstone areas. Post-Triassic (presumably Miocene) movements along faults in the Lancashire Coalfield were considerable. Thus of the 1050 ft (320 m) displacement along the Great Haig Fault, 725 ft (221 m) represents post-Triassic movement (Trotter, in Trueman 1954), whilst the 3000 ft (915 m) displacement along the western boundary fault of the Coalfield is completely referable to post-Triassic movements.

Earthquakes at Bolton and Manchester appear to represent modern movements along the Irwell Valley Fault, whilst the 1907 earthquake at Malvern shows that slip is still occurring along the post-Triassic eastern boundary fault of the Malvern Hills.

It has therefore been considered more convenient to delay the more detailed description of the main tectonic elements in the region until this late stage. After all, these elements, though primarily of Hercynian moulding, have undergone considerable Alpine "trimming" and many of their fractures are more worthy of the latter designation. These major elements include:

The Lancashire Coalfield
The East Pennine Coalfield
The North Staffordshire Coalfield
The South Staffordshire Coalfield
The Warwickshire Coalfield
The South Derbyshire and Leicestershire Coalfield
The Abberley-Trimpley Inliers (3 in fig. 7.16)
The Malvern Hills (2)
The Woolhope-Mayhill Inliers (1).

The Lancashire Coalfield is exposed over a total area of 500 miles and is "axe-shaped" (Trotter, 1954) with its handle projecting southwards to Macclesfield. The main interruption within the Coalfield is the caledonoid Rossendale Anticline exposing Millstone Grit over its core. The Coal Measure succession amounts to some 6400 ft (1952 m), the coals being of bituminous type. Redbeds occur, but these sub-Permian reddened horizons range from Millstone Grit to high Morganian. The reddening is therefore secondary and has depths of up to 1000 ft (3o5 m).

To the east of the Coalfield lies the broad and partly Alpine N.-S. arch of the Pennines, with again Millstone Grit over its surface core between the Lancashire and Yorkshire coalfields. This anticlinal area is, as far as Hercynian structure is concerned, much more complex and includes the N.-S. aligned synclines of Goyt andPemberton and the Edgehill Anticline. The westward dip of the Pennine Anticline is up to 1 in 1½ (Trotter, 1954). Within the Coalfield, more caledonoid folds include the Burnley Syncline and the Knowsley Anticline, whilst more acute flexures occur beyond the northern rim of the Burnley fold. The major faults trend north-west or north-north-west and are developed in two belts, through Manchester and Wigan respectively. In the former zone, the Irwell Valley Fault has a north-easterly downthrow of 3000 ft (915 m). Along the eastern margin of the Skelmesdale Syncline is the Upholland fracture, with a throw of 4000 ft (1220 m). The fractures are of "scissor-type" (Trotter, 1954) with often rapid diminuitions of throw (1220 m to 60 m) in 3½ miles along the Upholland Fault). All appear to be tension fractures.

The East Pennine Coalfield extends for over 60 miles (96 km) from Leeds to Nottingham. The exposed western portion covers some 900 square miles, whilst at least 2000 square miles of Coal Measures occur buried eastwards beneath the Permo-Triassic and later strata as far as the East Coast. Beyond Lincoln the base of the Permian lies more than 3000 ft (915 m) below sea-level. The southern limit of the buried coalfield cannot be a great distance beyond Grantham, whilst the proved occurrence of Precambrian beneath the Triassic at North Creeke (Norfolk) precludes any southern extension of the Coalfield beyond the Wash. The Ammanian-Morganian succession in the coalfield amounts to 5600 ft with red beds occurring lower in the Upper Coal Measures of Nottingham than in the Leeds district.

Structurally the Coalfield is a shallow basin which Edwards thought could be a western embayment in a still greater basin (Edwards, in Trueman, 1954).

The search for gas in the Southern North Sea has shown that a vast area of Coal Measures, including younger (Stephanian) measures, stretches from Yorkshire and Lincolnshire across to the Netherlands (see Ch. 10).

A number of NW.-SE. upfolds, including the Nocton-Blankney, Eakring, Brimington and Kiveton anticlines, interrupt the general basin-like structure. The Don Monocline trends NE.-SW. Some of the folds exhibit isolated elevations along their course. Fracturing is often intensely concentrated along these folds. Normal and reverse faults trend north-west and north-east.

The North Staffordshire Coalfields include the larger Potteries Coalfield and the less important Cheadle, Shaffalong and Goldsitch Moss basins. The Cheadle and Goldsitch downfolds lie along the N.-S. Goyt trough. The Potteries Coalfield is exposed over a triangular area, and is a deep syncline, axis NNE.-SSW., and with an appreciable southward pitch. The fold is assymetrical, with a steeper western limb. The succession of Coal Measures reaches over 6000 ft (1830 m) and a large area of Coal Measures must lie southwards beneath the Trias (Cope, 1954). To the west the syncline is bounded by the sharp, complex "Western Anticline" exposing upper Dinantian and Millstone Grit. Ammanian rocks on the western flank of this fold frequently dip at 70°. the steeply dipping coals being known to miners as "rearers". The main fractures trend NNW.-SSE.; NNE.-SSW. and WNW.-ESE. and include normal, reverse and wrench types. The most powerful fracture is the Apedale Fault with an easterly downthrow of up to 2000 ft (610 m). In the south of the Coalfield the (Tertiary) Butterton Dyke trends NNW.-SSE.

The Shaffalong downfold is a N.-S. aligned, flat-bottomed, steepsided flexure separated from the assymetrical Cheadle Syncline by the sharp, much faulted, Wetley Rocks Anticline.

The South Staffordshire Coalfield has a visible width of 9 miles (14 km). Precambrian and Cambrian rocks break the surface in the extreme south, but elsewhere the Coal Measures rest on a Silurian basement which appears on the surface near Walsall and Great Barr in the east, and in the cores of the very sharp NNW.-SSE. periclines of the Dudley-Sedgeley area in the west. Many smaller domes, which do not bring the Silurian to the surface, are known from underground working (Mitchell, in Trueman, 1954). Thin feather edges of Carboniferous Limestone and Millstone Grit are known to occur underground at the northern end of the Coalfield. Large intrusions of olivine basalt occur in the Coal Measures near Rowley Regis and Wednesfield.

The Cannock part of the Coalfield is separated from the Dudley trough by a broad area near Walsall. Complex folding occurs along the Lickey-Dudley-Sedgeley line and again from Netherton to Walsall (Mitchell, 1954). Extensive faulting occurs near the margins of the Coalfield. On the north-eastern side, the Eastern

Boundary, Clayhanger and Vigo fractures form a complex belt. The Birmingham Fault cuts along the Triassic towards the southern tip of the Coalfield, whilst on the western margin, extensive faulting defines the surface limit of the Carboniferous, particularly between Upper Penn and Pedmore. Important westward-throwing fractures occur a short distance west of the Coalfield in this Stourbridge area, and again (the Bushbury Fault) to the north of Wolverhampton. Within the Coalfield occur several N.-S. trending (e.g. Mitre), NNE.-SSW. (e.g. Hazelslade) and several WNW.-ESE. fractures.

The Warwickshire Coalfield lies some 15 miles (24 kms) further east, but extensive areas of productive Coal Measures probably underlie the Triassic over the intervening area. The exposed coalfield suggests a shallow steep-sided basin, aligned somewhat west of north and with a southward pitch. Dips within the coalfield are low except at Arley. High dips occur at the Cambrian-Westphalian junction both on the eastern (Nuneaton) and north-western (Dosthill) margins of the coalfield. A knot of faults marks the northern limit of the exposed coalfield, and both the north-eastern and western margins are bounded by important fractures.

The South Derbyshire and Leicestershire Coalfield is centred on Ashby de la Zouch and has a greatest width of 9 miles (14 kms). The productive Coal Measures total 2000 ft (610 m). The structure is dominated by the NNW.-SSE. Ashby Anticline which separates the more productive belts of South Derbyshire and Leicestershire. The former portion is much broken by subsidiary domes and basins and by important fractures (Moira, Overseal, etc.). To the west of this Derbyshire portion other large fractures are suspected (e.g. Coton Park Fault). To the west of the axis of the Ashby Anticline there occurs the parallel Boothorpe Fault with a westerly downthrow of 1000 ft (305 m). The eastern limit of the Leicestershire Coalfield is the reverse-type Thringstone Fault which (at depth) near Desford brings Cambrian against Ammanian. Later movements, affecting the Triassic, have been in an opposite direction between, for example, Coalville and Worthington.

The Abberley-Trimpley Inliers expose Silurian (Abberley Hills) or Lower Devonian strata (Trimpley) along complex faulted belts at the western margin of the Kidderminster-Worcester Triassic plain. An almost continuous line of fractures downthrowing eastwards extends from the upper reaches of the River Stour for nearly 30 miles (48 kms) to the river Teme. Because, however, of the continued unrest along this belt during Carboniferous and Permian times, upfoldings and horst-like movements being followed quickly by planation the throw along these faults is much less than the stratigraphic discrepancy.

The Salopian shales and limestones of the Abberley Hills are steeply folded and in places overfolded (see 7.17c). Reverse dips occur, for example, due east of Abberley village, and west of Woodbury Hill. These much-affected Silurian rocks are covered unconformably by gently dipping Upper Coal Measures and/or Clent-type breccias on Woodbury Hill, Ankerdine Hill and Abberley Hill. The gap represented by this unconformity is wide but it is narrowed somewhat in nearby areas, as, for example, the small Mamble Coalfield where the relatively low-dipping Highley Measures (presumably Halesowen type) rest on more tightly folded Downtonian-Dittonian strata. Mykura (1951) recognizes this earlier Morganian tectonic phase, but believes the main folding and overfolding along the Abberleys to be late-Stephanian. In this he seems to be supported by Butcher (1962). Hollingworth (in Butcher, 1962), voices a doubt about this late age for the more severe folding of the Abberley Salopian, and suggests that these rocks were deformed before the infolding of the Hillside Measures (strata which he equates with the Woodbury Measures beneath the Haffield Breccia).

The Malvern Hills are composed mainly of Precambrian metamorphic rocks with a small remnant of Uriconian-type volcanics, the Warren House Series. The Precambrian inliers extend for 7½ miles in a N.-S. direction from North Hill to Chase End Hill. On the whole the outcrop is narrow but widens on to Hereford Beacon, to the west of the Uriconian area. The structure of the Malvern Hills has aroused considerable controversy but an admirable summary of some of the various views has been given by Butcher (1962). He has also put forward his own picture of the structure suggesting that the main structural element is a N.-S. monoclinal fold whose steep western limb may have been about 1000 ft in length (see fig. 7.17b). This monocline is broken by flat and steep dip-slip strike-faults (which arose parallel to the fold limbs) and also by oblique faults. Butcher regards the Malverns as a spectacular example of the way in which an ancient Precambrian crystalline basement can bend and form the core of a much younger anticline. Butcher's structural interpretation is illustrated in his Fig. 1 (1962, p.108). The thrust western margin of Hereford Beacon would be one of the flat dip-slip strike-faults. This agrees with the earlier view of Groom (1900) regarding the thrust edge of the Beacon. Raw (1952) had earlier interpreted the Hereford Beacon thrust as a limb-replacing thrust of a gigantic overfold (see fig. 7.17a). In most interpretations, post-Triassic faulting has defined the eastern boundary of the Malverns, though ideas differ regarding the amount of the easterly downthrow.

Before leaving this area, it is as well to remember two other lines of evidence. Firstly, the demonstration by Reading and Poole (1961) of important pre-Llandovery uplift and erosion along part of the Malvern Axis. Secondly, geophysical work by Brooks suggests that the junction between the Lower Palaeozoic and Malvernian rocks on the west side of the Malverns must extend steeply to a considerable depth. This could support Butcher's interpretation. Brooks considers, however, that the presence of a pre-Llandovery fracture beneath the Silurian cover could have initiated and determined the site of the later monoclinal fold. Brook's work also indicates that to the north of North Hill the Precambrian surface begins to fall away northwards and that the eastern margin of the Precambrian outcrop is a series of step-faults, rather than one main fracture.

The Woolhope-Mayhill Inliers are important areas of Silurian just west of the Malvern axis. The Woolhope Inlier is a faulted, asymmetric pericline (its axis trending NW.-SE.), separated from a smaller Shucknall Inlier by a topographic depression, believed to mark a continuation of the Vale of Neath Disturbance (Squirrell and Tucker, 1960). The Mayhill Inlier is a complex periclinal structure in which a N.-S. axis branches into two fold directions, NNW. and NNE. It is situated at the convergence of the Woolhope and Malvern axes of folding (Lawson, 1955). The south-eastern boundary of the inlier is marked by the Blaisdon Fault. Longmyndian rocks occur near Huntley.

Having now looked at the various major elements in more detail, it only remains to survey the net pattern made by these elements in Central England. They appear to fall into three major trends: (a) NE.-SW. or NNE.-SSW (Lancashire structures, Cheshire Synclinal, North Staffordshire Coalfield, Clee Synclines, Trimpley); (b) N.-S. (Abberleys, Malverns, west Pennine-Goyt structures); (c) NW.-SE. or NNW.-SSE. (Leicestershire Coalfield structures, Charnwood Forest, Nuneaton Core, Warwickshire Syncline, Sedgley-Dudley-Lickey line). These major trends have been noted previously by Rastall (1925, Fig. 3). One interesting feature is that, on the whole, the caledonoid structures occur on the west side of the N.-S. structures whilst the charnoid axes lie to the east (see fig. 7.16). This feature is perhaps best brought out and recognized on the Geological Survey Aeromagnetic maps of Great Britain (Sheet 5, 1964). This shows the importance of two broad charnoid lines, one running through Charnwood Forest north-westwards to the Peak District (i.e. to the southern end of the Pennine chain) and the second from the Vale of Moreton area through the Nuneaton-Sedgeley area to the North Staffordshire

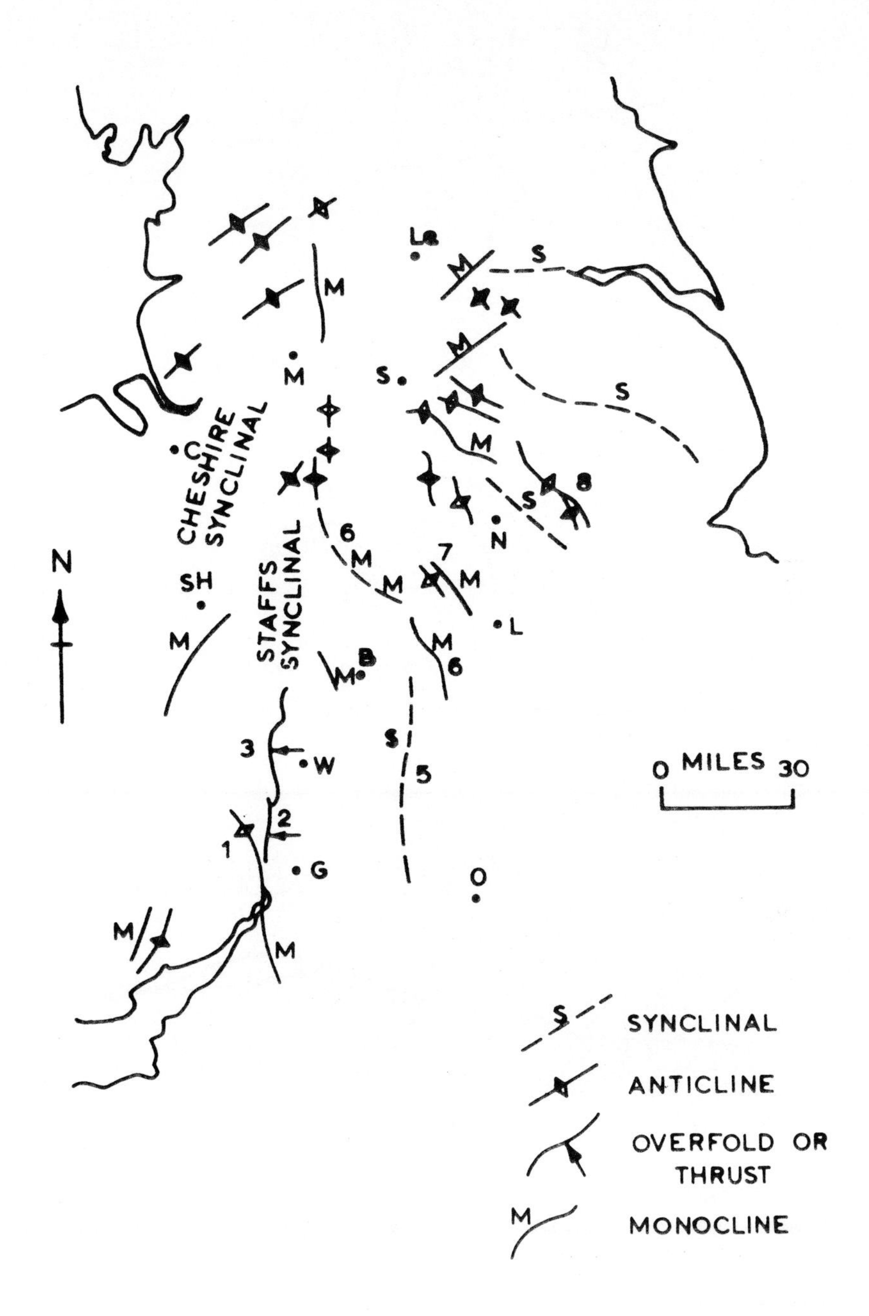

Fig. 7.16 Tectonic Map of Central England (partly after Wills). Key: B. Birmingham; C. Chester; G. Gloucester; L. Leicester; Le. Leeds; M. Manchester; N. Nottingham; O. Oxford; S. Sheffield; Sh. Shrewsbury; W. Worcester.

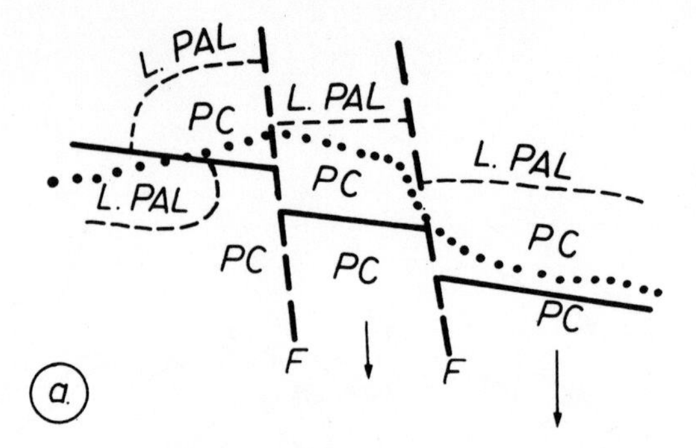

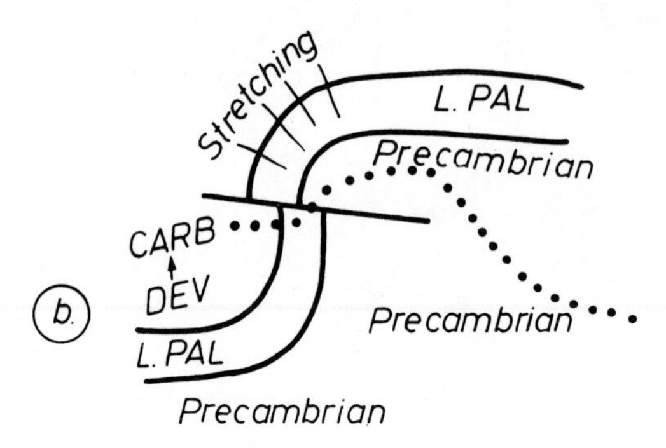

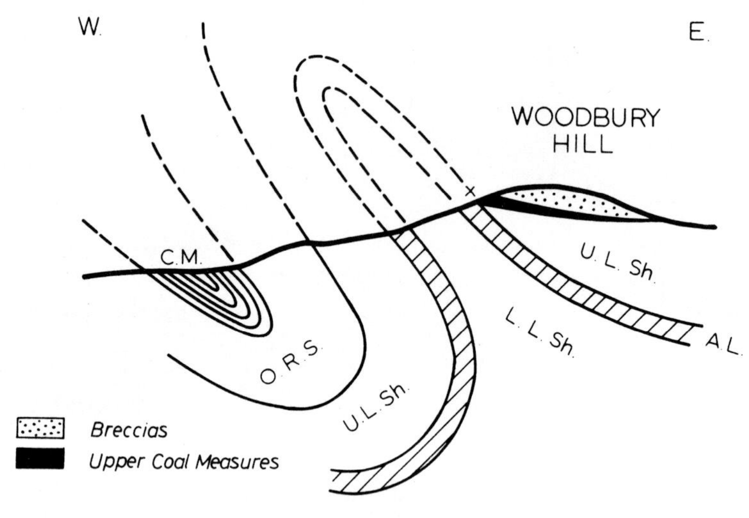

Fig. 7.17 Malvern-Abberley Structure.
Two Versions of the Origin of the Malverns Structure;
(a) after Raw; (b) after Butcher.
(c) Section Across the Abberley Hills (after Mykura).
A.L. Aymestry Limestone.

Coalfield. This major pattern differs slightly therefore from that outlined by Rastall. It is interesting to note that the second (wide) belt, when continued south-eastwards, also takes in the Cambrian occurrence at Calvert and the Silurian rocks near Missenden, beneath the Mesozoic cover.

One problem remains. The N.-S. aligned Armorican structures extend from the Bristol-Severn area through the Malverns and Abberleys to the western Pennines. Folding along this line is often intense and any overfolding or thrusting is usually directed westwards. Some of the elements appear to have been affected by at least two phases of the Hercynian Orogeny. Thus local E.-W. compression was active over a long period of time, and the difficulty is to account for this when the overall regional compressive stress was in a N.-S. direction. One explanation involving the inward movement of a St. George's and an East Anglian block, producing intervening E.-W. compression has been put forward by Owen (1958). The tectonic history of the Woolhope area, on the other hand (Squirrell and Tucker, 1960) suggests an earlier E.-W. compression followed by a later (main) N.-S. one. This makes critical the precise dating of the main buckling in the Malvern-Abberley area, since here the region was certainly affected by a powerful E.-W. compression. Hollingworth's impression of a slightly earlier date for the major effects in this region may therefore be correct.

7.4 North-East England

North-east England is defined for present purposes as the region bounded to the W. by the Pennine and Dent faults, to the S. by the North Craven Fault and to the E. by the curving fault running from Hartlepool (Fig. 7.18). It contains the northern part of the Pennines rising to 2930 ft (893 m) at Cross Fell and the high hills on the Scottish Border, including the Cheviot, 2676 ft (815 m). Near the North Sea, however, there is mainly low-lying ground, heavily covered with glacial drift, which includes the important North-East Coalfield.

The succession is as follows:

Permian
(unconformity)
Coal Measures
Millstone Grit
Carboniferous Limestone
Upper Old Red Sandstone
(unconformity)
Lower Old Red Sandstone
(unconformity)
Silurian
Ordovician
(unconformity)
Ingleton = Precambrian (?)

The Ingletonian rocks, consisting of slates, cleaved siltstones, grits and coarse arkoses, are seen only in two small inliers just N. of the North Craven Fault. These rocks are pre-Ashgillian and may be Precambrian.

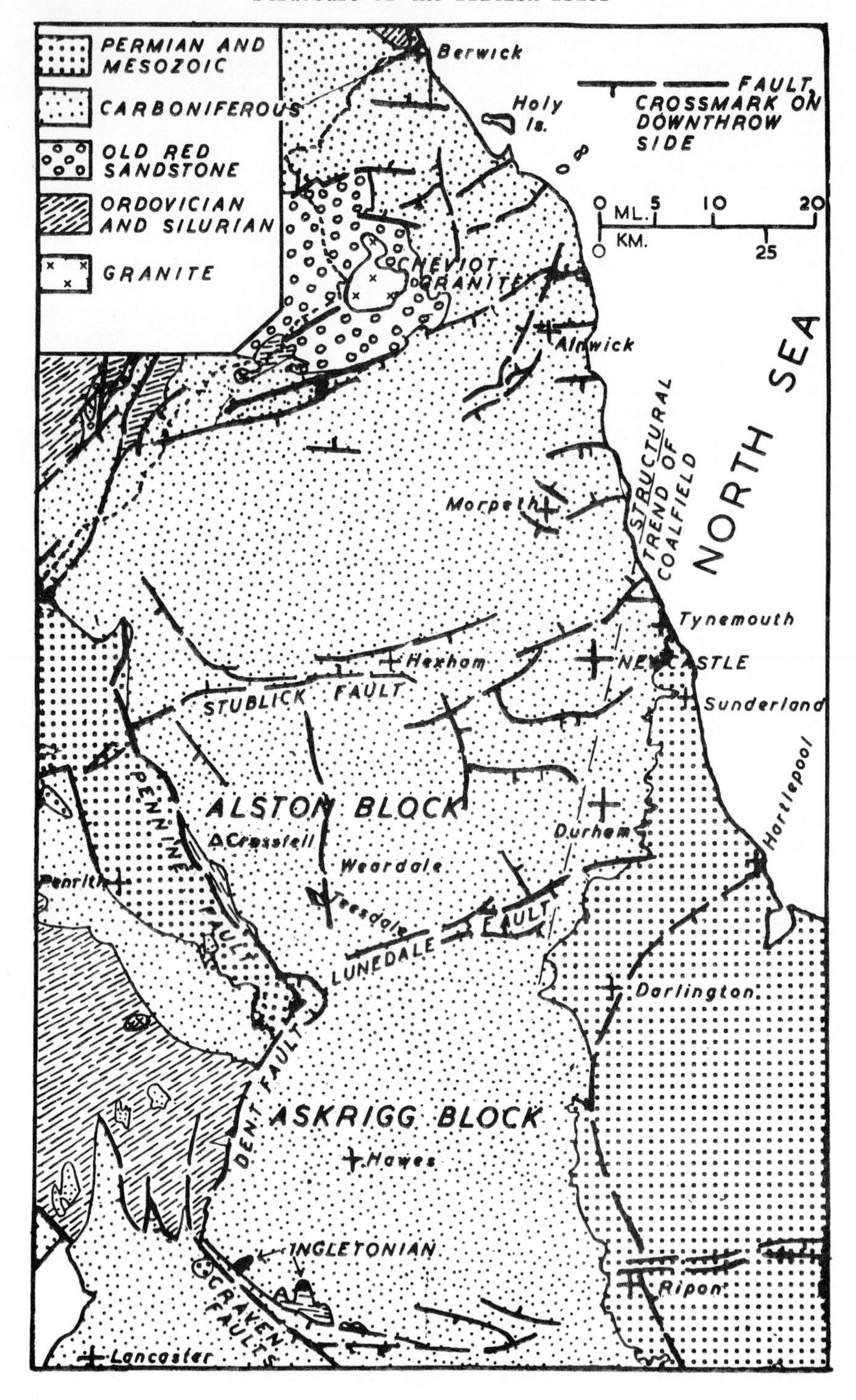

Fig. 7.18 Structural Map of North-East England.

Lower Palaeozoic rocks occur as small outcrops immediately N. of the North Craven Fault in the Ingleton area and E. of the Pennine Fault, near Cross Fell. The succession is similar to that in the Lake District (6.2). An inlier of Lower Ordovician slates is seen in upper Teesdale and inliers of Wenlock and Ludlow grits, greywackes and shales occur on the S. side of the Cheviot.

Lower Old Red Sandstone andesite and rhyolite lavas, agglomerates and sparse red sandstones or marls form a large outcrop around the Cheviot. They are intruded by a mass of granite and by numerous dykes. On the W. side these rocks are unconformably overlain by Upper Old Red Sandstone, conglomerates and sandstones, which spread across the border on to the Lower Palaeozoic strata of the Southern Uplands (6.3).

In the northern part of the region the Lower Carboniferous starts with a basal conglomerate succeeded by shales and thin, impure limestones - the Cementstone Group: basalt lavas occur along the Scottish Border. The Cementstone Group and the overlying Fell Sandstone Group are also termed the Tuedian. The succeeding Scremerston Coal Group, with workable coals, and an overlying succession of sandstones, shales and limestones are sometimes referred to as the Bernician.

Towards the S.W., i.e. at the N. end of the Pennines, marked changes in lithology occur with the appearance of more abundant marine limestones. The Scremerston Coal Group, as such, disappears and at about the same horizon the strata exhibit the typical Yoredale rhythm with a repetition of limestone, shale and sandstone, frequently topped by a thin coal seam. Further S., limestones become more and more dominant and spread to the base of the sequence.

Shales, siltstones and fine-grained sandstones, identified on maps as Millstone Grit, and lying between the highest limestone and the lowest coal of the Coal Measures, form large outcrops in the region but until more palaeontological evidence is available, it is not certain that these are equivalent to the established Millstone Grit Series of the Southern Pennines.

The Coal Measures in Northumberland and Durham are divided into a Lower Group, a Middle or Main Productive Group, with many workable seams, and an Upper Group, mostly sandstones with a few coals. Rocks up to the Coal Measures are penetrated by intrusions of quartz-dolerite, including the extensively quarried Great Whin Sill, and a number of dykes.

The Coal Measures are succeeded unconformably by the Permian, consisting of a thin sequence of yellow sands and the Marl Slate, followed by the Magnesian Limestone, up to about 800 ft (244 m) thick.

A number of west-north-westerly tholeiite dykes cut the region and are considered to be a continuation of the Mull Tertiary Swarm (Ch. 9). Structural events were:

	possibly Precambrian
Caledonian	Mid-Ordovician Late Silurian Structures due to post-tectonic Devonian intrusion Mid-Devonian
Hercynian	Mid-Carboniferous (Sudetic) Post-Westphalian (Asturian)
	Alpine (*sensu lato*)

Events of possibly Precambrian age:

The occurrence of north-westerly isoclinal folds in the Ingletonian has been demonstrated, by means of current and graded bedding, by Leedal and Walker (1950). The structures contrast with the post-Ashgill structures and are evidence for possibly late Precambrian events. Moreover, excavation exposed an erosional unconformity between the Ingletonian and overlying Ashgillian quartzites. According to some radiometric dates, however, the Ingletonian is probably not earlier than the Cambrian and could be early Ordovician (O'Nions *et al* 1973).

Mid-Ordovician events:

The unconformity, in the Cross Fell Inlier, between the Borrowdale Volcanic rocks and the equivalent of the Coniston Limestone shows that the mid-Ordovician movements, well known in the Lake District (6.2), also occurred considerably further to the E.

Late Silurian events:

The folding of the Lower Palaeozoic strata of the Cross Fell, Ingleton, Horton and other inliers is evidence that the late Silurian folding, which strongly affected the Southern Uplands (6.3), extended into the present area and was probably an early but important factor in the structural establishment of the Alston-Askrigg Block extending under the younger rocks as far E. as Newcastle and as far S. as the North Craven Fault. The north by westerly strike in the Cross Fell Inlier and the west-north-westerly strike in the Ingleton and Horton inliers are, however, anomalous in a Caledonian context. These strikes are nevertheless parallel to the adjacent faults and suggest the early establishment of a local tectonic style which was an influence on later sedimentation and tectonics (George, 1963a, p.34).

Structures due to post-tectonic Devonian intrusion:

Relief of the late Silurian compression was accompanied by the outpouring of volcanic material in Lower Old Red Sandstone times. The volcanic pile around the Cheviot is invaded by the large Cheviot granite, the intrusion of which may have been accompanied by uplift. Circumferential tension is shown by the presence of a surrounding dyke-swarm on a roughly radial pattern.

Mid-Devonian events:

Mid-Devonian movement, as in the Midland Valley of Scotland (7.5), is evident from the unconformable overlap of the Upper Old Red Sandstone on the W. side of the Cheviot.

Further S., the Alston and Askrigg blocks were probably further shaped by Devonian movements and by the intrusion of granites below the present erosion surface.

The Weardale Granite, in the Alston Block, was proved by the Rookhope bore (Dunham *et al.*, 1965) to lie unconformably beneath Visean at 1381 ft (390 m) and the Wensleydale Granite, in the Askrigg Block, was shown by the Raydale bore (Dunham, 1974) to occur at 1625 ft (495 m) beneath Visean. Both have ages around 400 m.y.

Mid-Carboniferous events:

Dinantian and Namurian sedimentation provide evidence of repeated distortion of the Pre-Carboniferous floor. Thus the Lower Carboniferous sediments, up to 8000 ft (2420 m) N. of Newcastle, thin to well under 1700 ft (520 m) over the Alston and Askrigg Blocks and thicken to 6000 ft (1828 m) beyond the Craven Faults S. of the Askrigg Block.

Movements of the Alston and Askrigg Blocks continued into the Namurian because over these structures the Millstone Grit is thin and interrupted by internal unconformities. There is no evidence, however, that the Alston-Askrigg Block had marked effects on Westphalian sedimentation, nor is there any sign in the stratigraphy of the Coal Measures of an anticlinal Pennine Axis (George, 1963, 46).

Further N.W., in the Bewcastle district, close to the Scottish border, the Upper Border group (part of the Carboniferous Limestone succession of this district) is about 7000 ft (2135 m) thick between the Antonstown and Harrett's Linn Faults compared with 2000 ft (610 m) to the S. and 4000 (1220 m) to the N. These faults, therefore, approximately follow older Carboniferous hinge-lines (Day 1970, 232).

Post-Westphalian events:

The main deformation was probably late-Westphalian and certainly pre-Permian. These movements had less effect on the Carboniferous of the Alston and Askrigg Blocks. The Coalfield Basin (fig. 7.20) has a N. by E. trend with the western limb dipping fairly gently off the Pennine anticlinal axis. The eastern limb rises under the Permian with, in Durham, flanking folds of a smaller order. On the E. side of the Durham coalfield, in fact, the Coal Measures are cut out by the sub-Permian unconformity and the Magnesian Limestone rests on the Carboniferous Limestone.

The Coalfield is traversed by numerous faults, nearly all of "normal" type. The major faults range within 25° of E. and W.; there is a second set trending N. by W. which is of less significance. Important Hercynian movements also took place along the faults bounding and cutting the Alston-Askrigg Block. The Pennine Fault (6.2) was initiated by Hercynian movement.

The N. by E. trend of the coalfield is evidence of roughly E.-W. Hercynian compression. Similar compression affected the region further N., around the Cheviot igneous mass, and towards Berwick-on-Tweed (Shiels 1964, Robson, 1977). The influence on the development of the folds of both the Cheviot igneous mass and the Lower Palaeozoic of the Southern Uplands can also be demonstrated. Near Berwick Shiels (1964) has described a major east-facing isocline and has pointed out that in this district folds generated by E.-W. compression are deflected round the Cheviot and Southern Uplands blocks. Robson (1977, 258) has noted that the Holburn Anticline has undergone a clockwise rotation, and the Lemmington Anticline the same type rotation, on the N.E. and S.E. margins of the Cheviot mass respectively.

The same author has shown that the first phase of post-Westphalian folding was followed by a second phase of Hercynian movements due to E.-W. compression and resulting mainly in faulting. Thus the axis of the Holburn anticline has been shifted in a dextral sense by at least four cross-faults. The Cheviot mass probably remained unfractured during the first phase but later it began to break against the Lower Palaeozoic to the W. through the development of tear-faults which also affected the adjoining sedimentary rocks.

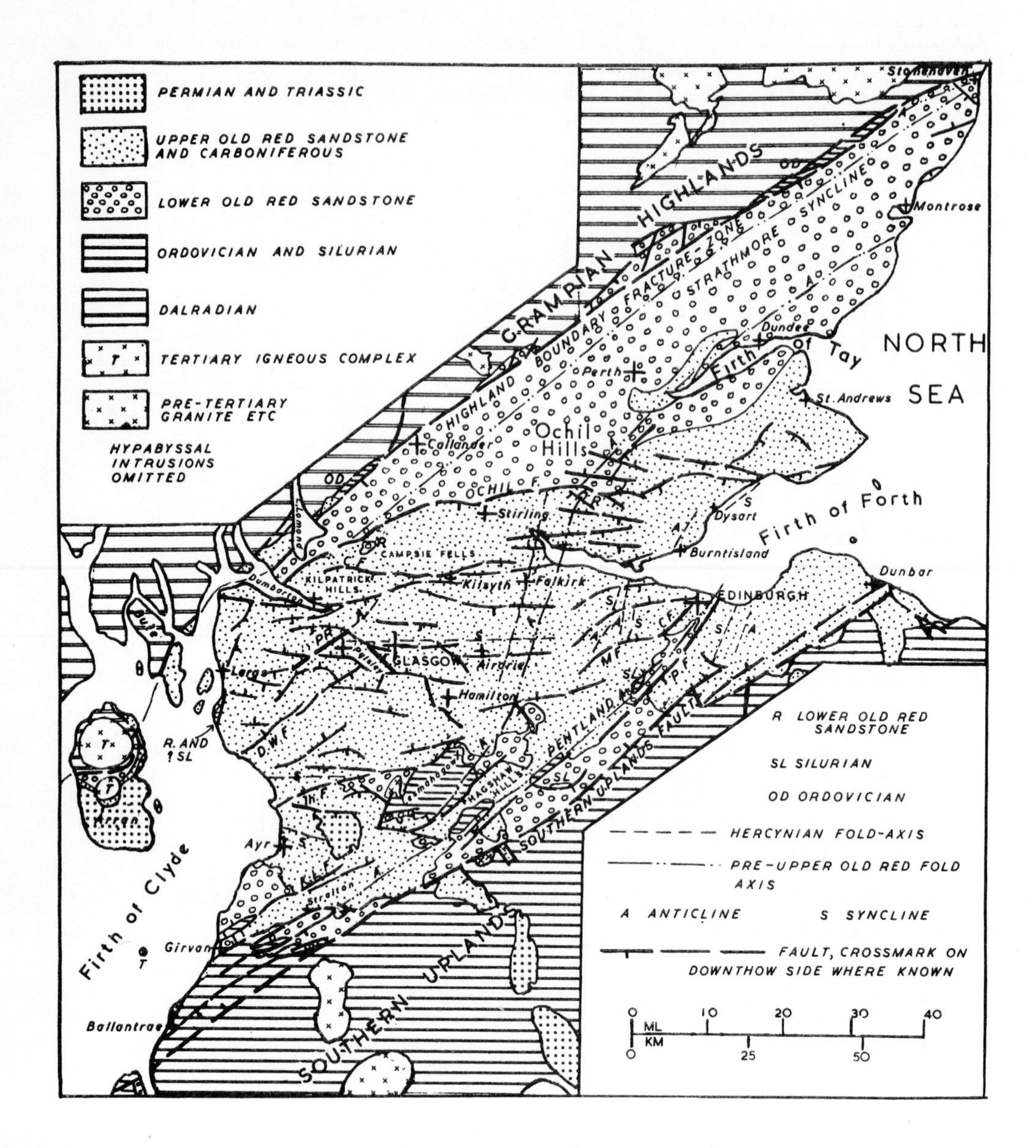

Fig. 7.19 Structural Map of the Midland Valley of Scotland
C.F. = Campsie Fault; D.W.F. = Dusk Water Fault;
In.F. = Inchgotrick Fault; K.L.F. = Kerse Loch Fault;
MF/CF = Murieston/Colinton Fault; P.F. = Pentland
Fault; P.R. = Paisley Ruck

Dolerite dykes with an E. by N. to E.N.E. trend arranged in an echelon near Holy Island and in another echelon in the High Green district near Otterburn post-date the folding but may occupy tension gashes associated with shear-faulting.

Further S.W., in the Bewcastle district, Day (1969, 232) has recognised a phase of N.-S. compression, producing gentle folding on E.-W. axes, before more powerful E.-W. compression. The folds produced by this later compression are deflected N.-N.E. by the Southern Uplands to the N. The Lower Carboniferous has undergone epigenetic lead - zinc - fluorine - barium mineralisation, and the region was at one time an important lead/zinc mining field. Significant mineralisation is concentrated in the relatively thin sediments of the Alston and Askrigg Blocks (Dunham, 1967).

Alpine events (sensu-lato):

Post-Liassic, probably Tertiary, movement led to the final development of the Pennine arch and the Pennine Fault (6.2). The fault cutting off the Magnesian Limestone is post-Triassic and almost certainly Tertiary, and it is possible that further faulting in the Coalfield and elsewhere took place. Older faults were reactivated and the movements were of both "normal" and transcurrent type. The occurrence of Tertiary dykes with a W.N.W. trend is evidence of Tertiary tension.

7.5 Midland Valley of Scotland

The Midland Valley (Fig. 7.19) is essentially a graben with a broad synclinal structure, but in detail it is complex, as its strata are affected by strong folding of both Devonian and Hercynian dates and by numerous faults of Devonian, Hercynian and probably Tertiary ages.

The Highland Boundary Fracture-zone (5.4) forms the N.W. margin of the Midland Valley and the Southern Uplands Fault (4.3) the S.E.

The term Midland Valley is justified in so far as it forms low ground relative to the Highlands and Southern Uplands but the topography is varied. Although the greater part of the valley lies below the 500 ft (152 m) level it is not a valley in the ordinary sense of the term, but consists of several low-lying areas, diversified by belts of hilly ground and isolated hills mostly corresponding to igneous outcrops. Several of these hilly regions exceed 1000 ft (305 m) and some 2000 ft (610 m). The highest point is Ben Cleugh, 2363 ft (720 m) in the Ochil Hills.

The succession in the Midland Valley is as follows:

		Triassic
		Permian (unconformity)
Carboniferous		Barren Red Measures
		Productive Coal Measures
		Millstone Grit
	Scottish Carboniferous Limestone Series	Upper Limestone Group
		Limestone Coal Group
		Lower Limestone Group
		Calciferous Sandstone Series

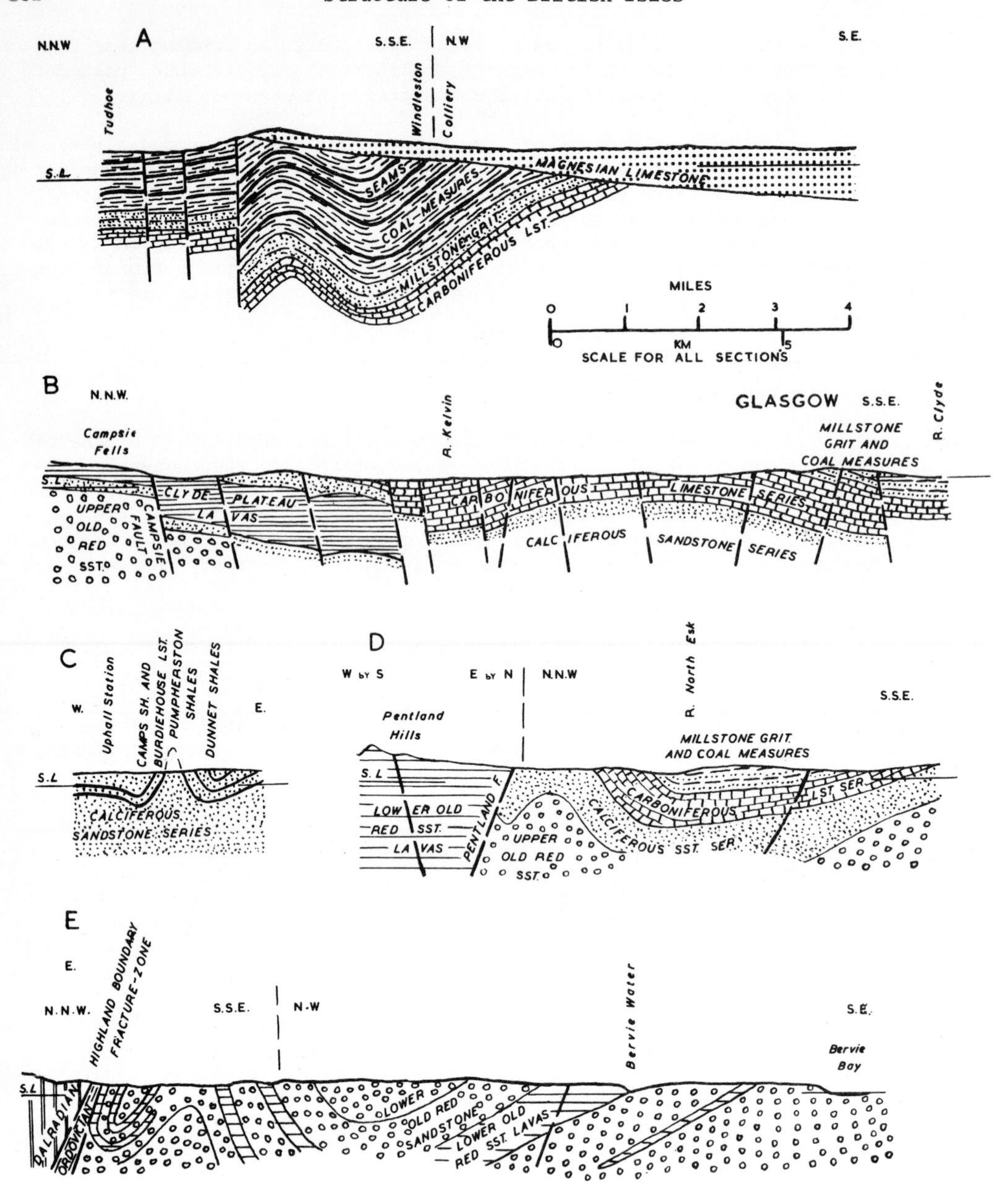

Fig. 7.20 Sections of North-East England and of the Midland Valley of Scotland
A. Durham Coalfield
B. Campsie Fells to Glasgow
C. Pumpherston District, Lothians Oil-Shale Field
D. East side of Pentland Hills
E. North-East of Midland Valley.

Devonian	Upper Old Red Sandstone (unconformity) Lower Old Red Sandstone (unconformity in places)
Silurian	Ludlow Series Wenlock Series Llandovery Series (unconformity)
Ordovician	Ashgill Series Caradoc Series (unconformity) Arenig Series

The most extensive outcrop of the Ordovician in the Midland Valley occurs in the S.W. in the Girvan-Ballantrae area: narrow outcrops also occur along the Highland Boundary Fracture-zone.

The Arenig consists of an ophiolite association of spilitic pillow lavas, pyroclastics, black shales, cherts and ultrabasics, including serpentinites, probably in an obduction zone (Church and Gayer, 1973). The Upper Ordovician consists of a shelf facies of mudstones, grits, conglomerates and limestones. In the extreme S.W., however, near the Southern Uplands Fault, grits and greywackes of geosynclinal facies are present. The Silurian makes a considerable outcrop in the Girvan area and is also exposed in several upfaulted and upfolded blocks well inside the Midland Valley. It consists entirely of sedimentary rocks, including grits, conglomerates, shelly sandstones, flagstones, limestones, shales and mudstones.

The Lower Old Red Sandstone forms extensive outcrops along both margins of the Midland Valley. It is essentially a fluviatile formation consisting of conglomerates, sandstones, flagstones and marls with abundant volcanic rocks. There are also numerous acid and basic hypabyssal intrusions of the same age, but compared with the Scottish Highlands plutonic rocks of the same date are rare, probably because erosion has not cut nearly as deeply. The Upper Old Red Sandstone forms rather less extensive and more disconnected outcrops and is a fluviatile or lacustrine formation of sandstones, conglomerates and marls.

The Scottish Lower Carboniferous contrasts strongly with that of the rest of Britain, and there are major facies variations within the Midland Valley itself related to older structures. The Calciferous Sandstone Series starts with a lagoonal succession of alternating shales and ferro-dolomites, followed in the W. by thick basalt lavas - the Clyde Plateau Lavas - and in the E. by thinner lavas of both basaltic and acid character. High up in the series the workable Oil-Shale Groups occur in the E., whereas in the W. there are varied sediments including thin marine limestones. In the Scottish Carboniferous Limestone Series the thick carbonate rocks of much of the Western European Dinantian are absent; only thin limestones occur, along with shales, sandstones and a deltaic coal-bearing succession which is extensively worked.

In fossil evidence the highest beds of the Scottish Carboniferous Limestone series belong to the Namurian. There is nevertheless, an unconformity or disconformity below the strata mapped as Millstone Grit, consisting of sandstones, shales and fireclays. A faunal and floral break one-third of the way up the Millstone Grit sequence is regarded as the base of the Scottish Upper Carboniferous.

The Coal Measures do not range above the Westphalian. Valuable seams occur in the lower part; the upper part, consisting of red sandstones and shales, is termed the Barren Red Measures.

The Permian, restricted to the Mauchline Basin in the S.W. and to the island of Arran, is formed of red desert sandstone and breccias and of basic volcanics. The lower beds of the sequence may be Stephanian (Mykura 1965).

Hypabyssal intrusive rocks and volcanic necks of Lower Carboniferous and Permian age are abundant, and there is an important suite of quartz-dolerite dykes and sills of Permo-Carboniferous age. North-west Tertiary dykes cut the western part of the region.

Structural events were:

Caledonian { pre-Devonian / Mid-Devonian

Hercynian { pre-Namurian (Sudetic) / post-Westphalian (Asturian)

Alpine (*sensu lato*)

Caledonian events of pre-Devonian age

On the S.W. of the Midland Valley, unconformity between the Ballantrae Igneous Series (Arenig) and a conglomerate at the base of the Barr Series (generally accepted as Bala) is evidence of important post-Arenig but pre-Caradoc movements. Folding took place predominantly along north-easterly axes. The complex and powerful late-Silurian folding and faulting (see below) have to some extent obscured the earlier structures.

On the opposite margin of the Midlands Valley Cambro-Arenig rocks, within or immediately S.E. of the Highland Boundary Fracture-zone, are unconformably succeeded by later-Palaeozoic formations. The most convincing evidence for post-Arenig - pre-Caradoc movements is in the Aberfoyle area. Here the Arenig is succeeded unconformably by grits and limestone containing poorly preserved fossils which are probably Caradocian. The same series, although without fossils, unconformably succeeds the Arenig near Loch Lomond, and in Bute the Loch Fad conglomerate, almost certainly Lower Palaeozoic, contains Arenig fragments. At Stonehaven the Cambro-Arenig is strongly folded but here one can only be certain that the movements were pre-Downtonian. There is, therefore, some evidence that intra-Ordovician structures were developed along the N.W. margin of the Midland Valley. The Border Fracture-zone itself (Anderson, 1947) may well have been initiated as a thrust during Arenig times and have provided a channel for the obduction of the ultrabasic and basic Arenig intrusions which occur at intervals along the fracture-zone (see also 5.4).

The main folding of the Southern Uplands was late-Silurian (6.3); there also is clear evidence of late-Silurian/pre-Devonian folding within the Midland Valley in the Girvan and Pentland Hills areas. On the other hand, in the Lesmahagow district there is no break between the Silurian and Devonian, and it may be concluded that these overlie a block which resisted this compression.

Following the work of Lapworth (1882) and Peach and Horne (1899), folding in the Girvan district, affecting strata up to the Wenlock, was regarded as largely isoclinal on north-easterly axes. Williams (1959, 1962), however, put forward the view that isoclinal folding is unimportant and that the succession is much thicker than previously believed. Five phases of deformation are recognized, when maximum pressures were horizontal. These were: firstly a main fold-phase consisting of overfolding due to overriding maximum pressure from the S.S.E.; secondly, thrusting with maximum stress roughly from S.E. to N.W.; thirdly a fold and reverse fault-phase with maximum stress from E.S.E.; fourthly wrench-faulting due to stress from N.E. to S.W.; and fifthly, wrench-faulting due to stress from N.N.W. to S.S.E.

That movements took place in the Girvan district during the deposition of the Upper Ordovician sediments is evident from unconformities within the succession and at the top. Williams (1962) further suggests that members of the Barr Series were deposited in basins defined by step-faults with continual slip below wave-base allowing deposition of the greywacke facies.

In the Pentland Hills, Silurian strata strike roughly N. 30°E. and are generally vertical; evidence points to an almost consistent upward succession to the W.N.W. Near North Esk Cottage a recumbent fold trending N.20°E. and closing E.N.E. has been recognized. One of the broad results of late-Silurian or immediate post-Silurian movements must have been the production of an intra-montaine hollow approximately coinciding with the Midland Valley. Rivers brought sediment into this depression from both N.W. and S.E. for the Lower Old Red Sandstone conglomerates of the N.W. outcrop contain boulders derived from the Highlands, and those of the S.E. outcrop boulders of Southern Uplands origin.

The Highland Boundary Fracture-zone was in existence by this time (late-Silurian movements along it have been recognized in Ireland). According to George (1960) the N.W. margin of the Midland Valley was not an important factor in Lower Devonian sedimentation which spread continuously and far across the Highlands. Some overlap on to the Highlands is obvious from the presence of outliers in Kintyre, near Crieff and near Blairgowrie, but these were probably limited embayments because the character of the conglomerates near the border suggests they were deposited close to the mountainous margin of a basin of deposition; moreover, conglomerates well above the base contain boulders of rocks derived from the border region which would have been covered had there been a spread of sediments far on to the Highlands. On the other hand, although there must have been downwarp along the S.E. margin, it is unlikely, as both Kennedy (1958) and George (1960) have also stated, that the Southern Uplands Fault was in existence at the end of the Silurian.

Events of mid-Devonian Age

The Midland Valley provides the most striking expression in the British Isles of mid-Devonian deformation, for throughout the region and its margins powerful folding and faulting took place on north-easterly axes and the Upper Old Red Sandstone rests with strong unconformity on the Lower or on older strata. The Strathmore Syncline, affecting the great thickness of Lower Old Red strata outcropping in the N.E., has an amplitude of at least 15,000 ft (4575 m) and is traceable from the east coast for 110 miles to Loch Lomond. The N.W. limb is steeper and in places vertical or slightly overturned.

South-east of the syncline is the Tay Anticline continued S.W. as the Ochil Anticline, partly cut off by the Ochil Fault (see below).

The Pentland Hills Anticline (Fig. 7.20) extends south-westwards from Edinburgh for at least 25 miles (40 km) and on the continuation of the same axis to the S.W., for a total distance of 80 miles (128 km), there are the Carmichael, Hagshaw Hills and Straiton upfolds. To the N.W. of this line lies the Lesmahagow Inlier. All the structures just mentioned bring Lower Palaeozoic to the surface. Mainly north-easterly faults, such as the Pentland Fault, the Straiton Fault and faults in the Hagshaw Hills, associated with these folds, are probably also mid-Devonian in origin, although in some cases reactivated by Hercynian movements (see below). Rolfe (1961) has suggested that bases of conglomerates acted as detachment horizons during tectonism in the Hagshaw Hills.

In mid-Devonian times the main displacement along the Highland Boundary Fault in Scotland took place, the Highland Block being steeply thrust over the Midland Valley (the fault appears normal near the Firth of Clyde but this may be due to later distortion). The throw cannot be estimated accurately both owing to the difficulty of comparing horizons and to the steep dip of the Lower Old Red to the south-east; Allan (1928) gives a downthrow to the S.E. of at least 10,000 ft (305 m) E. of the Tay. Study of the structural pattern in the Lower Old Red conglomerates by Ramsay (1962) supports the view that the main movement was due to N.W.-S.E. horizontal compression. The pre-Upper Old Red Sandstone date of the main movements is proved by outcrops of this formation near Loch Lomond.

The principal displacement along the Southern Uplands Fault was also mid-Devonian. The downthrow to the N.W., although clearly very considerable, cannot be determined accurately, nor is it known with certainty if the fault is reversed. By the beginning of the Upper Devonian the Midland Valley was thus a true graben bounded by powerful faults.

Hercynian events of pre-Namurian age

During the Carboniferous the graben floor remained active and influenced sedimentation, as shown by minor unconformities. Thus uplift and denudation followed the out-pouring of the Calciferous Sandstone Series lavas, and the Upper Sediments of this Group are transgressive: as a result of widespread subsidence at the beginning of the Scottish Carboniferous Limestone, the basal limestone of this series rests on a variety of older strata, and there is also a marked break at the beginning of the Millstone Grit. Kennedy (1958) has pointed out that during Carboniferous times the Caledonian folds became accentuated and that there was also general downward movement, greater towards the northeast and hinging upon a north-west line in the vicinity of Arran. George (1960a) has suggested that the Carboniferous sedimentation which spread far over the north-west margin of the Midland Valley in Ireland also extended well over the western part of the Highlands. During the Carboniferous several north-easterly faults in the Midland Valley, especially the Inchgotrick and Kerse Loch faults in the W., were active, as they separate belts of contrasted sequence.

In the Girvan area Williams (1959) distinguished three structural phases (following the five Caledonian phases mentioned above) during which the relief of pressure was horizontal; these phases were probably not later than Hercynian.

Hercynian events of post-Westphalian age

Although the framework of the Midland Valley is Caledonian the most important internal structures are post-Westphalian (Asturian) and were formed in phases of maximum pressure and relief of pressure acting mainly in a N.-S. direction towards, or at, the close of Carboniferous times. Nevertheless, the Caledonian borders and the buried Caledonian structures continued to exert a strong influence. Consequently there are two main sets of Hercynian structures, those with a "normal" E.-W. trend and those with a N.E. to N.N.E. or Caledonoid trend. In certain areas, however, including eastern Fife (see below), there may have been a phase of E.-W. compression.

The interlocking nature of the E.-W. and N.N.E. structures is epitomised by the complex structural depression of the Central Coalfield which has both an E.-W. axis and a N.N.E. axis. East of the latter, in fact, Caledonoid structures predominate in the Carboniferous rocks: to the west, on the other hand, E.-W. structures are more common. The Carboniferous strata of the west central Midland Valley, including the Glasgow district, form a broad syncline, pitching E. This structure is emphasized in the topography by the Clyde Plateau Lavas (Lower Carboniferous) which form a horse-shoe of hills around Glasgow rising north and south of the Clyde to some 1500 ft (458 m). The syncline is greatly complicated by many E.-W. folds and faults; (most of the faults step down to the south, north of the Clyde (Fig. 41B), and north, south of the river. The Campsie Fault has a downthrow to the S. which reaches a maximum of nearly 3000 ft (915 m). Along the Campsie Fells the lavas are on the N. side of the fault and here the high ground coincides with the upthrow. Further W. however, where the lavas of the Kilpatrick Hills are on the S. side the high ground is on the downthrow side. Further S. the Ayrshire coal-basin also shows a dominance of E.-W. faults and minor folds. The amplitude of the structures can be judged from the fact that the Lower Carboniferous coals, exposed N.E. of Glasgow are about 4000 ft (1220 m) deep under the centre of the syncline. The Coal Measures of the Upper Carboniferous extend N.N.E. to beyond Alloa where they form a centre of the Clackmannan Syncline initiated at least as early as Lower Limestone group times. This is cut off to the N. by the Ochil Fault, bringing up Lower Devonian lavas, a displacement of at least 10,000 ft (3050 m).

To the S. of the Ochil Fault the Coal Measures and older strata are cut by the subparallel Abbey Craig Fault. The Ochil Fault system continues W. then S.W. to beyond Loch Lomond where it separates Lower Old Red Sandstone sediments from Upper Old Red Sandstone sediments to the S.E. However, the Ochil Fault *sensu stricto* ends about 8 miles (13 km) W. of Stirling (Francis *et al* 1970, 246).

Caledonoid folding (and also faulting, see below) is present in the westerly district, but east of the Central Coalfield is much more important. The structure is known in considerable detail from the oil-shale workings in the Lower Carboniferous W. of the Pentland Hills.

These strata are disposed in numerous sharp folds with axes striking from N.N.E. to a few degrees E. of N. Generally, but not invariably, the eastern limbs are steeper, and the axial planes inclined westwards. The most striking fold, situated on a culmination, is the Pumpherston Anticline (Fig. 7.20), the E. limb of which is locally vertical. The folds are broken by west to west-north-west and north-east faults with inclinations generally about 45°. In some cases it can be shown that westerly faults are deflected into a south-westerly direction (see also below). In the N. most of the faults downthrow to the S., counteracting the general northerly pitch in this part of the field.

The Pentland Hills Anticline is primarily a mid-Devonian structure (see above) but it was reactivated by Hercynian compression, as the Upper Old Red Sandstone and Lower Carboniferous are folded round its northern end. The Pentland Fault, with displacement of several thousand feet, can be observed to be reversed and, in fact, has been shown to have a reversed hade of about 22° in a bore (Lees and Taitt, 1946, p.275). The fault separates the anticline from a zone of marked N.N.E. folding in the Carboniferous strata to the S.E. The most important of these folds is the syncline of the Midlothian coal-basin. The Lower Carboniferous coals (locally termed the Edge Coals) dip at angles of 60° or more on its N.W. limb, are some 5000 ft (1525 m) deep along the axis and rise at angles of 25° to 30° on the S.E. limb.

In West Fife north-north-easterly structures, of which the most important are the Burntisland Anticline and the Wemyss-Dysart Syncline (with amplitude of at least 6000 ft (1830 m) predominate. East-west faults are common. In East Fife an important structure is the large, northerly-trending Largo Syncline (Forsyth and Chisholm, 1977, 22).

The Midland Valley provides a good example of "folding within a frame", and the production of north-easterly folds near its margins by Hercynian N.-S. compression can be ascribed to the influences of the already up-faulted block flanking the graben. The north-easterly to north by easterly folds of the central areas, however, require further discussion. It seems probable that the Caledonoid structures exercised basement control, the anticlines not only being accentuated but being forced closer together and thus producing from the primary N.-S. Hercynian compression a resolved compression which extended upwards through overlying Carboniferous sediments. In the case of the folds E. of Edinburgh the controlling structures were probably the Southern Uplands and the Pentland Hills Anticline. For those W. of Edinburgh the controls may have been the Pentland Hills Anticline and a Caledonian anticline now completely buried by the sediments of the eastern part of the Central Coalfield. The existence of such a structure is supported by isopachyte studies by Kennedy (1958, 121-4), and by the presence of igneous rocks of Lower Old Red Sandstone type at from 3990 to 4267 ft (1217-1300 m) in a hole at Salsburgh, east of Aidrie (Falcon and Kent, 1960). Such an upfold, in fact, may be the continuation of the Ochil Anticline dropped down by the Ochil Fault. Just as the Hagshaw Hills Inlier lies on the continuation of the Pentland Hills Anticline, so the Lesmahagow Anticline may mark the further prolongation of the Ochil Anticline.

Movement which was probably due to N.-S. Hercynian compression took place along the graben margins. Horizontal slickensides and variations in the apparent throw of the Upper Old Red Sandstone and Lower Carboniferous suggest that the Highland Boundary Fault behaved as a sinistral transcurrent fault during the Hercynian deformation (Anderson, 1947; George, 1960a). There is evidence of similar displacement along the Southern Uplands Fault (George, 1960a).

A number of powerful north-easterly faults cut the interior of the Midland Valley and while these can be interpreted as normal (or in the case of the Pentland, reversed) faults, they may well be, as pointed out by E. M. Anderson (1951), transcurrent faults or at any rate have an important strike-slip component. Under N.-S. compression the relatively rigid basement would tend to yield by shearing rather than by folding and because of the Caledonoid grain there would be a preferential development of north-easterly shears. Once initiated in the pre-Upper Devonian rocks, these shears would spread upwards into the overlying sediments resulting both in faulting and the development of sharp folding within a narrow zone. Basement shears would, of course, develop most readily along pre-existing Caledonian faults. Among faults which may be of this type, are the Lochwinnoch, Dusk Water and Inchgotrick in the west. The first two are strongly expressed topographically as they determine erosion gaps through the Lower Carboniferous lavas S.W. of Glasgow. The Lochwinnoch Fault is continued by the "Paisley Ruck", a narrow zone of rucked strata and sharp folding which continues N. of the Clyde. The abrupt north-easterly folds known as the "Riggin" near Kilsyth may be also due to basement distortion. A north-north-easterly ruck and shear-belt S. of Largs (Patterson, 1949) brings up red sandstones and grey quartz siltstones with Lower Devonian miospores (Downie and Lister, 1969).

West of Edinburgh the swing of several westerly faults into a south-westerly course may be due to the influence of sub-surface structures, and it is perhaps significant that some change direction close to a line coinciding with the axis of the

Pumpherston Anticline (see above). The likely influence of Caledonian structures is emphasized by the manner in which one of these fractures, the Murieston Fault, turns S.E. as the Colinton Fault, on reaching the Pentland Hills Anticline about 5 miles (8 km) S.W. of Edinburgh. East of the anticline three curving faults cutting the Midlothian Coalfield provide evidence of lateral shift in a sinistral sense (although this is inconclusive) where they cut fold structures (Tulloch and Walton, 1958, 124).

In East Fife, two north-easterly belts of complex structures, the Ceres-Maiden Rock Fault-Zone and the Ardross Fault cut the Carboniferous rocks which are thrown into tight folds; along both there is evidence of dextral transcurrent movement (Forsyth and Chisholm, 1977, 227). This sense of movement and the purely northerly axis of the Largo Syncline (see above) are evidence of E.-W. Hercynian compression in this part of the Midland Valley, similar to a phase well-authenticated in North-East England (7.4).

Alpine *(sensu lato)* events

In Arran folding and faulting affect Triassic sediments and Tertiary igneous rocks. Some of this is due to Tertiary igneous events (Ch. 9) but structures in the southern part of the island, including a number of N.N.W. faults, are more likely to be of normal tectonic origin. Further S.W. Tertiary movement is well authenticated in Ireland.

In the absence of strata younger than the Permian, Tertiary movements cannot be proved with certainty in the mainland portion of the Midland Valley. The Upper Carboniferous to Permian of the Mauchline Basin is folded into a gentle syncline (which has a north-westerly) trend marked by a gravity low (McLean, 1966).

Many north-westerly faults, particularly in the W. part of the Midland Valley, have been ascribed to the Tertiary, although they may be reactivated older fractures. Tertiary dyke-intrusion is evidence of N.E.-S.W. tensional fissuring, and the formation of north-westerly Tertiary joints also took place.

7.6 Central Ireland

The north-eastern limit of this area is taken at the Carboniferous edge along the sides of "the Southern Uplands in Ireland". The south-eastern edge is defined by the Leinster Massif, whilst the southern limit is a line running from Dungarvan along the River Blackwater to Killarney and Tralee Bay (Fig. 6.11). The north-western edge is perhaps the most arbitrary to define but in this present account is taken at a line following the Highland Boundary Fracture-zone, much of which is hidden under younger strata.

The central region thus delineated appears from several considerations to fall into two (unequal) segments, the separating line running from the south side of Galway Bay across to Skerries or Loughshinny (on the east coast). In the first place, the northern of these two portions has a low relief (rarely rising above 400 ft (122 m) O.D.) whilst the southern, larger, area is of more varied height with some fifteen to twenty isolated hill masses or ridges rising out of a plain which is again between 200 (60 m) and 400 ft (120 m) above sea-level. These higher masses include Slieve Bernagh, the Galtee and Knockmealdown mountains and the Comeraghs. Some reach 1000 ft (305 m) O.D., some rise to over 2000 ft (610 m), whilst the Galtees have one summit of just over 3000 ft (915 m). The majority of these higher regions have trends which are either NE.-SW. or E.-W., and these reflect the structural trend, too.

Geologically, too, there is a distinction between these two portions. The northern region is surfaced almost entirely by Carboniferous Limestone, the only exceptions being the anticlinal Lower Palaeozoic tract of Balbriggan in the east and some very small inliers near the north-west boundary. The southern portion on the other hand is as varied geologically as it is topographically and surface rocks range from Ordovician to Ammanian Coal Measures (there may even be Morganian measures in the complex Kanturk Coalfield along the southern margin). Lower Carboniferous volcanics occur fairly extensively in County Limerick.

Thirdly, there is a marked difference in fold intensity between the two portions. This difference has already been pointed out by Gill (1962, Fig. 3/1), the concentric-type (and occasionally box-type) folds being less numerous and much more gentle in the northern portion, but becoming more closely packed, and accompanied by some thrusting, in a southerly direction, particularly towards the Kerry-Dungarvan line (see Fig. 6.10).

The general succession for the whole region is as follows:

Coal Measures
Millstone Grit
Carboniferous Limestone
Old Red Sandstone
(unconformity)
Dingle Beds (Downtonian?)
(unconformity)
Ludlow Series
Wenlock Series
Llandovery Series
(unconformity in places)
Ordovician
? Dalradian or older Precambrian

Ordovician rocks occur in a broad anticlinorium on the south-west side of Slieve Bernagh, from about Ballyenllen Castle to Croaghuan Peak near Broadford, and comprise shales and cherts with Caradocian graptolites. The cherts are highly folded but the shales display little cleavage. Caradocian shales, cherts and greywackes occur in the Slieve Aughty Inliers.

In the Balbriggan area, the Bala comprises 1700 ft (518 m) of brown and black slates alternating with green greywackes, whilst at Portrane and on Lambay Island, volcanics and limestones alternate with the Bala slates. Slump-produced breccias are a feature of the Portrane Ordovician. The only other proved occurrence of Ordovician in these Central Ireland inliers is that in the Chair of Kildare where Llanvirn gritty shales are overlain by Caradocian ashes, thick andesites and basalts, shales, greywackes and limestones, some of which could be Ashgillian. These Ordovician rocks of the Kildare Inlier dip south-eastwards at steep angles.

Ordovician also occurs in the Tyrone Inlier surrounding, above a thrust, metasediments. These could be Dalradian but they are of higher regional grade than nearby Dalradian rocks N. of the Highland Boundary fracture-zone. The schists of the inlier may, therefore, be older Precambrian.

The major parts of the Silurian outcrops in Central Ireland comprise rocks of Llandovery age (but often of Valentian facies). They occur in the inliers of Balbriggan, the Galtees, Slieve Bernagh and Slieve Bloom. The rocks include red, grey and green greywackes, mudstones, slates, flags and shelly conglomerates. The unfossiliferous slates (good enough, in such places as the Lingham Valley, to be used for roofing), grits and greywackes of Slievenaman, the Comeragh Mountains and

Slieve Aughty are probably also of Llandoverian age. In the Chair of Kildare Inlier, Valentian (?) red and black shales may follow Ashgillian Beds with little stratigraphical break. This may be true also at Balbriggan and Portrane (McKerrow, 1962). Wenlock strata are found in the Slieve Felim and Silvermines areas, the Keeper Hills and on Devil's Bit Mountain. The rocks consist of dark grey and blue mudstones, siltstones and interbedded greywackes. Wenlock flags occur in the Cratloe Hills, whilst highly cleaved slates with greywackes in the Galtee Mountains are probably of basal Wenlock age. The Wenlock-Ludlow succession on Devil's Bit Mountain is over 8000 ft (2440 m) thick, whilst that of the Slieve Bernagh Inlier includes spectacular slump-conglomerates, one formation being over 700 ft (214 m) thick.

A well-marked unconformity separates the Old Red Sandstone and the underlying Lower Palaeozoics in Central Ireland. The unconformity does, of course, reflect the Caledonian Orogeny, but these earth movements, noted earlier (Ch. 6), may have been quite complex and prolonged. The Downtonian (?) Dingle Beds of the Dingle Peninsula (see Ch. 7, sect. 7.7) and probably also of the Croughaun Hill Outlier in the Comeraghs rest unconformably on older strata but are in turn followed with great unconformity by the Old Red Sandstone. The Devonian succession in Central and Southern Ireland may hide important stratigraphical breaks. MxKerrow (1962) hints that, except in Dingle, there may be no Lower Old Red Sandstone in the southern half of Ireland.

The highest mountains in Central (as also in Southern) Ireland are formed of Old Red Sandstone, and include Galtymore Mountain (3015 ft, 919 m), the Knockmealdown Mountains and the nameless summit (2597 ft (793 m) behind Coumshingaun in the Comeraghs. The Old Red Sandstone frequently forms broad outer ringed outcrops around central Lower Palaeozoic areas in Central Ireland, and invariably the harder conglomerates and sandstones of the Devonian System stand higher than the more weathered, older, central slates, shales and greywackes. The Lower Palaeozoics in fact often form farmland "oases" within the more desolate encircling moorlands of the Old Red Sandstone.

The Devonian rocks comprise red, purple, green and grey sandstones, shales, flags, and occasionally thick conglomerates (even of cobble grade) and breccias. The succession is extremely thick (over 10,000 ft, 3050 m) in the Comeragh-Knockmealdown boundary region and a sub-caledonoid trough of maximum thickness probably then extends across County Cork towards Kenmare, near which locality the Old Red Sandstone sequence reaches over 15,000 ft (4575 m) in thickness. A marked thinning takes place, however, both northwards and north-eastwards of this south Comeragh-Kerry line (see Capewell, 1957a, Fig. 4). In the Galtee Mountains the Old Red Sandstone has decreased in total thickness to 4000 ft (1220 m) whilst around the Slieve Felim and Slieve Bernagh inliers only about 1000 ft (305 m) of rocks represent the Devonian System. In the south-east, the feather edge of the Old Red Sandstone beneath the overlapping Carboniferous runs from near Bannow Bay to Bagenalstown on the western margin of the Leinster granite. This feather edge of the Devonian beneath first Tournaisian, and further north Visean, limestones probably runs across to near the western half of our Galway Bay-Loughshinny line (cf. George, 1962a, Fig. 2/IV). This marked thinning away northwards of what in the south amounts to many thousands of feet of sedimentary cover has played an important part in the differing response to Hercynian compression of our two portions of Central Ireland. It is almost a similar "tectonic" environment to that existing in the Lower Carboniferous across the southern edge of the Pennine Block.

The Carboniferous Limestone covers at least two-thirds of the surface of the region described in this chapter. Its thickness over Central Ireland is about 3000 ft (915 m), thinning towards the margins of the Leinster and Longford-Down massifs but swelling to almost 4000 ft (1220 m) in the intervening Dublin Basin. Herein, again, lies an explanation for the buckled character of the Carboniferous rocks in that basin. The most important single element of the Dinantian succession in Central Ireland, at least as far as the southern of our two portions is concerned, is undoubtedly the great sheet reef, which extends as far north as County Clare in the west and County Kildare in the east, an area of nearly 3000 square miles. This massive reef, nearly 2000 ft (610 m) thick in the Cork district, belongs to the lowest Visean and consists of polyzoan and algal fronds, often in position of growth. Other important occurrences in the Dinantian of Central Ireland are the two basaltic groups (S_1 and early D_1) of the Limerick district, and the striking Rush and Lane conglomerate groups. These coarse rudites of the Loughshinny-Rush coastal region probably represent submarine faulting and slumping during the Nassauian orogenic phase of mid-Dinantian times. Fragments of Leinster granite and mica-schist in the Upper limestone of the Clondalkin district point to active fault scarps along the northern edge of the Leinster Massif in the Selke (mid-Visean) orogenic phase. In the Carrick-on-Shannon Syncline, beyond the northern limit of our definition of "Central Ireland", basal conglomerates and pale calcareous sandstones of C_2S_1 age have been shown to occur beneath the limestone sequence (Caldwell, 1959). This raises the possibility that other small inliers of arenites further south may be of basal Carboniferous formations and not Old Red Sandstone as previously thought.

At the top of the Dinantian sequences occur black Pendleside-type shales and thin limestones, and on the south side of the Balbriggan Massif these P_2 beds are followed by E_2 sandstones and shales (Smyth, 1950). The absence of the E_1 stage represents slight movements during the Sudetic orogenic phase. Namurian sandstones and shales up to R_1 in age are preserved in the nearby Summerhill Basin (Nevill, 1957). In South Clare, in the vicinity of the Shannon estuary, the Namurian is 3500 ft (1067 m) thick, comprising shales, flags, siltstones and sandstones and with many slump sheets and even sand volcanoes (Brindley and Gill, 1959). In the upper cyclothemic portions of the Namurian sequences of Clare, Rider (1969) has shown the existence of "birdsfoot type" deltas, like the modern mouth of the Mississippi. In West Clare, the E_1-R_2 Namurian sequences is 5000 ft (1525m) thick. The palaeogeography has been determined by Rider (1974). Delta slips along "growth faults" have recently been described by Rider (1978). In North Clare, the sequence thins appreciably and, on Slieve Elba, the E and H stages are missing. Similar thinning occurs southwards from the Shannon estuary (Hodson and Lawrence, 1961).

Coal Measures occur over an area of 150 square miles in the Leinster or Castlecomer Coalfield. This is an oval-shaped plateau with a central depression so that the Coalfield is a basin both structurally and topographically, the outer rim comprising tough Millstone grits, rising up to 1000 ft (305 m) O.D. Three workable seams of anthracite occur in the lowest 800 ft (244 m) of the Westphalian succession. Theother coalfields of central Ireland are very much smaller, but again form ridges above the central plain. The Slieve Ardagh Coalfield, 16 miles (26 km) south-west of the Leinster Coalfield, is more sharply folded, but even more intensely buckled is the Kanturk Coalfield along the River Blackwater. This area of Coal Measures lies near to the Hercynian front of Southern Ireland. Exposures are scanty but are enough to reveal the great complexity of structure. Generally though, the succession "youngs" towards the south, i.e. through a maze of numerous repetitions by folding and strike-faulting.

Whilst intervening earth movements must have occurred in Central Ireland in mid-Ordovician, late-Ordovician and mid-Carboniferous times, the major structural effects were the results of the Caledonian and Hercynian climaxes. The timing of these peaks of compression can only be broadly placed at the Ludlow-Upper Old Red Sandstone and Morganian-Permian intervals, respectively, and it is not as yet possible to date the main movements more exactly. McKerrow (1962, Fig. 1) appears to favour a Middle Old Red Sandstone age for the Caledonian climax in Central Ireland whilst Cole (1922) concluded that the Hercynian "crust waves reached Ireland during the Stephanian age".

The Caledonian structures of Central Ireland are, of course, only glimpsed here and there where the later (eroded) Hercynian upfolds allow, but one can see that the Lower Palaeozoics were thrown into NE.-SW. or ENE.-WSW. (further west) trending anticlines and synclines, which frequently combine into anticlinoria and synclinoria. On Slieve Bernagh a broad, open anticlinorium in the Broadford area is followed northwards by a synclinorium with strong axial plane cleavage on its northward-dipping limb. Strike-faults follow the axis of the southern upfold and there are wrench-faults trending WNW.-ESE. and also just east of N.-S. lines. It is sometimes difficult to differentiate some of the Caledonian fractures from later Armorican effects.

To understand the effects of the Hercynian Orogeny in Central Ireland one has first to bear in mind a number of contributing factors: (a) the form of the sub-Old Red Sandstone floor; (b) the presence of hard cores such as granites, thick volcanic sheets, even thicknesses of tough sedimentary formations both within the basement and the overlying younger blanket; (c) the trend and form of these bulwarks, whether they be in the basement (or as buried topographies on its surface) or in the blanket; (d) the variation in thickness of the blanket, more especially in the direction of major compression. One can use an analogy to illustrate the importance of the above factors. Imagine an empty room with layers of carpet on the floor. If two people are asked to push at either end of the room, they can (if they are strong!) fold the carpets into long uniform and parallel folds, all at right angles to the direction of compression. If, however, there are odd items of furniture in the room and the same two people try now to push, the carpets will twist into variable warps and folds, more especially in the vicinity of the furniture. The same would happen if the floor beneath the carpets had a very irregular surface with deep holes or upstanding mounds or ridges here and there. The analogy of the furniture is not an entirely valid one because in our consideration of the structural behaviour of rock layers, the items of "furniture" are within the "carpet" layers or within the floor and influence the contortion of the blanket in this internal way.

In the case of Central Ireland, the main internal obstacle to the S.-N. Hercynian compression of the Old Red Sandstone and Carboniferous sedimentary sheets was undoubtedly the great Leinster Massif with its hard cores of granites, diorites and Bala volcanics. Other contributing influences were the Connemara mass in the north-west, the Longford-Down Massif in the north-east, the caledonoid trend of the Caledonian structures in the Lower Palaeozoic strata, irregularities in the sub-Devonian floor (the biggest being the Leinster "headland") and lastly the way in which the Old Red Sandstone wedges out in thickness from its great 10,000 (3050 m) to 17,000 ft (5185 m) successions in the south to virtually nothing in the Dublin-Galway Bay region. Further hidden influences have been hinted at by other authors. Capewell (1957a) believes that the broad, tectonically simple, flat-crested arch of the Comeraghs surrounded by more severe folding immediately to the south, west and north, was due to the presence of a deeper, possibly plutonic, mass as yet unexposed. This could be of granite as the area lies on an almost direct prolongation south-westwards of the Leinster Granite (Capewell, 1957). A similar granite core could occur in the Slievenaman Inlier to the north (Hollingworth, in Capewell, 1957a).

When one looks at the pattern of the Hercynian folds in Central Ireland the most outstanding feature is the way in which the fold axes turn gradually north-eastwards or even north-north-eastwards towards the western side of the Leinster

Massif (Fig. 6.10). In the Leinster Coalfield almost N.-S. folds occur especially on its eastern margin. Traced through to the Dublin-Loughshinny coast, the folds then gradually turn back to an easterly direction. The folds have been "slewed" round because of the position and resistance of the Leinster Massif, acting both as a resistance within the basement and as a marked raised irregularity in the sub-Devonian floor. The numerous folds of the Dublin district are no doubt due to the coming together (with continued compression) of the Leinster and the Longford-Down-Balbriggan massifs. The folds include the Swords-Lambay Anticline (the Chair of Kildare upfold may be a south-westward continuation), the Loughshinny Syncline, the Skerries Anticline and the Summerhill Syncline (see Smyth, 1950). The latter downfold has a steeper south-eastern limb.

The way in which the Hercynian folds are more numerous and more intense in the southern portion (than in the northern part) of Central Ireland is due mainly to the thicker affected blanket of Upper Palaeozoics and to the resistance presented by the Leinster "headland". Further north, in the broad region around Athlone, the massifs recede and the mainly Carboniferous blanket was here only very gently rippled. Moreover, a certain amount of protection to compression was afforded by the shield of Lower Palaeozoic kratons just to the south, i.e. Slieve Aughty, Slieve Bloom, etc. The northward diminition in the folding is well seen even across one uninterrupted belt of Upper Carboniferous sandstones and shales in County Clare. In South-west Clare the folds are sharp and close and dips moderate to steep (see Gill, 1962, Fig. 3/11, sects. C-D). Further north, dips become very gentle indeed. It is important to remember also that different formations may have responded differently to compression, and planes of decollement may occur along some bedding surfaces, more particularly at shale junctions.

Faulting in Central Ireland followed a number of directions. The longest faults trend ENE.-WSW. and include important fractures separating the Cratloe and Slieve Bernagh inliers, crossing the Slieve Aughty Inlier and the Castlecomer district of the Leinster Coalfield and the lengthy fault-zone which appears to run from Charleville along the northern fringe of the Slievenamuck rim of the Galtee Inlier. One is tempted to recall the great curving caledonoid fractures of South and East-central Wales and suggest that this Tipperary fracture belt runs on to form the north-western edge of the Slieve Ardagh (Upper Carboniferous) Outlier (it could even be the fault through Castlecomer). It might then be that the basement marks the site of long controlling deep-seated fractures which have determined the trend and psoition of later folding and fracturing in an overlying cover. Other important east-north-east fractures occur in the north of County Dublin and in the Comeraghs where the Croughan Hill Fault has a northerly downthrow of at least 1000 ft (305 m).

North to south aligned fractures also occur widely and are a particular feature of the coalfields and the Comeraghs. In the Leinster Coalfield the Luggacurren Fault has an easterly downthrow of 700 ft (214 m). In the Comeraghs, some of these fractures are dip-slips, others are tears. In the Kanturk Coalfield, the nearly vertical coals often bulge to enormous proportions adjacent to the many N.-S. faults (and of course on the fold axes).

Important WNW.-ESE. faults occur in the Comeraghs, the Maum Fault having an appreciable northerly downthrow. An E.-W. fracture, with a northerly downthrow cutting out 2000 ft (610 m) of E_1 and E_2 shales, defines the southern margin of the Summerhill Basin, north-west of Dublin. Interesting sharp E.-W. monoclines in the Comeraghs could represent a recrudescence of movement along similarly aligned fractures in the sub-Old Red Sandstone basement.

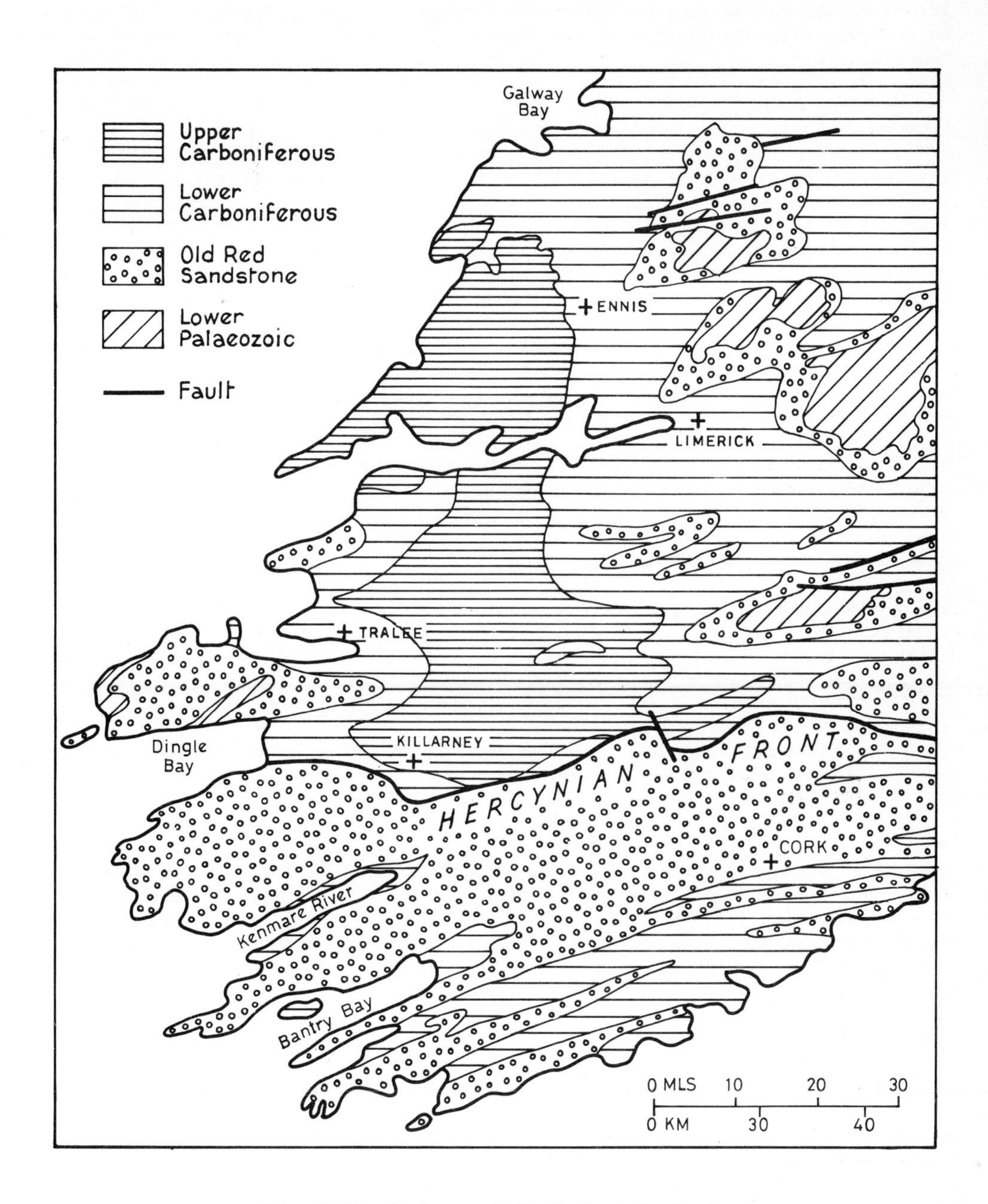

Fig. 7.21 Geology of South-West Ireland.

The Reef Knolls in Central and Eastern Ireland are often closely associated with lead-zinc mineralisation. Russell (1972) suggests that ascending hot brines occurred over hotspots located at the intersection of NE.-SW. faults with N,-S. "geofractures".

7.7 Southern Ireland

This region lies south of a line linking Tralee, Killarney and Dungarvan Harbour (Fig. 6.11). The western portion of this area has a rugged relief with many hills reaching to over 2000 ft (610 m) O.D. and two summits, Brandon (in the Dingle Peninsula) and Carrantuohill (in Macgillicuddy's Reeks) reaching over 3000 ft (915 m) O.D. In this western region rivers are short and flow fairly swiftly to the sea. This western coastline is a deeply indented one and represents one of the finest examples of a drained "ria" or Atlantic-type coast in the British Isles (Fig. 7.21). The projecting form of the headlands reflects the trend of the major folds, with the promontories marking the major antiforms and the bays the major synforms. The former expose the Old Red Sandstone (and in the case of the Dingle Peninsula, also the Silurian) whilst the latter preserve the Carboniferous (fig. 7.21). The eastern half of the area is of much lower relief, rarely rising above 800 ft (244 m) O.D. and with large portions being below 400 ft (122 m). Three major W.-E. flowing rivers, the Blackwater, the Lee and the Bandon, drain the region. The coast is of a more "Pacific-type" but a number of inlets breach outer structure to cut deep back into more inland folds, particularly so in the case of the harbours of Cork and Youghal.

The geological succession in the region is as follows:

Bollandian and Namurian Shales
(unconformity)
"Carboniferous Slate" or Carboniferous Limestone
Old Red Sandstone
(unconformity)
Dingle Beds (Siluro-Devonian)
(unconformity)
Silurian

Holland (1969) has described the Silurian succession in the Dingle Peninsula. This sequence includes rhyolites, agglomerates, tuffs and ignimbrites. This Wenlock-Ludlow succession (1700 m thick) passes up conformably into the Siluro-Devonian Dingle Group (almost 3000 m thick) - grey to purple sandstones, mudstones and locally thick conglomerates. Both these thick sequences were folded during a late Caledonian episode. Gigantic inversions and overfolds can be seen, as for example, on Clogher Head. Much of the steepening and inversion of the limbs of isoclinal folds occurred however during later (Hercynian) movements, which also caused extensive faulting and zones of cleavage (see Holland, 1969, p.306).

The succeeding Old Red Sandstone includes several incursions of very coarse-grained material, attributed to alluvial fan environments by Horne (1975). The Middle O.R.S. Inch Conglomerate Formation (600 m thick) is a thick immature conglomerate or breccia of metamorphic provenance with an imbrication indicating a southerly source. An elevated ridge along the site of the present site of Dingle Bay may have separated a Dingle sedimentary basin from a more southerly Munster Basin. The Old Red Sandstone succession thickens rapidly southwards into McGillicuddy's Reeks (Walsh, 1968) and the Caha Mountain region (7000 m thick). The thick piles of purple red, grey and green sandstones, siltstones, slates, cornstones and conglomerates are largely of fluvio or fluvio-lacustrine origin. Their pre-Devonian base is not seen. Sparse plant evidence suggests that this great thickness is

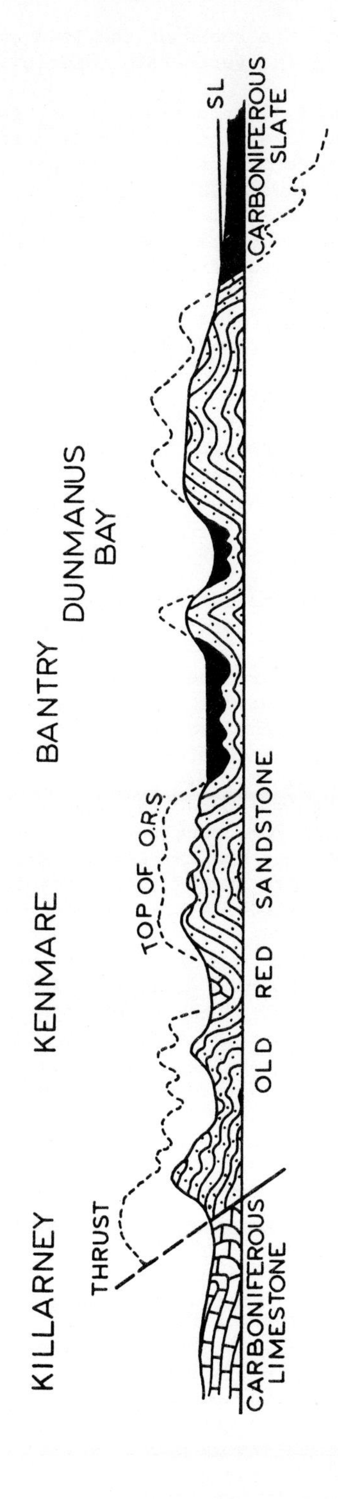

Fig. 7.22 Diagrammatic Section to Illustrate the General Structure of South-West Ireland (based on Geological Survey).

largely of Upper Devonian age. The rocks of the Inveragh Peninsula have yielded late Middle to Early Upper Devonian vertebrate material (Russell, 1978).

The lowest Carboniferous rocks fall into two distinct facies within the area of Southern Ireland, the dividing line running from Kenmare to Cork Harbour. North of this line, the Lower Limestone Shales (1000 ft) pass up into the Main Limestone as, for example, near Ardmore (Dawson-Grove, 1955). In the vicinity of Cork, the thick Waulsortian sheet-reef of Central Ireland makes its appearance. To the south of the line deltaic type sediments appear, to reach a maximum total thickness of 8000 ft (2440 m) at the head of Bantry Bay. They are generally referred to as the "Carboniferous Slate" but can be subdivided into the more arenaceous Coomhoola Grits below and the Ringabella argillites (with thin limestones) above (Turner, 1939). Their deposition appears to be the result of the rapid erosion of an uplifted folded area to the south or south-west of Ireland. Bouguer anomalies suggest a southward thinning of the Old Red Sandstone, in the extreme south of Ireland, against a rising Lower Palaeozoic floor (Thirlaway, 1951). These Lower Carboniferous rocks appear to be overlain unconformably by Bollandian-type highest Visean and/or Namurian cherty shales in a few small outliers in Southern Ireland, the main one occurring near Minane (George, 1962a, Fig. 2/V). Jackson (see Gill, 1962) has proved an unconformity between low Namurian and basal Visean at Ballyheada, near Ballinshassig.

The effects of pre-Old Red Sandstone earth movements in Southern Ireland can be seen only in the Dingle Peninsula. Here the Dingle Beds and the Silurian are folded on ENE.-WSW. lines, parallel in fact to the trend of the later Hercynian folds. On Clogher Head the folds are nearly isoclinal, the axial planes dipping south-south-east at 60°. Further east the folds are a little more open. Younging structures such as graded and current bedding, ripple marks, etc., prove overturning, e.g. in Ventry Harbour (Shackleton, 1940). The regional cleavage has been shown by Shackleton to be of the same intensity in the older rocks as in the Old Red Sandstone, and is probably, therefore, an Hercynian cleavage. It is inclined south-south-east at about 70°, parallel to the axial planes of the more open folds in the Old Red Sandstone and the Carboniferous. That, however, the Caledonian folds must have been overturned northwards before the Old Red Sandstone was deposited is shown, e.g. near Bull's Head, where the Old Red Sandstone lies with low dip on inverted Dingle Beds.

South of the so-called "Armorican Front" - i.e. a line running west-east from Dingle Bay to Dungarvan (see fig. 7.21) - the thick Upper Palaeozoics are intensely folded into large-scale anticlinoria and synclinoria of cleavage or parallel type flexuring (fig. 7.22). This area was the northernmost fringe of the great "Armorican Arc" that curved westward through Cornubia and Brittany, turning slightly south of west in West Cork and in Kerry. The complicated upfolds (bringing up the thick Devonian sequences) include the Mangerton, Clashmore and Kilcrohane (amplitude 4000 m) anticlines. The synclinoria preserve Carboniferous successions and include the Bantry Bay, Blarney, Ardmore, Cork and Minane downfolds. The famous ria inlets of Kerry mark the downfolds while the intervening massive headlands are dominated by the Devonian cores of the major upfolds (fig. 7.22). A large number of minor flexures (ranging in amplitude from hundreds of feet to mere inches) are superimposed on all the major folds. Structural zones, characterised by different fold types have been made out in several areas, e.g. in the Sneam area (Capewell, 1957), the Beara peninsula (Coe and Selwood, 1963) and Mizen Head (see fig. 7.23). Coe (1969) has described a complex pattern of deformation and igneous intrusion in southwest Cork. Intrusions were directed where dips were steep or along steep strike faults. Coe thinks the tectonic style of that area suggests movements on block faults in an underlying rigid basement and that the igneous activity was probably connected with fault block movement at depth rather than with fold development at high crustal levels. Coe and

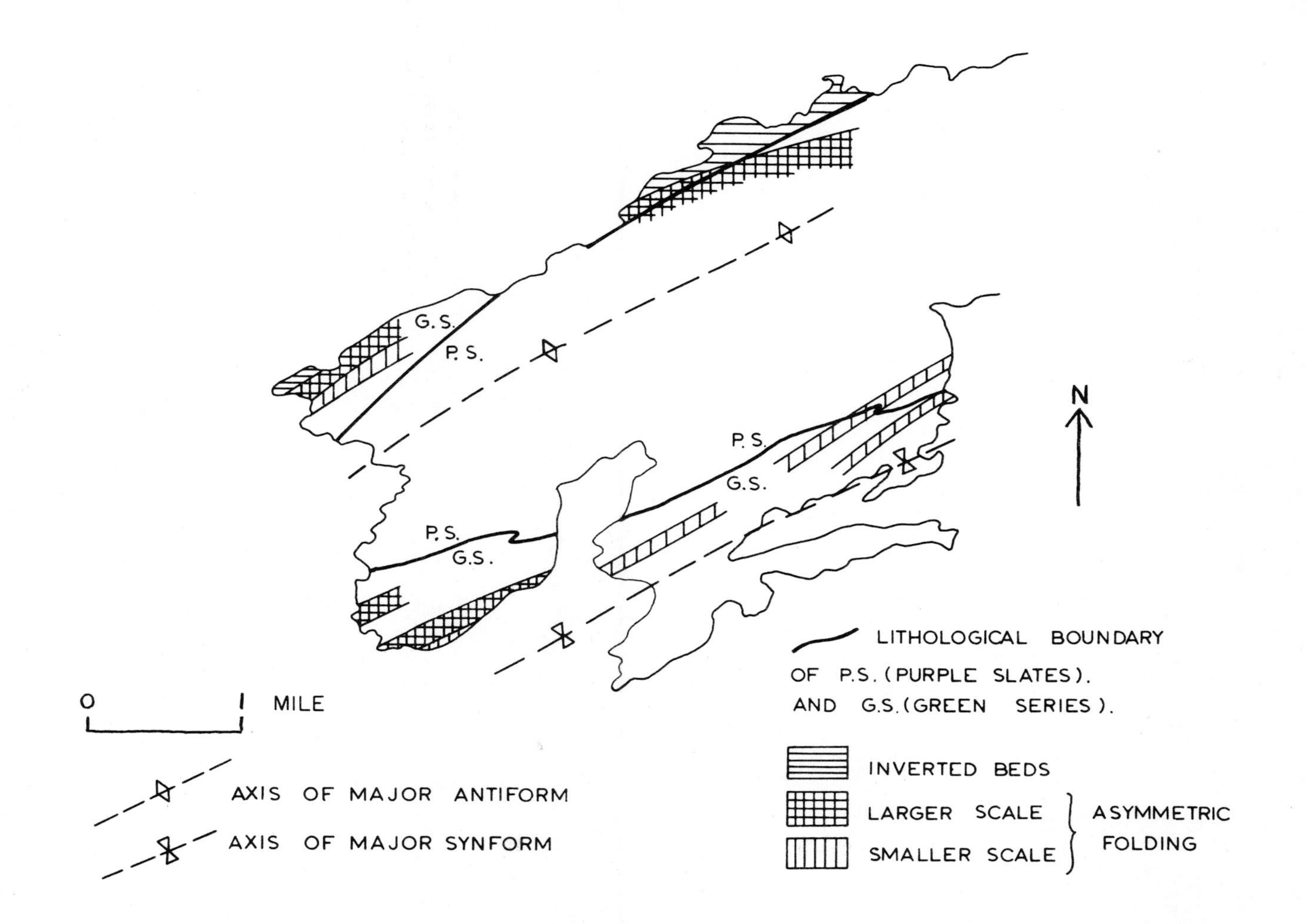

Fig. 7.23 The Structure of Mizen Head, South-West Ireland (after Reilly and Gill).

Selwood (1963, p.58), in their study of the Beara peninsula in County Cork, state that the final uplift of the central zone of that complex diapiric fold could be related to the emplacement at great depth of a granite pluton.

The fault pattern in southernmost Ireland includes major ENE.-WSW. thrusts, E-W trending wrench faults (see Capewell, 1957) and a large number of cross faults trending from NW.-SE. through to NNE.-SSW. N.-S. faults in the Reeks appear to have a massive (accumulative) easterly downthrow in order to cancel out the persistent westerly plunge (about 15-20 degrees) of the folds in that region (Walsh, 1968, p.19).

Walsh, in his study (1968) of the area west of Killarney, describes the "Armorican Front" as a line marking a sharp contrast in structural elevation and fold style. The site of the "Front" may have been decided well before the Hercynian climax. The line is for example near to the Upper Devonian barrier postulated by Horne (1970). It lies almost near to the southern positive area of Namurian times, noted by Hodson and Lewarne (1961). Again it lies along a region where the very thick Devonian-Carboniferous sediments were rapidly thinning northwards. Such northward restriction within the Upper Palaeozoic cover could result in (a) a marked piling upward of the southern folds and (b) eventual giving way northward of slices along powerful thrusts (Owen, 1974, p.239).

The nature of the "Amorican Front" changes eastwards across Southern Ireland. Near Killarney, powerful, northward-downthrowing, faults such as the Muckross-Millstreet system (3000 m throw), the Benson's Point fracture (1600 m throw) and the Black Lake Fault (3300 m throw) may together represent a broad zone of upthrusting stretching all the way to the west coast at Rossbeigh. Walsh (1968) notes high fault plane inclinations of between 60 and 90 degrees. In more central parts of the "Front", a zone of lower angled "schuppen" occurs (Gill, in Coe, 1962, p.54) involving slices of Devonian and Carboniferous rocks all essentially the right way up. Philcox (1963) also describes changes in the frontal zone, this time from Mallow eastwards to Dungarvan. No major thrusting occurs in this eastern area and "fanning out" of the structures here may in some way be related to caledonoid influences and/or to the presence of the Leinster pluton.

Gardiner (1978) believes that the "Front" in Ireland has no direct link with South Wales, being separated by a major basement block, and may well be a localised internal foreland feature. In both South-West Wales and in Southern Ireland, the location and trend of the Hercynian Front is a function of basement control, according to Gardiner and simple metallogenic correlations are therefore improbable.

CHAPTER 8

Alpine Terrains

8.1 Eastern England

The western margin of this area extends along the base of the Lias from Northallerton, in the north, through Scunthorpe and Leicester to Gloucester, whilst the southern margin runs from Gloucester to Swindon and then eastwards through Maidenhead to London and the Thames estuary. The area has a long eastern coast formed for the major part of mainly Chalk or the overlying Tertiary-Pleistocene sediments (Fig. 8.1). Jurassic rocks form the coast around the Wash and again along the Yorkshire coast from Filey to Redcar.

Topographically, the region is one of diverse relief with long hill scarps and intervening clay vales, but apart from the Cleveland Hills, in the north, the surface height rarely exceeds 1000 ft (305 m) O.D. The higher ground tends to follow two main belts, corresponding to the Middle Jurassic limestones and the Chalk respectively. The western line of hills comprises the Cotswolds and the Northamptonshire Heights and Lincoln Cliff, but the summit heights diminish appreciably northwards. The eastern line extends from the Marlborough and Berkshire Downs along the Chilterns to the lower rolling wolds of Suffolk and Norfolk. Beyond the Wash, the hills continue as the Lincolnshire Wolds. The two hill lines then merge in east Yorkshire as the Yorkshire Wolds, separated by the broad Vale of Pickering from the higher, more desolate Cleveland Hills.

The area is drained mainly towards the North Sea by such rivers as the Tees, Derwent, Humber, Welland, Yare, Stour and Thames.

The rock succession in the area is as follows:

Pliocene-Pleistocene
(unconformity)

Eocene
(unconformity)

Cretaceous
(unconformity in most places)

Jurassic
Triassic (in the extreme north of this area)

The structure of the region is relatively simple by comparison with the other British and Irish areas. Nevertheless there are several interesting facets of the area's structure. These include the sub-Mesozoic geology, the Market Weighton Block, the Alpine structures and striking superficial structures, more particularly in the Jurassic strata of Northamptonshire, Rutland, Gloucestershire and Oxfordshire. Earth movements, on a regional scale, occurred during the early part of Middle Jurassic times, in late Jurassic and Lower Cretaceous times, at the close of the Cretaceous Period and more especially in the Miocene.

Our knowledge of the geology and form of the sub-Permian or sub-Mesozoic foundation of Eastern England is based on a large number of boreholes drilled for oil prospecting or water supply and in the search for concealed coalfields. Geophysical

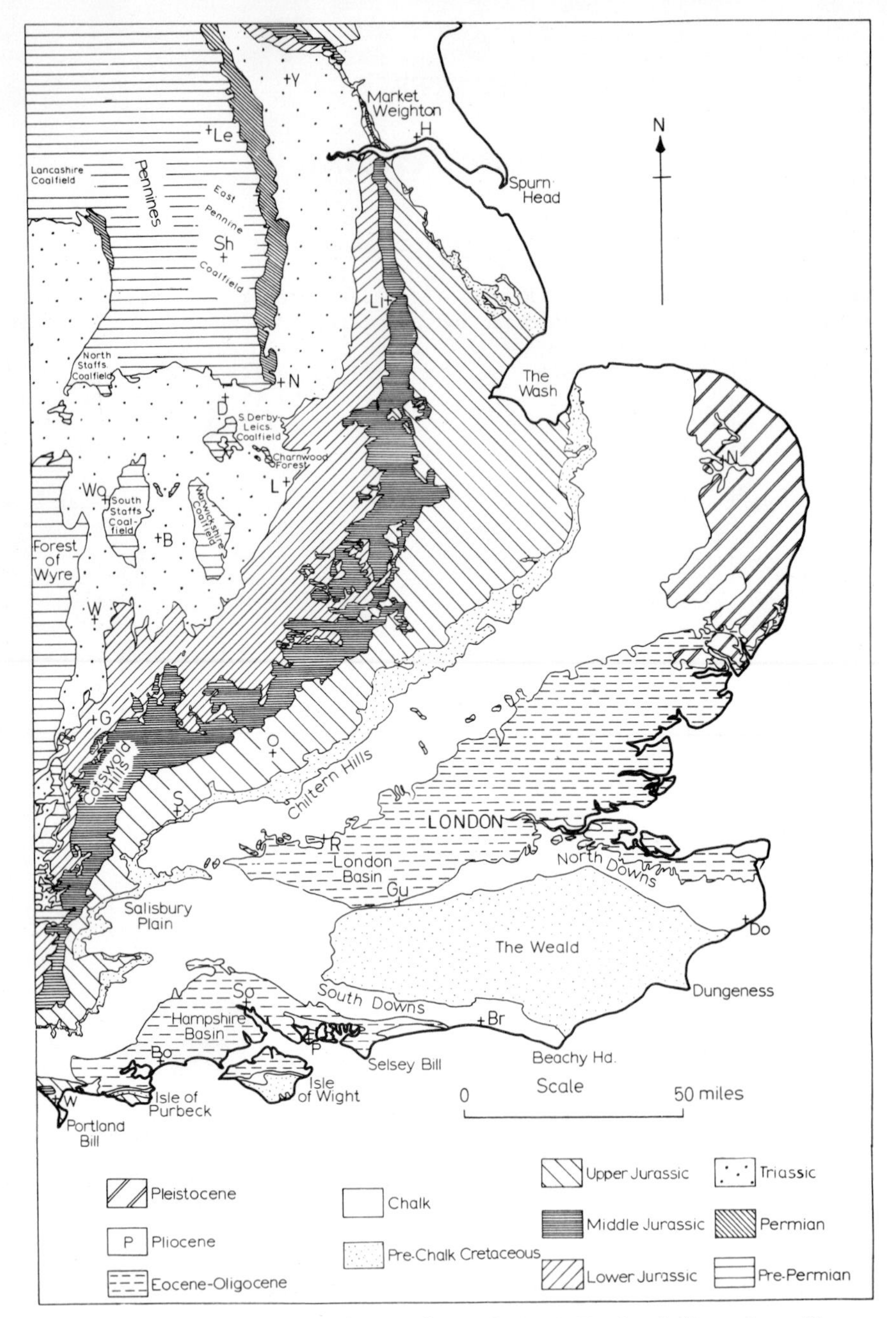

Fig. 8.1. The General Geology of South-East England (based on the Geological Survey's 10 miles to the inch map). Key: B, Birmingham; Bo, Bournemouth; Br, Brighton; C, Cambridge; D, Derby; Do, Dover; G, Gloucester; Gu, Guildford; H, Hull; L, Leicester; Le, Leeds; Li, Lincoln; N, Norwich; No, Nottingham; O, Oxford; P, Portsmouth; R, Reading; S, Swindon; Sh, Sheffield; So, Southampton; W, Worcester; We, Weymouth; Wo, Wolverhampton; Y, York.

methods have also substantially contributed to our knowledge of the basement. The results of these different methods have been published by Lees and Taitt (1946), Kent (1949), Falcon and Tarrant (1951) and Falcon and Kent (1960). Comprehensive discussions of the results, and full bibliographies are given by George (1962a, 1963a).

The depth to, and surface form of, the pre-Permian surface of Eastern England has been discussed by Kent (1949), and are shown in his Plate 2. In Yorkshire and Lincolnshire, this floor falls steadily eastwards to some 6000 ft (1830 m) below sea-level at Flamborough Head. This fall is an expression of the N.-S. Miocene arching over the Pennines. South of the Wash, the main feature is the broad London Platform, "a remarkably flat-topped area of shallow ancient rocks over which there are no transgressions associated with major downwarping" (Kent, 1949). The shallowest part of the Platform appears to be an oval rise (-360 ft or -49 m O.D.) under Cambridge and there is a gentle fall towards the Harwich coast (-1023 ft or -312 m O.D.). The Platform falls much more quickly alongs its southern edge (south of London).

The geology of the sub-Permian (or, south of the Wash, the sub-Mesozoic) floor is very varied and is conveniently summed up in George's Fig. 3 (1963a) and Fig. 1 (1962a). Carboniferous rocks occur beneath the Permian over the region north of the Wash, with an anticlinal area of Carboniferous Limestone and Millstone Grit near Bridlington but Coal Measures elsewhere. The occurrence of Carboniferous Limestone beneath the Permian at Foston near Grantham marks the southern edge of the East Pennine Coalfield whilst further east the occurrence of Precambrian rocks on the sub-Mesozoic surface at Peterborough (Fig. 8.3) and North Creake (Norfolk) sets the southern limit of Coal Measures in that direction.

Silurian (mainly Salopian) rocks must occupy large spreads over the surface of the London Platform, being proved at Harwich, Lowestoft, Ware, Little Missenden and south of the Thames mouth (Fig. 8.2b). Devonian rocks (not always of O.R.S. facies) form the surface of the buried Platform around London, near Southend and near Oxford (Witney). Carboniferous Limestone occurs south-west of both Northampton and Cambridge. The complex sub-Mesozoic geology of the Oxford area is shown by the occurrence of Cambrian at Calvert and of Coal Measures in the Burford-Bourton area. These Morganian measures are over 4000 ft (1220 m) thick, of Pennant aspect, but with several coal seams (Stubblefield and Trotter, 1961). The unconformable contact of these Morganian Beds with the underlying Devonian points perhaps to movements contemporaneous with the "Symon Fault" of Coalbrookdale.

The chances of locating other areas of Coal Measures over the more eastern portion of the London Platform do not appear to be good. For example, though the form of the sub-Mesozoic surface does not indicate any south-eastward prolongations of the Leicester and Nuneaton axes of uplift, the occurrence of Silurian at Ware and of Cambrian at Calvert seems to support the idea of pre-Mesozoic uplift and erosion along these extensions. Further north, of course, lies the buried eastern extension of the Yorkshire Coalfield. Permo-Triassic salt deposits (like those in the Eskdale borehole) must also occur widely beneath East Yorkshire and Lincolnshire and out under the North Sea. The discovery of gas at Eskdale in 1937 and Aislaby in 1963 made the proving of the Permian beyond the Yorkshire-Lincolnshire coastline an important matter.

Today Triassic deposits have a south-eastern limit (running from near Great Yarmouth through Huntingdonshire to Oxford) beneath the post-Triassic Mesozoic cover of East Anglia (Fig. 8.2a). The narrow overlapping or overstepping fringes of Lias, Middle-Upper Jurassic and Lower Cretaceous sediments on to the London Platform point to the long-continued positive behaviour of this block during Mesozoic times. The greatest stratigraphical break occurs over a large area from about

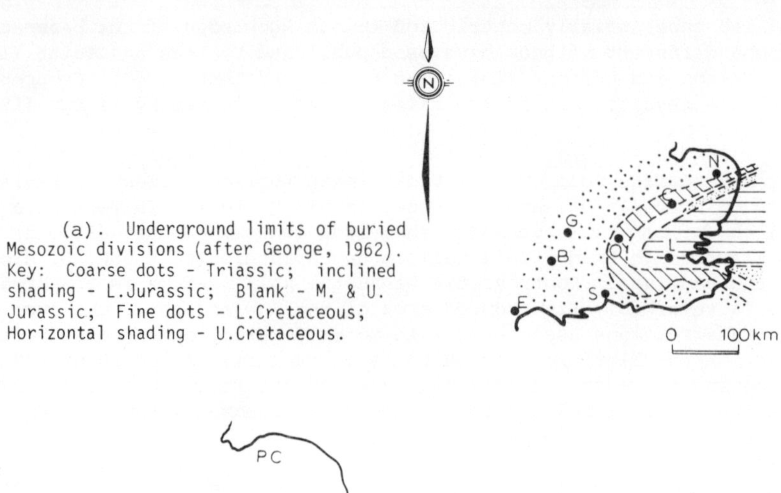

(a). Underground limits of buried Mesozoic divisions (after George, 1962). Key: Coarse dots - Triassic; inclined shading - L.Jurassic; Blank - M. & U. Jurassic; Fine dots - L.Cretaceous; Horizontal shading - U.Cretaceous.

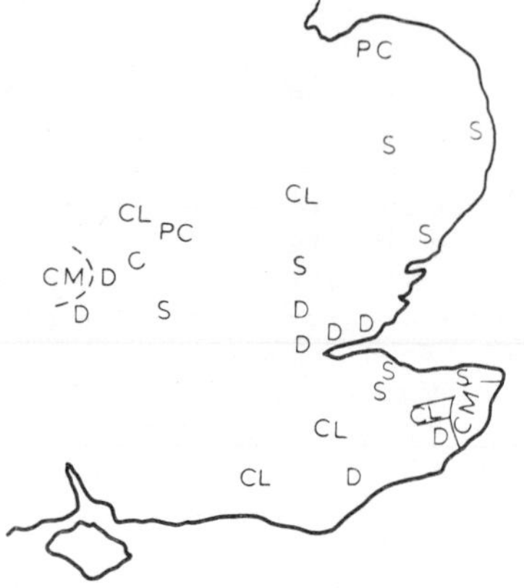

(b). Sub-Mesozoic geology of S. E. England (after George, 1962). Key: P.C., Precambrian; C., Cambrian; S., Silurian; D., Devonian; C. L., Carb. Limestone; C. M., Coal Measures.

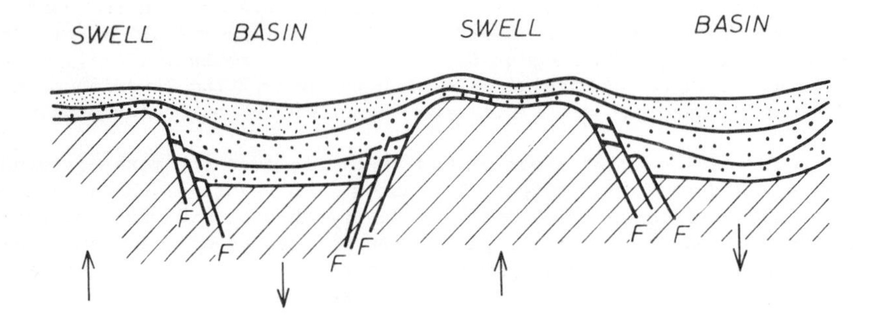

(c). Effect of basement fracturing on later sedimentation producing basins and swells (after Whittaker, 1975).

Fig. 8.2. Foundation of S.E. England.

Luton to the Suffolk-Essex coast, with the Gault resting directly on the Palaeozoic floor.

Over the area, Jurassic rocks,1100 ft (308 m) thick in the North Norfolk borehole, have an S-shaped surface outcrop with an extremely thin mid-portion between the Humber and the Vale of Pickering, i.e. in the Market Weighton area. The widest outcrop occurs in the counties of Northampton, Huntingdon, Rutland and Cambridge that of the Oxford Clay being particularly broad. The Jurassic succession over the whole region is characterized by the lateral and vertical variations which occur in lithology and thickness. The Lower Lias, up to 900 ft (275 m) thick in areas like Evesham, Northamptonshire and north-easternmost Yorkshire, thins to less than 200 ft (60 m) at Moreton and to 100 ft (30 m) near Market Weighton. The Middle Lias over the Moreton area is reduced to 30 ft (9 m) and the Upper Lias to some 12 ft (4 m) at nearby Fawler. Similar thinning occurs on either flank of the Market Weighton "Block", over the centre of which Upper Cretaceous rests on Lower Lias. Similar thickness variations along the outcrop can be demonstrated for the Inferior Oolite and Great Oolite Series and it is clear that areas of shallows were dominant and concentrated in the Moreton-Oxford and Market Weighton areas. Hallam (1958) doubts the existence of narrow "axes" of uplift and prefers to think of them as "swells", "zones of relative uplift existing nearly always as submarine shallows or land". They give the false impression of narrow axes merely because these areas of shallows are now crossed by narrow Jurassic surface outcrops.

Whittaker (1975) attributes the changes in thickness in the cover Mesozoics to fracturing positive and negative blocks in the Palaeozoic basement (see fig. 8.2c). Sellwood and Jenkyns (1975) ascribe the Market Weighton swell to a relative buoyant rise of a salt pillow or granite body whose movement was again triggered by faulting in the basement.

Kent (1956) considered the Market Weighton intra-Jurassic variations as being due more to a broad non-subsiding rigid block than to an anticline. Faunal connection existed across and past the area in Jurassic times. Facies changes, too, are not as marked as was once thought, for the Middle Jurassic, the Oxfordian or the Corallian. The Market Weighton Block probably had a NNW.-SSE. orientation with troughs of deposition on either side (Kent 1958). Both Hemingway and Neale, discussing Kent's views, accepted a Jurassic basin to the east of the block but not on its western flank. Hemingway further thought that "the block was corrugated to form a complex of intra-Jurassic and late Jurassic folds of small amplitude and length, modified and partly obscured by pre-Albian erosion and by Chalk deposition".

Some intra-Jurassic folding and erosion occurred in the south-west of the region - in the vicinity of Birdlip and Painswick, Gloucestershire - in the middle of the deposition of the Inferior Oolite. The (Vesulian) Upper Trigonia and Clypeus "grits" transgress across Lower and Middle Inferior Oolite, folded along sinuous N.-S. lines (see Arkell, 1947, fig. 2).

These minor earth movements died down over most of the area (excepting perhaps the Market Weighton Block) during the Kimmeridgian but were resumed again towards the latter part of the Jurassic and during the Lower Cretaceous. One outstanding feature of the geological map of Eastern England is the unconformity (or unconformities) developed at the base of the Cretaceous System. The Lower Greensand rests on Kimmeridge Clay in many areas and in Bedfordshire and Buckinghamshire the Corallian, and even the Oxford clay, is overstepped by these Cretaceous sands. Uplift along a Bedford axis has therefore been involked to account for this appreciable stratigraphical break.

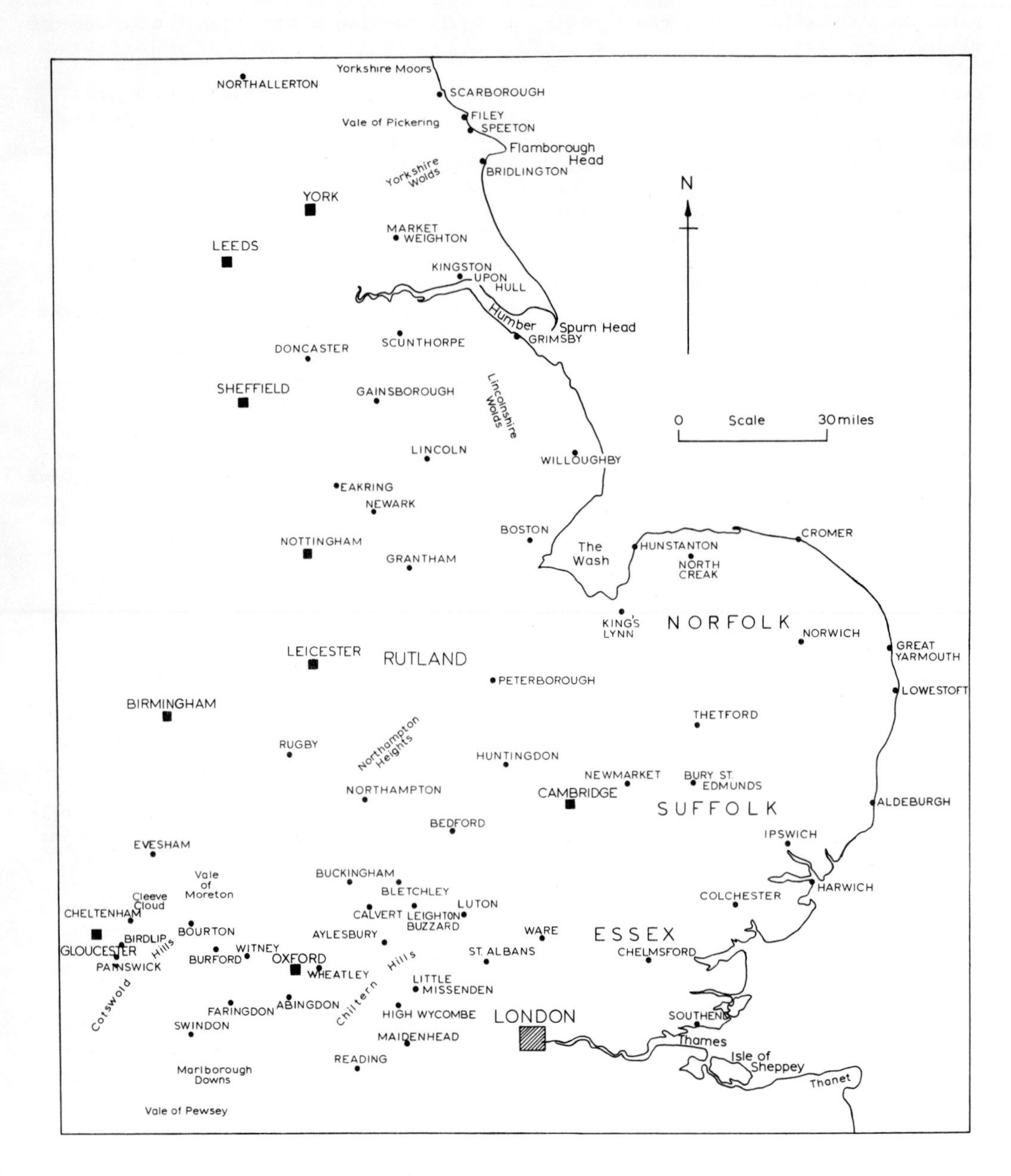

Fig. 8.3. Locality Map of East England.

In the Howardian Hills of Northern Yorkshire an intricate system of WNW.-ESE. and NE.-SW. faults is of post-Jurassic and pre-Cretaceous age. Further south, a series of NW.-SE. aligned faulted synclines in the Upper Jurassic rocks runs out from beneath the Gault along the foot of the Marlborough and Berkshire downs and of the South Chilterns (Arkell, 1947). Wealden Beds are involved in at least one instance whilst the Lower Greensand oversteps at least two of the synclines and is itself unaffected. At Wheatley, a belt of shallow folds and vertical faults trends NW.-SE. for 12 miles (19 km) with a width of up to 1 mile (1.6 km). Directions of movement appear to have been reversed, in some cases, in pre-Greensand and post-Gault times. Earthquakes occurred near Wheatley in 1666 and 1683. Other similarly aligned folds occur at Faringdon, Bourton and Swindon. The north-eastern limb of the former is steep and probably faulted.

It would appear then that important movements, involving shallow folding, faulting and uplift took places, probably intermittently, from uppermost Jurassic through to early-Albian times. Some high Jurassic and Lower Cretaceous deposits are missing through subsequent pre-Greensand or pre-Gault removal, others through non-deposition. According to Rastall (1925) the greatest breaks occur at south-eastern prolongations of the Charnwood and Nuneaton "axes", but this has been doubted by later workers. In any case the whole queston of this sub-Cretaceous unconformity, complex enough as it is, has been rendered even more complicated by the more recent work of Casey (1961-3) who has shown that our previous identification of post-Kimmeridge to Aptian horizons over this area has to be radically revised. He has shown that at about the Jurassic-Cretaceous junction, there existed three separate basins of deposition, (1) a Southern Basin, embracing Kent, Dorset and Buckinghamshire, (2) a Spilsby Basin, Lincolnshire and Norfolk with relics preserved in Cambridgeshire and Bedfordshire, and (3) a Speeton Basin, beyond the Market Weighton "swell". The Spilsby Sandstone, previously ascribed to the Neocomian now correlates with the Portland-Purbeck of Dorset, as also do the Sandringham Sands. Part of these formations are still lowest Cretaceous in Casey's new sense in view of the defining of the Purbeck Cinder Bed in Dorset and the mid-Spilsby Nodule Bed of South Lincolnshire as the basal horizon of the Cretaceous (1963). In view of these revisions, the nature and intensity of the sub-Cretaceous and intra-Lower Cretaceous oversteps may, in places at least, be less intense than previously thought. Nevertheless, a tendency for "shallows" or emergence at times still applies to the Bedford area at least during this Upper Jurassic-Lower Cretaceous span.

The Albian Gault and the succeeding Chalk together mark a great marine transgression over the whole region, even the London Platform being submerged for good. The Chalk is about 1200 ft (366 m) in Norfolk and Yorkshire. The highest zones are seen in Norfolk. The close of Cretaceous times was marked, as in the case of the succeeding period, by a major regression due to another broad uplift of much of the British area. The Tertiary Eocene sands and clays therefore rest unconformably on the Cretaceous. The regional extent of the break can be made out only by zonal mapping within the Chalk, and is illustrated by Wooldridge and Linton (1955, Fig. 4). The greatest gap occurs along the base of the Eocene from about the Colne to Reading. A broad E.-W. anticlinal axis trends from the Thames mouth towards Wallingford and Oxford (Arkell, 1947). It has been suggested that some lines of crumpling within the Chalk of Yorkshire may represent renewed movements along Howardian faults in post-Chalk times, but when an incompletely indurated Chalk would crumple rather than fracture. One such line of crumple occurs at Foxholes, 8 miles (13 km) south-west of Filey.

The numerous transgressions and regressions of the Eocene sea over the south of the region represent continued unrest (unless the changes in sea-level were purely eustatic) and are a prelude to the Alpine earth movements of Miocene times. The major effect of these movements in Eastern England was of course to impose a gentle easterly tilt over the northern portion and a small south-easterly slope to the southern part, i.e. along the northern flank of the London Basin.

This simple picture is, however, complicated in several areas by minor Tertiary folds and by fractures. Faulting on a small scale affects the Middle Jurassic rock to the north and south of Lincoln and a NW.-SE. trough occurs near Scunthorpe. A similarly aligned upfold has been inferred to account for the presence of Lower Cretaceous rocks beneath the glacial drifts of Louth and Willoughby, with anomalous dips at the latter locality. A major WNW.-ESE. fault with a northerly downthrow extends through Stamford, Rutlandshire.

In northern Yorkshire, the eastward regional dip off the Pennine uplift is upset in the Cleveland Hills where the Lias and the Middle Jurassic rocks were domed along an axis trending south of east. To the east two subsidiary domes are centred on Sleight's Moor and Robin's Hood Bay (Wilson, 1948). A basin structure is centred on Whitby. Lesser warps are present in north Cleveland. The Vale of Pickering is a broad trough structure. At the western end of the Vale occurs the Gilling Gap, a narrow faulted trough with inward throws of up to 1000 ft (305 m). On the coast occur a number of NNW.-SSE. faults, but one of these, the Peak Fault, had a complex monoclinal history in Upper Lias times. A dome exposes a small patch of Trias within a Jurassic area on the bed of the North Sea at 54°22'N., 0°7'W., (Donovan and Dingle, 1965).

The London Basin too comprises a number of minor folds and several fractures (see Fig. 8.5a). The gentle flexures partly represent the foreland ripples beyond the main belt of folding, but also represent some incompetent adjustment of the thin Mesozoic-Tertiary cover to deformation of the thick and rigid undermass (Wooldridge and Linton, 1955). The larger flexures have amplitudes ranging up to 200 ft (60 m). The folds fall into a number of trends, viz. E.-W., N.-S., NE.-SW. and NW.-SE. A series of elongated domes extend eastwards along the Lower Thames Valley, and include the Thanet, Cliffe and Purfleet-Grays elements. They may represent a posthumous expression of an anticlinal structure in the basement. North-south monoclinal flexures, especially present on the south-eastern rim of the London Basin, may be a cover readjustment to fracturing in the basement. The main fault belt trends NE.-SW. across the Thames near Greenwich and continues (not as one continuous line but as impersistent elements) into Essex. The East Anglia earthquake of 1883 suggests repeated instability along this line.

One last structural feature of Eastern England is worthy of mention. In dissected areas involving Lower and Middle Jurassic strata, structures of superficial origin are particularly marked and locally intense. These structures include cambers, gulls, dip and fault elements and valley bulges. Cambers are due to the sub-surface erosion and valleyward outflow of Lias beneath the Middle Jurassic limestones. Gulls are widened joints in the camber filled in with material from above. In the ironstone field of Northamptonshire they can be 40 ft wide (Hollingworth, Taylor and Kellaway, 1944). Step-faulting represents advance cambering. Valley bulges are due to differential loading of incompetent Lias. In Northamptonshire they are typified by those in Bytham Brook, and the Glen Valley. Dips of 40° occur in valley bulges along the Evenlode Valley, in the Oxford district (Arkell, 1947). Cambers and escalator structures occur also in the Corallian on the Wytham Hills, near Oxford, owing to the sapping of the sands of the Lower Calcareous Grit at the springline.

An interesting study of the detailed joint patterns in Cotswold Jurassic limestones has been made by Hancock (1969). See for example his figs. 3 and 8.

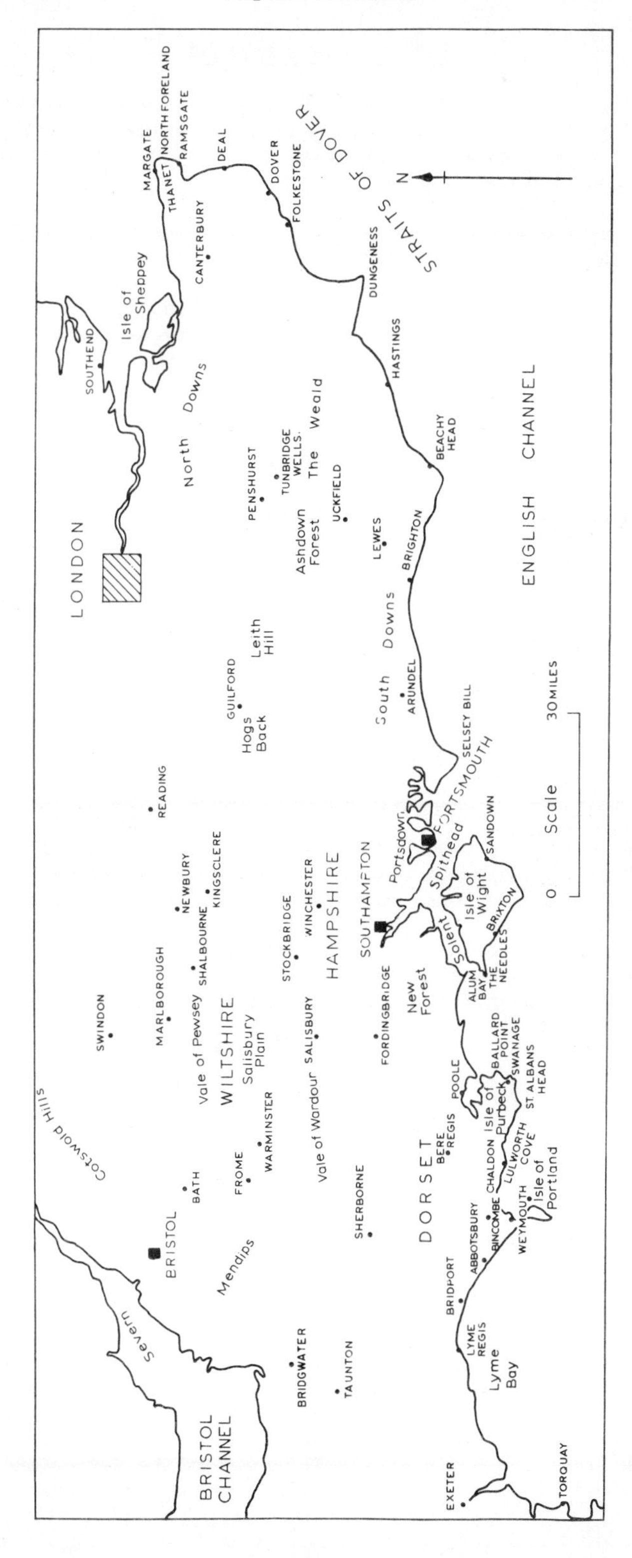

Fig. 8.4. Locality Map for Southern England.

8.2 Southern England

The northern limit of this area is taken as a line from Gloucester to Swindon and thence eastwards to the Thames estuary, whilst the western boundary is chosen at the base of the Middle Jurassic, approximately through Frome and Sherborne to Bridport on the Dorset coast.

Cretaceous rocks occupy much of the surface of the area. The major regional structure affecting these Cretaceous areas is the Wealden Dome exposing a large tract of Lower Cretaceous sands and clays in its core and ringed on three sides by Chalk downlands. The Chalk then extends westwards to Salisbury Plain and south-westwards as the Dorset Downs. Two extensive areas of Tertiary sediments occur within the area, a south-western portion of the London Basin, in the north, and the markedly assymetrical Hampshire Basin in the south. The latter has a steep southern flank of Cretaceous rocks across the Isle of Purbeck and the Isle of Wight. Jurassic rocks form the western fringe of the region and also reach the surface in the Weymouth Peninsula and the core of the Purbeck Anticline.

Like Eastern England, the region is one of undulating relief, the highest point, Leith Hill on the Lower Greensand in the north-west of the Weald, being 965 ft (304 m) O.D. The Chalk reaches heights of over 800 ft (244 m) in several places, especially on the Marlborough Downs and the South Downs. The Wealden sands also give rise to high ground reaching to 700 ft (214 m) in Ashdown Forest. The main drainage is to the Thames, or to the Wealden and Southampton-Poole coastlines. The varied coastline includes the classic areas of Purbeck, Lulworth and Chesil Bank, with spectacular examples of both coastal erosion and deposition.

Palaeozoic rocks form the deeper foundation of the area. They slope away southwards and south-westwards from the shallow London Platform and were met at depths (below O.D.) of 3250 ft (990 m) at Hellingley (near Eastbourne), 4550 ft (1387 m) at Penshurst and 5707 ft (1739 m) at the Ashdown No. 2 borehole. The Shalford borehole (near Guildford) proved Rhaetic rocks resting on steeply dipping Silurian but further west at Kingsclere a borehole stopped in Keuper Marl at a depth of over 5000 ft (1524 m). Silurian rocks form the surface of the buried Palaeozoic floor under the north-eastern part of the Weald, overlain in places by Carboniferous Limestone, or, as on the north flank of the Kent Coalfield, by Coal Measures. Lower Carboniferous rocks form the sub-Mesozoic floor over parts of the inner Weald, whilst Devonian has been encountered between Brighton and the western edge of the Kent Coalfield.

The two dominant units influencing the deposition (or non-deposition) of the Triassic-mid-Cretaceous sediments in the area were the positive London Platform in the north-east and the subsiding Wessex Basin in the south and west. The former was more extensive in earlier-Mesozoic times and then gradually retreated away north-eastwards during Jurassic and early-Cretaceous times to reach a southern limit from London to Sheppey and Pegwell Bay by the beginning of Upper Cretaceous times. Triassic rocks, on the other hand, do not extend north-east of a line curving roughly through Oxford, Newbury, Petersfield, Arundel, Wadhurst and Folkestone.

Evidence of intra-Jurassic movement is confined to the western Jurassic fringe of the area. The most active earth movements were in the vicinity of the Mendips, where marked thinning, or even non-deposition of Liassic and Middle Jurassic horizons was a feature of this stable block. Some early unrest along the line of the Weymouth and Purbeck anticlines may possibly be indicated by thickness variations in Dorset, as, for example, in the Junction Bed (the boundary of the Middle and Upper Lias) and again in the Inferior Oolite Series and in the Kimmeridge Clay (1650 ft (502 m) thick in Kimmeridge Bay, only half that thickness

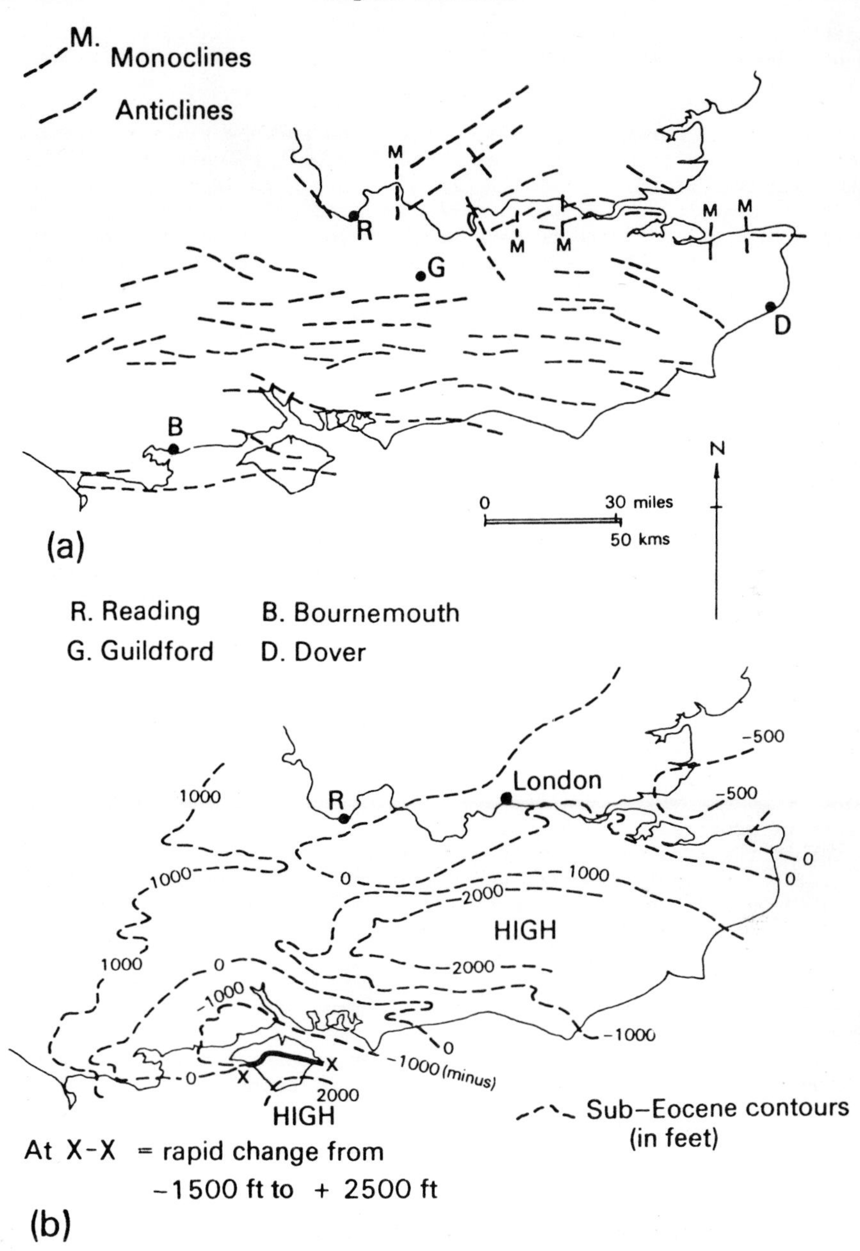

Fig. 8.5. Aspects of the Tertiary Warping of South-East England (after Wooldridge and Linton).

at Ringstead. Intra-Jurassic movements in Kent are suggested by the disappearance of the Lias beneath over-stepping Bathonian at about the latitude of Deal (George, 1962a).

Widespread regional movements during Portland and Purbeck times resulted in the restriction of the deposition of these Upper Jurassic members and uplift occurred over the London Platform, especially along its southern (Kent) border. As a result those Jurassic members which were deposited against the flank of the Platform were eroded prior to the transgressions which occurred over their eroded edges during Wealden or later-Cretaceous times. The north-eastern limits of the Purbeck, Middle Jurassic and Corallian beneath the Cretaceous run through Folkestone, Dover and Snowdown respectively, the Corallian edge running west-north-westwards to Chatham (see Edmunds, 1948, Fig. 2). The thickness of Jurassic rocks preserved beneath the Cretaceous of South-east England therefore varies appreciably, being nil at Deal and Whitstable, 635 ft (194 m) at Dover and 4305 ft (1312 m) at Portsdown. These variations must be explained by the differential subsidence of Jurassic times and the amount of erosion prior to the deposition of overlying Cretaceous Beds.

The base of the Cretaceous presents, in fact, as complicated a picture, in terms of overstep and overlap, in this area as in Eastern England, and in Dorset the complexity was accentuated by the occurrence of locally intense movements during the post-Wealden to pre-Albian interval. In Wiltshire, the Gault oversteps the Lower Greensand which in turn cuts across the outcrops of the Kimmeridge and Corallian. In the vales of Pewsey and Wardour, the Purbeck and/or Portland emerges from Beneath the Albian. Pre-Lower Greensand NE.-SW. faulting occurred near Calne whilst in the Bridport area, pre-Albian faults affect the Lias and Lower Oolites. Westwards across Dorset and East Devon the Albian transgresses across already (eastward) tilted Jurassic and Triassic formations.

It follows then that late-Jurassic and Lower Cretaceous times in southern England were times of continued crustal unrest but it is sometimes difficult to allot precise dates to the movements. It is also necessary to remember once more the new age allocations to those deposits which span the Jurassic-Cretaceous boundary (Casey, 1961-3). In this way, the timing of some of the intra-Cretaceous movements may be brought forward, even perhaps to Purbeck-Wealden times, in areas like North Wiltshire.

In the Weymouth area, spectacular intra-Cretaceous structures can be shown to be of post-Wealden age (fig. 8.5d). Their upper age limit is less clear but they are probably of pre-Aptian origin, the structures probably standing high and being eroded when the Aptian sands were being deposited further east.

The Weymouth intra-Cretaceous structures (fig. 8.6d) comprise sharp E.-W. folds and faults. The folds include the Poxwell Pericline, the Upton Syncline, the Ringstead Anticline and the Chaldon Pericline. Dip faults of pre-Albian age also occur, as, for example, at Ringstead. A recent survey of the sea-floor south of Dorset by Donovan and Stride (1961) has detected a NW.-SE. aligned Shambles Syncline which is transgressed by the base of the Upper Cretaceous and is therefore another pre-Albian fold. This would suggest that the Purbeck Anticline, to the north, had at least in part, a pre-Albian ancestry. The northern limb of the Weymouth Anticline is also partly of intra-Cretaceous age (House, 1961), as also may be some of the southward-downthrowing faults in the core of the Weymouth Anticline.

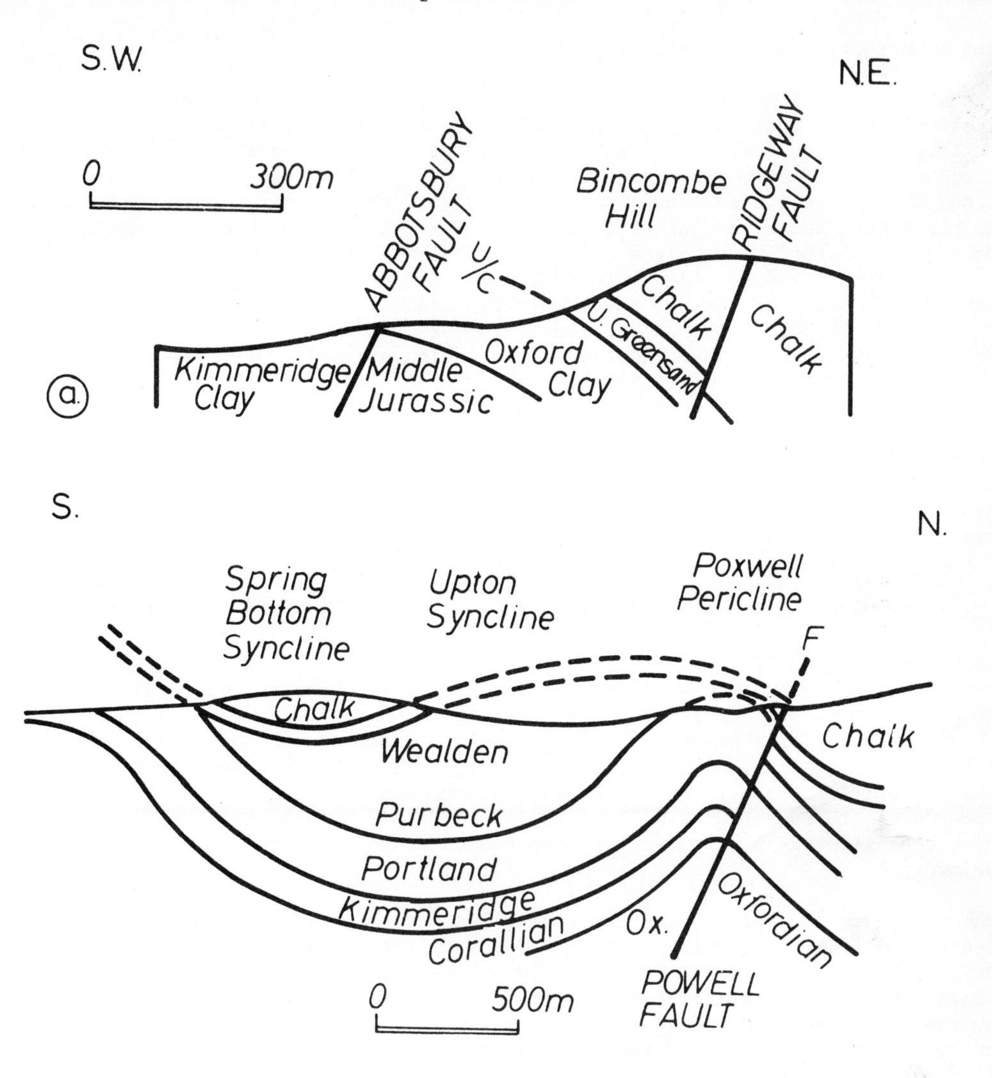

Fig. 8.6. Mid-Cretaceous Structures in Dorset (after Arkell, 1946).

The Abbotsbury Fault

The chief fault-line produced by these mid-Cretaceous movements was the Abbotsbury Fault system extending from the Abbotsbury coast eastwards to at least the Chaldon Pericline (House, 1961). The intra-Cretaceous fracture (or fractures) suffered a large southerly downthrow, as much as 1500 ft (457 m) in places. On the upthrow side, the higher Jurassic and Wealden rocks were eroded prior to the transgression by the Albian sea. Thus at Bincombe (Fig. 8.6c) the Upper Greensand rests unconformably on Oxford Clay to the north of the Abbotsbury Fault, whilst to the south of the fracture the Portland-Purbeck successions are preserved. Near Abbotsbury Church, the upthrown side of the Abbotsbury Fault was eroded down to the Forest Marble horizon before the deposition of the Greensand. All Arkell's sections (1947) across the fracture indicate it as having a fairly steep dip to the south. The Poxwell Fault System has now been shown by Mottram and House (1954) to comprise probably four post-Cretaceous fractures, two of them also of intra-Cretaceous age. These two fractures are 75 ft apart, dip southwards at 50°-60° and had an intra-Cretaceous southerly downthrow of 500-600 ft. An abnormal thickness of Upper Greensand encountered in the 1936-7 Poxwell borehole, and then described as steep fault-drag, is now explained by repetition through faulting. House (1961) has identified two post-Cretaceous fractures along the northern edge of the Chaldon Pericline, further east. One of the fractures had an intra-Cretaceous origin and a normal southerly downthrow and is therefore the Abbotsbury Fault. Its rather flat dip in places is ascribed by House to a mid-Tertiary folding of the fracture.

These tectonics of South Dorset are similar to those in the neighbourhood of Hanover (being there of Necomian age) but in Germany they are associated with salt domes. Lees and Cox (1937) believe that the Weymouth structures are attributable to the presence of a plastic series at depth. Under such conditions a moderate force of compression, normally producing only gentle anticlines, could give rise to steep local effects. The salt series could be Triassic or Permian. It is here interesting to recall that at Bere Regis and Fordingbridge, pre-Gault structures have been located at depth whilst thin bedded anhydrites are present in the Trias below.

The Albian transgression over the region was extensive and extended beyond the feather edges of the Wealden-Aptian of Kent on the London Platform. Pre-Albian Cretaceous rocks thin from 2000 ft (610 m) near Dover to nil near Ramsgate (George, 1962a), though the greater part of this reduction took place before any overstep by the Gault. Marine conditions continued with the deposition of the Chalk, up to 1000 ft thick (305 m) in the Marlborough-Reigate areas and to 1600 ft (488 m) in Sussex and Hampshire.

These quiet marine conditions were disturbed by regional uplift, with slight warping, at the close of Cretaceous times, and varying amounts of Chalk (and, further west, earlier-Cretaceous formations) were removed prior to the deposition of the cyclic Eocene and Oligocene sediments. Phillips (1964) hints that the presence of Upper Greensand and Purbeck pebbles in the Bagshot Beds of Dorset could mean that "the rising land to the south of the Lulworth fold and fault scarp had been eroded down into the Purbeck before the deposition of these Eocene gravels to the north" (p.401).

Eocene deposits occur within the area in the south-western portion of the London Basin, along the north coast of Kent and in the Hampshire Basin (including the thin strip of steeply dipping strata across the middle of the Isle of Wight). They comprise alternations of marine and non-marine sands, gravels and clays representing westward transgressions and eastward regressions of the Anglo-Gallic Sea. Oligocene strata are now confined to the Hampshire Basin (New Forest and

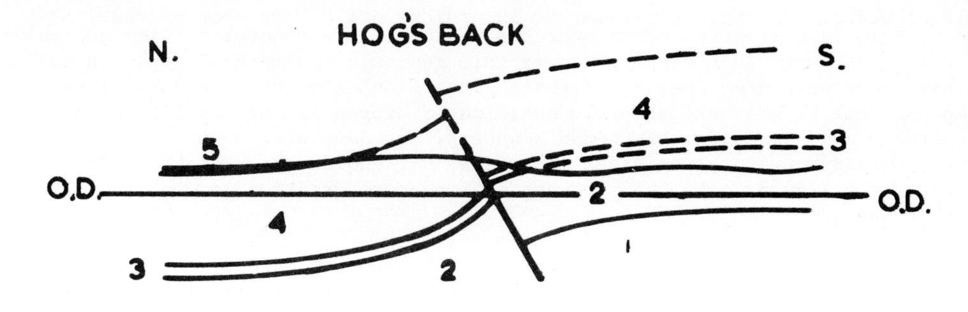

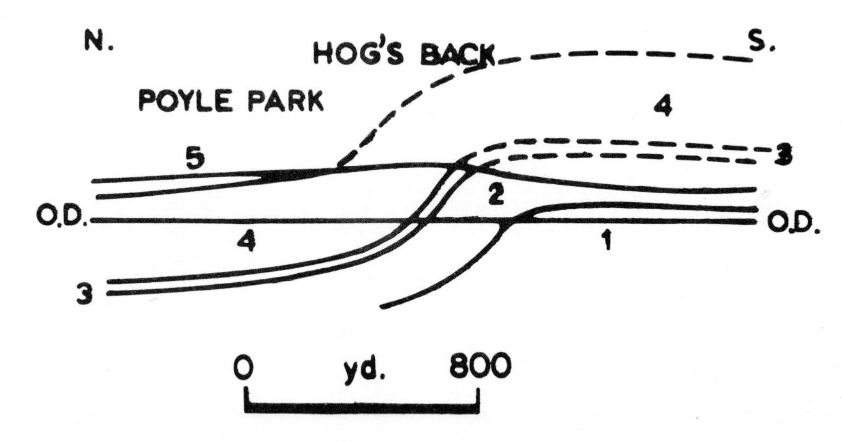

Fig. 8.7(a) Sections across the Hog's Back Disturbance, near Guildford (after Geological Survey). 1. Wealden; 2. L. Greensand; 3. Gault-U. Greensand; 4. Chalk.

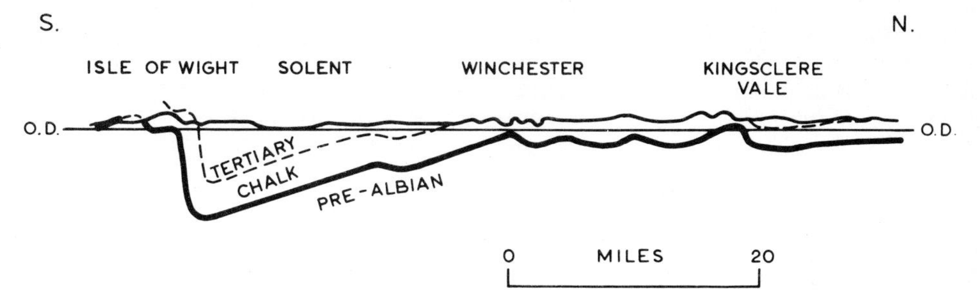

Fig. 8.7(b) Section across the Hampshire Basin (after Wooldridge and Linton, 1955).

Isle of Wight). By this time, marine incursions had become short-lived. No Eocene-Oligocene deposits occur over the Weald and the general opinion appears to be that they were never deposited over this region. The central Weald is believed to have been warped up into an elongate "inversion" structure in early-Tertiary times so that we see here an early anticipated expression of the late main Miocene structures which include the London downwarp, the Hampshire downwarp and the intervening Wealden upwarp. The Blackheath Beds, marking the base of the Ypresian Cycle, are believed to be the products of longshore drifting around the north coast of this Wealden Island.

The most marked structural effects in the area were produced by the Alpine earth movements in Miocene times. The overall effect of these movements is shown in Fig. 8.5b, which depicts the present height variation of the sub-Eocene surface. Over the western Weald this rises to well over 2000 ft (610 m) O.D., whereas in the Isle of Wight - in the base of the Hampshire Basin - the basal Eocene surface falls to over 2000 ft (610 m) below sea-level. The steep rise in height up the monoclinal fronts of the Hog's Back, Sandown, Brixton, Purbeck and Weymouth anticlines is also evident.

The Weald

The London Basin, the Wealden Dome and the Hampshire Basin are broad regional expressions of the Miocene structure. In more detail there are within each one of these regional units a considerable number of more impersistent folds, trending chiefly E.-W., and usually with steeper northern limbs to the anticlines. In the Weald, they trend a little south of east and are particularly frequent near Tunbridge Wells and Pullborough. In the central Weald none of the anticlinal lines is simple, the main folds being frequently accompanied by parallel minor rolls (Wooldridge and Linton, 1955). The main axes are often accompanied by powerful longitudinal faulting, usually along the northern flanks of the flextures. One such line from Lamberhurst to Appledore is 25 miles (40 km) long. Considerable faulting occurs again north of Chiddingley and near Uckfield, both in the Lewes district. A fracture between Henfield and the Brighton coast has a southerly downthrow of over 500 ft (152 m).

The most intense monoclinal warping occurs in the Guildford-Reigate district on the north-west of the Weald (Fig. 8.7b). The folding is most marked on the Hog's Back, the fold being accompanied in places by a reverse strike-fault which is displaced laterally by NW.-SE. or NNE.-SSW. cross-faults of wrench origin. The stratigraphical throw of one such cross-fault at Runfold is as much as 900 ft (274 m). The maximum throw of the Hog's Back reverse faults is 400 ft (122 m). These fractures may be positioned over the site of intra-Cretaceous normal faults with southerly downthrow.

Wessex

West of the Guildford folds lies the Kingsclere Pericline (Fig. 8.7a), previously thought to mark the eastward continuation of the Mendip Axis, but the great thickness of Jurassic rocks encountered in the borehole refutes this theory. On the other hand, the Miocene fold may owe its position to the nearby presence of a pre-Gault Fault. The northern limb of the pericline is dissected by swarms of small dip faults which shift both tilted strata and strike-faults. Other upfolds north of the Hampshire Syncline include the Portsdown, Dean Hill, Stockbridge, Winchester, Wardour, Warminster and Pewsey anticlines.

Isle of Wight

The most intense effects of the Alpine Orogeny in Britain were felt along the extreme southern fringe of the area, that is from the Isle of Wight to the Isle of Purbeck and the Weymouth Peninsula. Two major monoclines, the Sandown and Brixton folds, affect the Isle of Wight. They are arranged *en echelon*, and where each dies out (one westwards, the other eastwards) the outcrop of the Chalk widens appreciably (between Calbourne and Gatcombe). The northern limbs of both folds are almost vertical (Fig. 8.7a) as revealed by the striking Chalk pinnacles of The Needles and the highly coloured Eocene Beds in Alum Bay. Shallow folds such as the Bouldnor Syncline and the Porchfield Anticline affect the Oligocene Beds in the northern portion of the island.

Isle of Purbeck

The Purbeck Anticline is probably not continuous with the Brixton fold. Appreciable cross-faulting may occur between The Needles and Ballard Point, similar perhaps to that located by Donovan and Stride (1961) south of Ringstead Bay, further west. Their work shows that the Purbeck Anticline is not continuous with the Weymouth fold, the former trending well south of the latter's eastern extremity and having its axis about 1½ miles (2.4 km) south of the Lulworth coast (1961, Fig. 6). The Corallian core of the Purbeck fold has been shown, by oblique asdic methods, to be considerably faulted, as also are the north-eastern and southern flanks. The fractures trend NNE.-SSW. and N.-S. On the mainland the main fold axis must lie very close to St. Alban's Head. The steep northern limb affects particularly the Chalk, which is vertical at Ballard Point, and on Brenscombe Hill a very assymetrical "fore-syncline" lies immediately north of this vertical limb, in the Eocene sands and clays. No lengthy fractures affect the peninsula, but some minor faulting (again almost N.-S.) is visible around the southern coastline. Three important E.-W. fractures occur on the east coast, the most interesting being the curved Ballard Down Fault. Its origin has been discussed by Arkell (1947, pp. 304-11). The "onion-scale" adjustment hypothesis would seem to be the most convincing explanation. The new Wych oilfield has proved the presence of major pre-Albian faults with large southerly downthrows. The Purbeck Anticline has been probed for oil. Boreholes were drilled in a minor upfold at Kimmeridge Bay, after the discovery of oil in the Cornbrash of the Paris Basin, and oil was discovered at a depth of 1800 ft (548 m) in highly fissured Cornbrash.

The Lulworth Coast (Plate V).

The intense structures along the Lulworth coast between Bat's Head and Mupe Bay have been described in detail by Arkell (1947) and more recently by Phillips (1964). The latter believes these structures to be the localized accommodation of a blanket of largely unconsolidated sedimentary beds above a major thrust fault in the basement. White (1948) inferred from gravity data that the Isle of Wight monocline passed down into a fracture (a normal one?) in the Upper Palaeozoic basement.

On the Northern limb of the Purbeck fold between Mupe Bay and Durdle Door, the dip of the Portland Beds increases to 80° whilst the overlying and more incompotent Purbeck and Wealden are overturned to the north in many places. The axis of the adjacent Lulworth Syncline is well exposed in the cliffs just west of Bat's Head, the axial plane of this highly assymetrical downfold dipping south at 35°. At Lulworth a unique occurrence of the stratified Broken Beds, jumbled blocks of Purbeck limestone,is seen. Arkell (1938) interpreted this formation, the remarkable Lulworth Crumple (Plate V) and other minor structures in the Purbeck Beds, as being

A. Vertical Purbeck Beds. Man O'War Cove, Dorset, England (K.J.).

B. Stair Hole and Lulworth Cove, Dorset, England (K.J.).

formed by upward drag on the steep limb of a major fold. Phillips (1964) puts forward considerable evidence in favour of these structures being due to *down-dip* movement, i.e. down the northward-dipping limb of the developing Purbeck fold. The Broken Beds were formed by the detachment and partial crushing of joint-bounded blocks of laminated limestones, during this downward sliding. The drag-folds are associated with two sets of oblique shear-planes, with normal displacements on a gently dipping set and reverse movements on a steeply dipping set.

Phillips has put forward different interpretations for the complex groups of minor fractures established by Arkell (1938) for the Chalk area between Bat's Head and Warbarrow Bay. Phillips claims that bedding plane shears and shear-planes dipping northward at 50-70° (with upward displacements on the northern sides) were developed during the folding of the Lulworth Syncline, more especially in the axial zone. They were followed by somewhat flatter, southward-dipping, thrust planes whose consistent dip suggest formation after the completion of the syncline. Further southward-dipping fractures, almost parallel to the overturned Chalk on the northern side of Lulworth and Man o'War coves, are termed thrusts by Phillips and normal faults by Arkell. Phillips has presented (1964, Fig. 15) a convincing picture of the development of the Lulworth Syncline and its associated minor folds and faults. He claims that as the Chalk was folded up into a verticalposition so this formation began to slide down northwards under its own weight. The Wealdon and Purbeck sediments then collapsed downwards to fill the gap behind the Chalk, but at higher levels than those now exposed. After the completion of the synclinal closure, thrusting from the south occurred, with some further crumpling down-dip of Purbeck-Wealden during a final uplift of the southern side.

The Weymouth Anticline

The structural crest of this assymmetrical fold passes approximately due east from Clay Hard Point on the Fleet shore (House, 1961). The northern limb dips at 10° to 20° (except where the Abbotsbury and Upwey Synclines give southerly dips near the Ridgeway Fault). The long southern limb dips at 5° to 10° and as low as 2° at the southern tip of the Isle of Portland. The Weymouth fold pitches eastwards at 6° near Redlands, 2 miles (3 km) north of Weymouth. The core is affected by many E.-W. faults and House believes that the Radipole Fault is the westward continuation of a reversed fault that replaces the middle limb of the Purbeck Monocline near White Nothe. The extreme northern flank of the Weymouth Anticline is disturbed by the intra-Cretaceous Abbotsbury Disturbance and also by the post-Chalk, probably Miocene, Ridgeway Fault System, which attempted to follow the earlier-Abbotsbury line as closely as possible, but with a now northerly downthrow. A complex pattern of both strike and cross-faulting of both pre-Albian and post-Cretaceous age occurs between Sutton Poyntz and Poxwell.

CHAPTER 9

Tertiary Igneous Terrains

Spreads of Eocene basalts occur in Ardnamurchan, in the Hebridean islands of Skye and Mull and in N.E. Ireland; they overlie Mesozoic and, more locally, thin Eocene sediments. Large Tertiary igneous complexes occur in Skye, Rhum, Ardnamurchan, Mull, Arran and N.E. Ireland (Richey, 1961; Charlesworth, 1960; Charlesworth, 1963). Many contain ring-dykes and cone-sheets, the recognition of which was an important advance in the understanding of the mechanics of igneous intrusion (Fig. 9.1). The presence of an undersea Tertiary basic intrusive complex about 63 miles (100 km) N. of the N. tip of Shetland, is indicated by geophysical evidence (Chalmers and Western, 1979).

The Tertiary centres occur with a N.-S. zone, 250 miles (400 km) long and 40 miles (65 km) wide. They show few relationships to pre-Tertiary structures penetrating indifferently the N.W. Caledonian Front, the Northern Highlands, the Grampian Highlands, the Midland Valley of Scotland and the Irish continuation of the Southern Uplands. Moreover, the igneous rocks all pre-date the Neogene topography and have been deeply dissected by Tertiary erosion and by Pleistocene glaciation. They form some of the most spectacular scenery of the British Isles, rising to 3309 ft (1010 m) in the gabbro peaks of Skye. Coastal and mountain exposures are excellent.

The formations penetrated are:

Tertiary
(unconformity)
Cretaceous
(unconformity)
Jurassic
Triassic
Permian
(unconformity)
Carboniferous
Upper Old Red Sandstone
(unconformity)
Lower Old Red Sandstone
(unconformity)
Silurian
(unconformity)
Ordovician
Cambrian

Dalradian Metamorphic Assemblage or Supergroup (partly Cambrian)
(Moinian Metamorphic Assemblage
(Torridonian
(unconformity)
Lewisian Metamorphic Assemblage

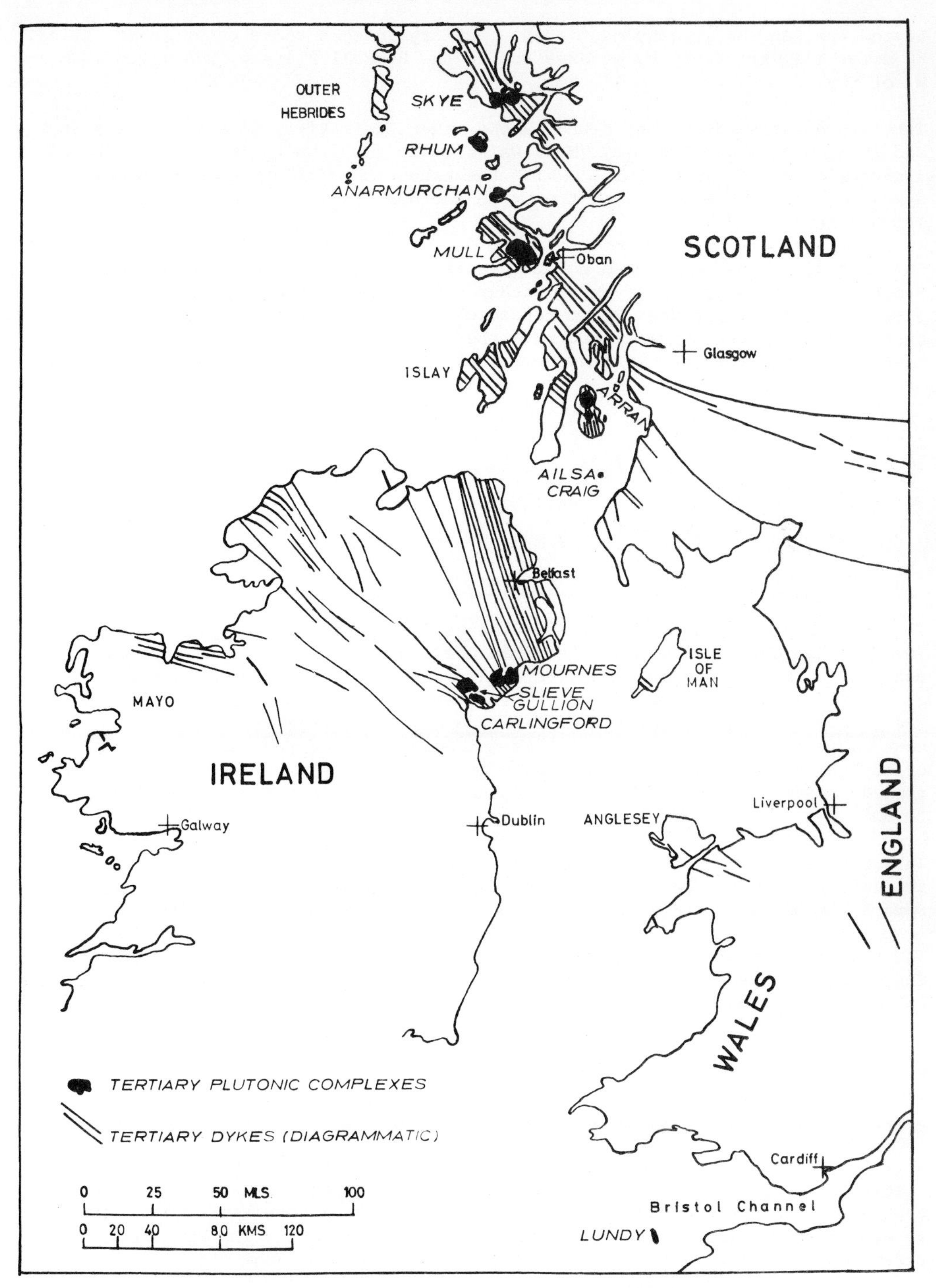

Fig. 9.1 Tertiary Igneous Districts of British Isles.

Beyond the centres already mentioned, Tertiary igneous rocks occur in St. Kilda, 88 miles (141 km) W. by N. of Skye and in the Rockall Plateau, 275 miles (440 km) W. of Skye.

Tertiary dolerite dykes are found, among other localities, in Mayo (W. Ireland), and in Snowdonia and Anglesey (N.W. Wales). Eocene granite forms Lundy Island (Bristol Channel), (Fig. 9.1). With the exception of the Mayo intrusions, the dykes and the granite occur on the broad southerly continuation of the Scottish/ Irish zone.

The intrusions and lavas need not be described: excellent general accounts of the Scottish occurrences are given by Richey (1961) and of the Irish by Charlesworth (1963). On the other hand, post-Mesozoic structural elements, some of distinctive character and closely associated with the emplacement of the intrusions, are features of the Tertiary Igneous terrains. These structures will be considered under the following headings:

Ring fractures associated with ring-dykes and cone-sheets

Domes and concentric folds due to magmatic intrusion

Dilation due to dyke intrusion

Mesozoic and Tertiary folds and faults.

Ring fractures associated with ring-dykes and cone-sheets

The intrusion of ring-dykes and cone-sheets is not only a magmatic phenomenon but involves the formation of arcuate or annular faults. In the Tertiary Igneous centres it is possible to demonstrate in a number of instances both the existence of such fractures and the amount of displacement involved. For example, in Mull the Loch Ba Ring-dyke, about 4½ miles (6.8 km) in average diameter, is one of the most perfect known (Fig. 9.2). It is intruded along a ring-fault, with downthrow of at least 3000 ft (915 m) bounding a late caldera of subsidence. The fault and ring-dyke are observed in a few places to be inclined outwards at angles of 70° to 80°.

In N.E. Ireland the Tertiary Ring-fracture of Slieve Gullion, up which has welled two acid ring-dykes, is about 7 miles (11.2 km) in diameter and conforms in a remarkable way with the margin of a lobe of the Caledonian Newry Granite.

The formation of some twenty concentric ring-dykes around the Beinn Chaisgidle Centre in Mull and of fifteen around Centre 3 in Ardnamurchan must have been accompanied in each case by successive phases of subsidence, amounting in all to several thousand feet.

Ring-fracturing resulted in the formation of volcanic calderas such as that of south-east Mull and that of the Central Ring-complex of Arran. In Mull the Central type of basalt lavas accumulated within the caldera to the extent of several thousand feet, at times (as shown by pillow-structure) flowing into a lake. In Arran the caldera is surrounded mainly by Old Red Sandstone but within there are blocks of Mesozoic sediments and Tertiary lavas interpreted as relics of a succession which foundered during the formation of the caldera. The main subsidence is estimated by King (1955) at 3000 ft. Sub-radial faults in the surrounding strata may be due to differential movement of peripheral segments.

The ring-fractures associated with cone-sheet formation are individually continuous for only part of a circle, although a zone of such fractures may make a complete ring. Similarly, the uplift accompanying the intrusion of a single cone-sheet is small but the formation of a whole cone-sheet complex must have been marked by a considerable uplift of the contemporary surface. For example on

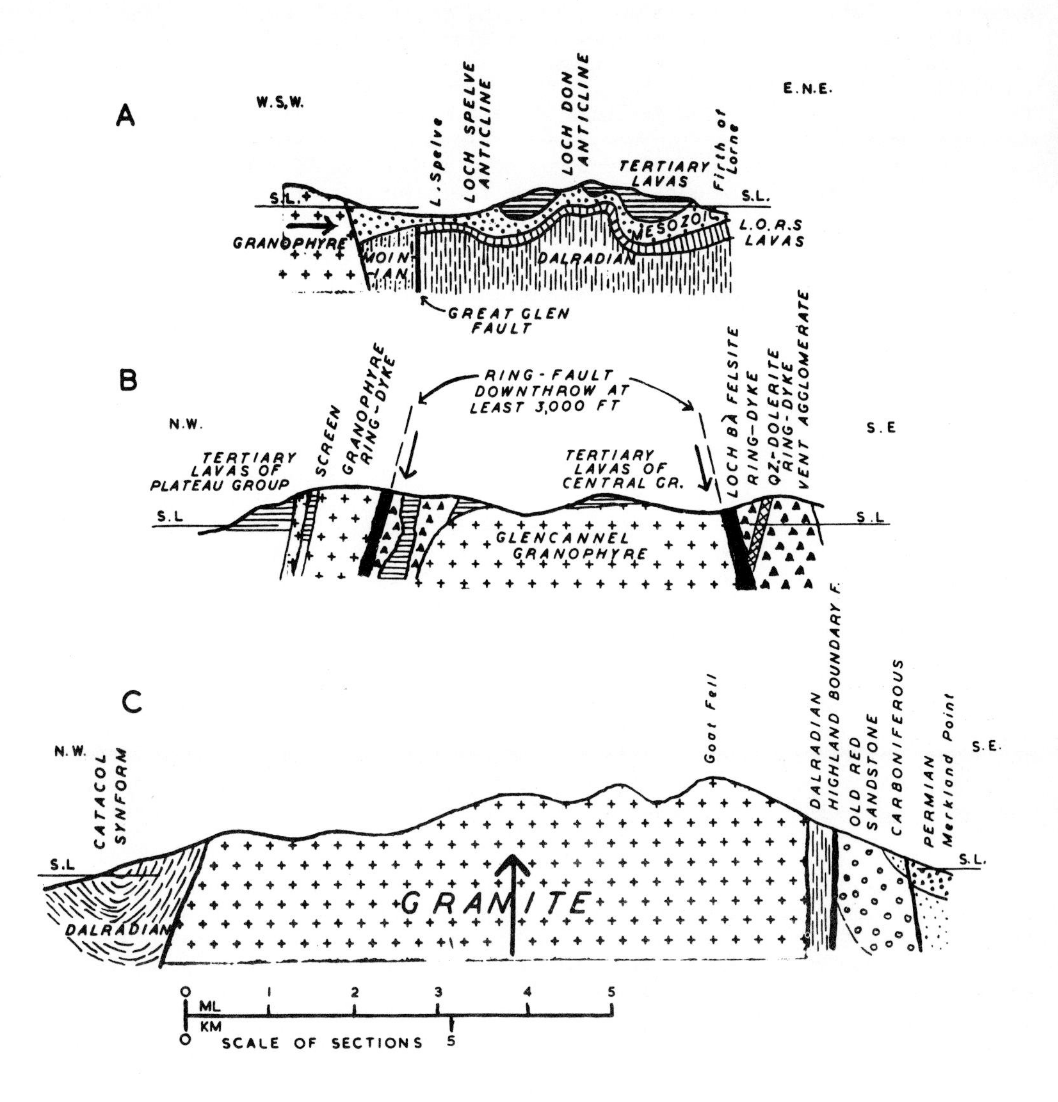

Fig. 9.2 Sections Showing Structural Effects of Tertiary Igneous Intrusions.

A. Lateral compression - the concentric folds of south-east Mull.

B. Subsidence - the Loch Ba Ring - dyke.

C. Uplift - the granite dome of Northern Arran.

Creach Bheinn in Mull the cone-sheets total 3000 ft (915 m) in thickness; as the average angle of inclination is about 45°, the total uplift must have been over 4200 ft (1280 m). Uplift of the same order also took place when the cone-sheets of Centre 2 in Ardnamurchan were intruded.

In Rhum, uplift within a ring-fracture about 6 miles (9.6 km) in diameter has been demonstrated (Bailey, 1945; Hughes, 1960). Immediately within the fracture, Lewisian Gneiss has been brought up by as much as 6000 to 7000 ft (1830 to 2135 m) against Torridonian on the outside. Inside the major ring-fracture there is a much smaller arcuate fault downthrowing centrally (Hughes, 1960).

Domes and concentric folds due to magmatic intrusion

Major structural effects accompanied the emplacement of the Tertiary Granite Complex of North Arran (Fig. 9.2). This has an almost perfect circular outcrop with a diameter of 8 miles (12.8 km), and everywhere, except on the N.E. margin, the surrounding schists dip steeply away from it, as do the Old Red Sandstone, Carboniferous, Permian and Triassic sediments, by which the schists are succeeded. Where the surrounding rocks originally dipped away from the granite, this dip has been merely accentuated and on the S.E. they have even been slightly overturned indicating outward pressure. On the north-west side, the schists dip towards the granite and have been bent into a sharp syncline, which is, in fact, a synform as it has been superimposed on Dalradian rocks inverted by Caledonian folding. It is only on the north-east side that the granite seems to have broken through the schists, instead of carrying them upwards with it, and here the schists strike sharply against the igneous rock, rather than striking parallel with its margin. It is only on this side too that the granite is in contact with rocks younger than the Dalradian; at such contacts mylonite is developed. The granite dome, the intrusion of which resulted in uplift of over 3000 ft (915 m) consists of two main components, an older, outer, coarse-grained granite and a younger, inner, fine-grained granite. Both granites are traversed by a zone of crushing which may be the result of further upward pushing, with differential movement, after solidification.

Another remarkable tectonic effect of Tertiary intrusion is the concentric folding peripheral to granophyres around the early caldera in Mull. The folds describe about 270° of arc, within most of which there are two or three well-marked anticlines with intervening synclines (Fig. 9.2). The most striking fold is the Loch Don Anticline, with a core of Dalradian schists, on the eastern or outer limb of which the beds are locally vertical. In Skye the arcuate Beinn an Dubhaich Granite occurs in the core of a curving anticline of Cambro-Ordovician limestone overlain by thrust Torridonian.

Dilation due to dyke intrusion

Most of the Tertiary complexes formed centres for the intrusion of great linear dyke-swarms. The dykes are so numerous that there was significant dilation of the continental crust at right angles to their general trend, i.e. in a NE.-SW. or ENE.-WSW. direction (Richey, 1939). In the case of the Arran swarm, for example, a belt of country which was originally 13.7 miles (21.9 km) across in a north-easterly direction became 14.8 miles (23.7 km) across after dyke intrusion.

Such stretching of the crust may have continued to the edge of the continental block concerned resulting in an overall addition to its width. For example, as continental drift took place the sialic blocks may be fissured as they move, and the fissures sealed up with basic magma rising from the sima, thus adding to the breadth of the block. Such may have been the case with the Tertiary dykes of the Western Highlands, formed near the edge of the European continent at the time that the great Atlantic rift was widening. Although the general movement was westerly it may have been south-westerly on the western seaboard of Scotland owing to the influence of Caledonian trend-lines.

Mesozoic and Tertiary folds and faults

Mesozoic folding and faulting are proved by the manner in which Upper Cretaceous rests on Jurassic and older formations in both Scotland and Ireland. In Skye, the Camusunary Fault with powerful downthrow to the south-east is post-Oxfordian (perhaps even earlier, e.g. post-Lias) but pre-Upper Cretaceous. In Ulster, a gentle anticline developed along the Highland Border ridge with a complimentary downfold to the south-east and pre-Upper Cretaceous faulting also took place. In both Scotland and Ireland warping and erosion occurred between the Upper Cretaceous and the outpouring of the Tertiary lavas.

Post-lava deformation on a considerable scale is evident in the Tertiary Igneous terrains. In Scotland the folding is gentle, but there are many post-lava faults, (Anderson and Dunham, 1966, p.76), the main directions being NW., NNW., N. and NE. Some of these faults follow pre-existing fractures. In Ireland the folding is more acute; it included the accentuation of the anticline along the Highland Border and the development of downfolds to NW. and SE. Faulting took place along the NW. and SE. margins of the Highland Border ridge and also at right angles. Some of the faults follow old Hercynian fractures. The scale of the deformation may be judged from the fact that combined folding and faulting has led to differences in level of the Chalk of over 3000 ft (915 m). The maximum known depth of the Chalk is 2200 ft (670 m) in the Lough Neagh Basin. In the north, near Ballycastle, the base of the Chalk is dropped about 700 ft (213 m) to the W. by the Tow Fault. North-north-westerly faults cutting the Slieve Gullion and Carlingford complexes show dextral transcurrent displacements of up to about 1½ miles (2.4 km).

CHAPTER 10

The Seas Around Britain

Until the 1950's, little was known about the sea-floor geology and deeper structure of the seas which surround the British Isles. Then came the results of the seismic prospecting in the English Channel (Hill and King, 1953) and the important paper by King (1954) on the geological history of the English Channel.

The first important breakthrough came however with the discovery of natural gas in 1959 by the Slochteren No. 1 Well of the Groningen Field in north-eastern Holland. The gas was found in Lower Permian (Rotliegendes) sandstones at depths of 9000 ft. Permian palaeogeographical reconstructions had, many years previously, postulated a major depositional basin (the Zechstein Sea) extending from N.W. Germany and the Netherlands across to N.E. England (Durham to Lincolnshire). The southern North Sea area thus immediately became the important target for seismic investigation. Geophysical research soon revealed the presence of widespread salt structures and associated diapiric folding in the Dutch and German portions of the southern North Sea. British seismic surveys too gradually spread northwards within our portions of the North Sea.

In 1964 applications for licences to explore and produce in British waters were invited and the detailed search really began. The first successful British gas discovery was made in December 1965 with the West Sole field (fig. 10.4) followed in 1966 by Hewett, Indefatigable and particularly Leman (the largest offshore gas field in the southern North Sea and proved to contain over 10 trillion cubic feet of gas).

The second breakthrough came in December, 1969, with the first major oil discovery at Ekofisk in Norwegian waters of the North Sea (fig. 10.4). The oil occurred in Palaeocene and Upper Cretaceous chalks. The discovery gave a tremendous impetus to the subsequent exploration of the central and northern areas of the North Sea. The first major British oil find (Forties) was made in September 1970, the reservoir this time being Middle to Upper Palaeocene sands (at a depth of 7000 ft). Many other important finds (Brent 1971, Auk and Argyll 1971, Piper 1972, etc.) have by now been made in the British and Norwegian waters of the North Sea.

Meanwhile the impetus of the early discoveries had spread to the more western seas of the British area. The important geophysical survey of the northern portion of Cardigan Bay (Griffiths, King and Wilson, 1961) opened the way for the inspired siting of the Mochras Borehole in 1967 on the coast of Merionethshire (see Wood and Woodland, 1968). Other important contributions have been made by Bullerwell and McQuillin (1968, South Irish Sea); Wilson (1968, Irish Sea); Blundell Davey and Graves (1970, South Irish Sea); Curry, Gray, Hamilton and Smith (1967, Celtic Sea and 1970, Western Approaches); Donovan, Lloyd and Stride, (1971, Bristol Channel); Dobson *et al* (1973, southern Irish Sea); Bailey (1974, Rockall Trough).

This mass of information about the geology of our sea areas has made an important contribution to the reconstruction of the geological evolution of the British area. Thick sequences of Mesozoic and Cenozoic sediments have now been located in important sedimentary troughs or grabens around Britain (fig. 10.1). Their presence (and their thick, young infillings) was not previously suspected. The whole area across Britain and its seas appears to have behaved in Mesozoic and

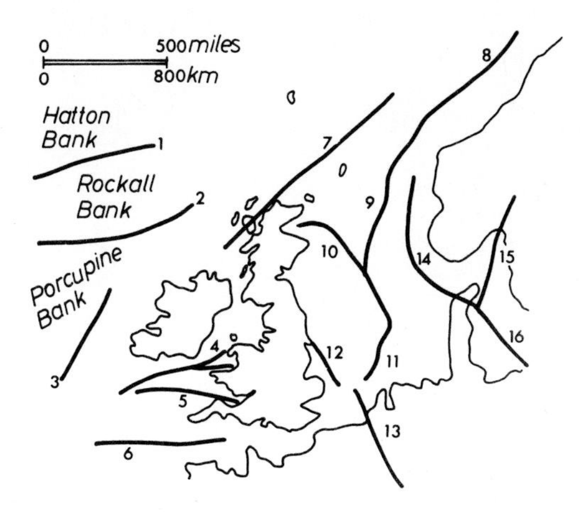

Fig. 10.1 Main Troughs around the British Isle (after Naylor and Mountenay, 1975). Key: 1. Rockall-Hatton Trough. 2. Rockall Trough. 3. Porcupine Trough. 4. N. Celtic Sea Trough. 5. S. Celtic Sea Trough. 6. W. Approaches Trough. 7. W. Shetland-Minch Trough. 8. Voring Basin. 9. N. North Sea Trough. 10. Forties Trough. 11. N. Netherlands Trough. 12. West Sole Trough. 13. Lower Rhine Trough. 14. W. Norway Trough. 15. Oslo Rift. 16. Danish-Polish Trough.

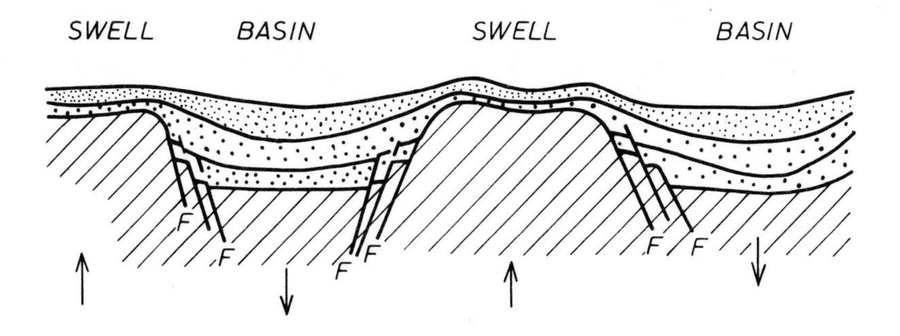

Fig. 10.2 Effect of Basement Fracturing on Later Sedimentation Producing Basins and Swells (after Whittaker, 1975).

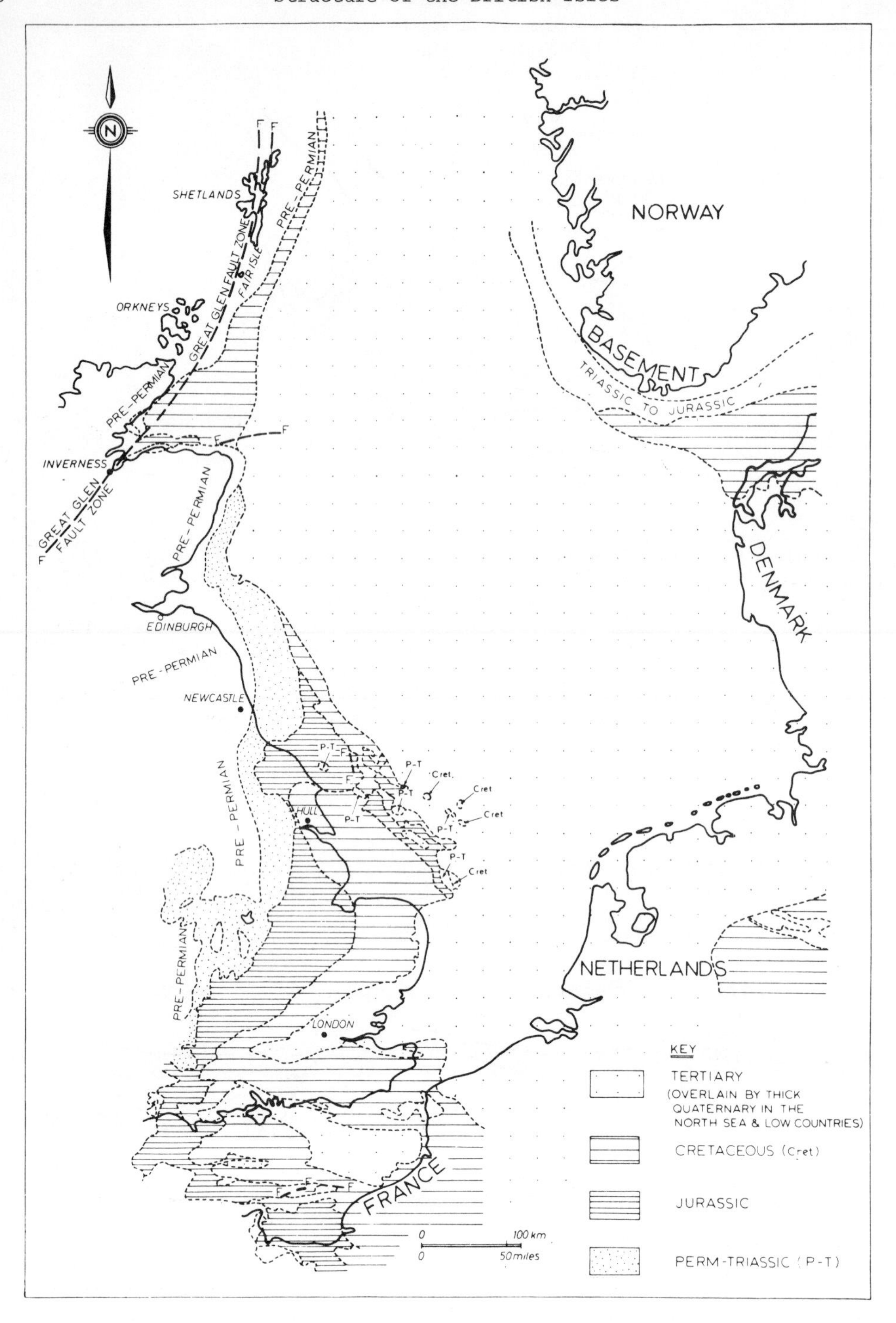

Fig. 10.3 The Post-Carboniferous Geology of the North Sea and Adjacent Areas (based on the I.G.S. map of 1972).

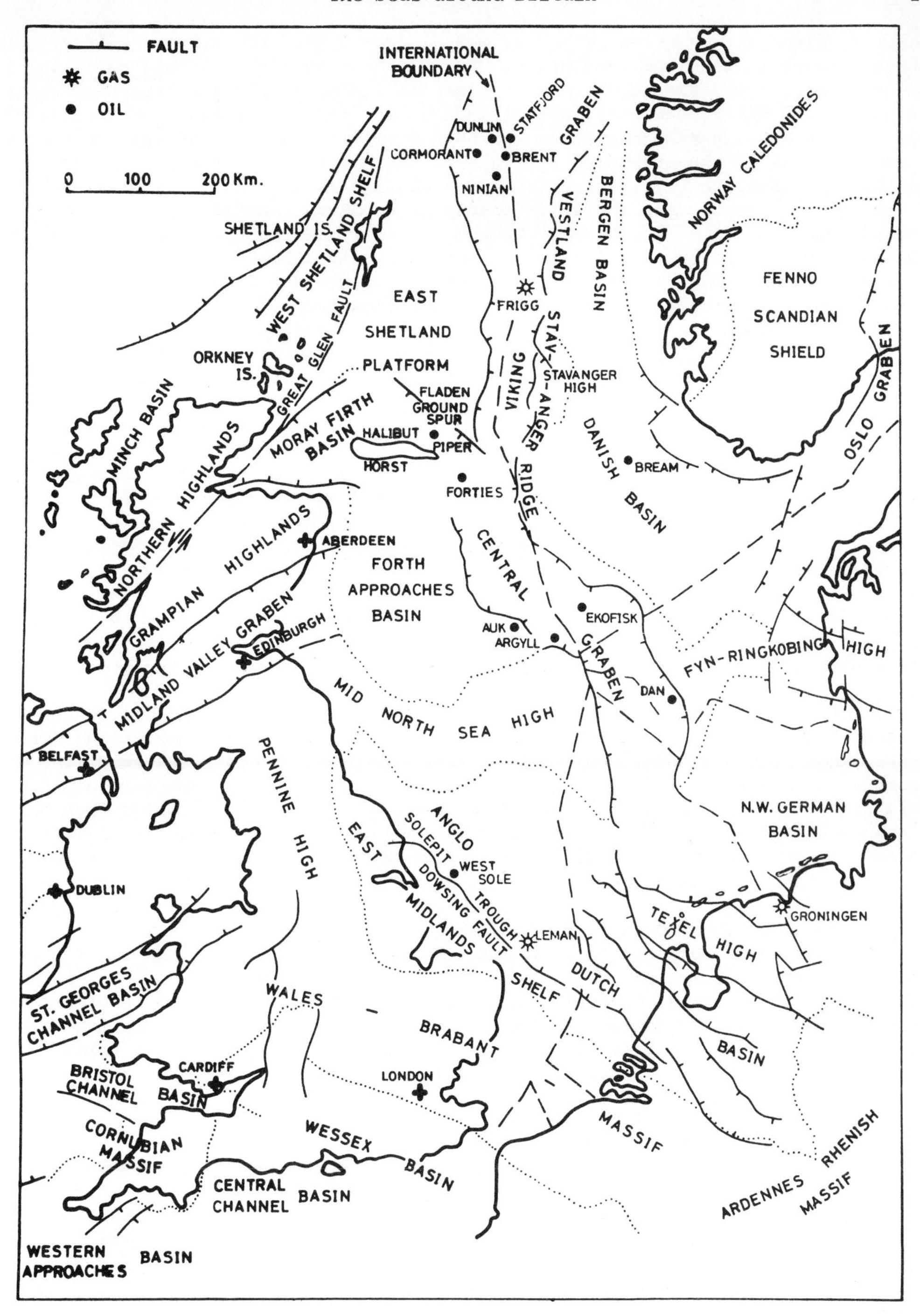

Fig. 10.4 Broad Structure of the North-West Continental Shelf (partly after Kent 1975).

Cenozoic times as a pattern of basins and swells. Whittaker (1975) has attributed these effects to basement fracturing. Such fracturing (in fact "geofracturing") suggests major, widespread, crustal influences and major plate rifting and separation must be the fundamental, underlying cause. In other words, the "cracking" of the British area in Mesozoic times (perhaps from even as early as late Palaeozoic times) was due to the attempted openings of the North Atlantic area. Troughs such as the North Sea, Rockall, Porcupine and Rockall-Hatton (fig. 10.1) were probably abortive attempts to open and which mostly never quite reached the final stage of ocean-crust injection (the Rockall Trough actually did reach this stage). All these troughs ("failed arms") were however to receive thick infillings of post-Palaeozoic sediments. The intervening blocks, on the other hand, were loathe to subside and as a result received much thinner (and frequently interrupted) sedimentary covers.

10.1 The North Sea

The floor of the North Sea is an extensive area of Tertiary and Quaternary sediments (see fig. 10.3). This Cenozoic blanket extends almost to the Norwegian and British coasts. Mesozoic and Permian rocks appear however along the western (i.e. British) margin of the North Sea. A particularly complex pattern of Permian-Cretaceous outcrops occurs to the east of Yorkshire and Lincolnshire (fig. 10.3) due largely to the salt tectonics of the gas field.

This wide extent of Cenozoic cover in the North Sea does however completely hide the very complex structures which exist at depth. Fig. 10.4 shows these deeper structures and is a tribute to the detailed studies of the North Sea area.

The major deep structure is the central rift made up of the Viking Graben and the Central Graben. The latter (with its southern splays, the West Danish and North Netherlands troughs) separates two shallower platforms of older (Palaeozoic) rocks - the Mid North Sea Platform and the Fyn-Ringkobing Platform (or "High"). This crossed pattern of grabens and highs results in the separation of four basins - the Forth Approaches, Bergen-Danish, Anglo-Dutch and N.W. German basins (fig. 10.4). The London-Brabant Platform marks the southern limit of the North Sea basins. The East Shetland Platform lies on the west side of the Viking Graben.

Fig. 10.5 illustrates the general (simplified) structure of the central and southern portions (respectively) of the North Sea. In the central North Sea (a), the three main features are (i) the shallow bounding platforms of pre-Permian rocks; (ii) the Central Graben; (iii) the very thick (over 3000 m in places) Tertiary infilling. Note how the fracturing tends to have died out by mid-Cretaceous times to be replaced by an overall sagging to accommodate the thick Tertiary infill. In the southern North Sea (b), the main feature is the halokinesis of Zechstein evaporites, forming numerous salt diapirs. There is often no Jurassic between the Trias and the Cretaceous, the absence of the system being attributable to Kimmerian uplifts and erosion over salt risings.

Broadly speaking, the gas fields are mainly in the southern North Sea (fig. 10.6) extending in a WNW.-ESE. orientated oval from E. Yorkshire and Lincolnshire out towards the Dutch coast. West Sole, Viking, Indefatigable and Leman are notable centres. The gas has migrated upwards from the underlying Coal Measures (now known to extend all the way across the southern North Sea) into the Rotliegende (Permian) sands which form a perfect reservoir very effectively capped and sealed by Zechstein dolomite and salt. The largest gas field is Leman, 18 miles long by 8 miles wide, a broad gentle anticline trending NW.-SE., the upwarp being accompanied by several faults. The recoverable reserves are some 10.5 trillion cubic feet (Pegrum *et al*, 1975). Hewett Field is an exception to the predominant

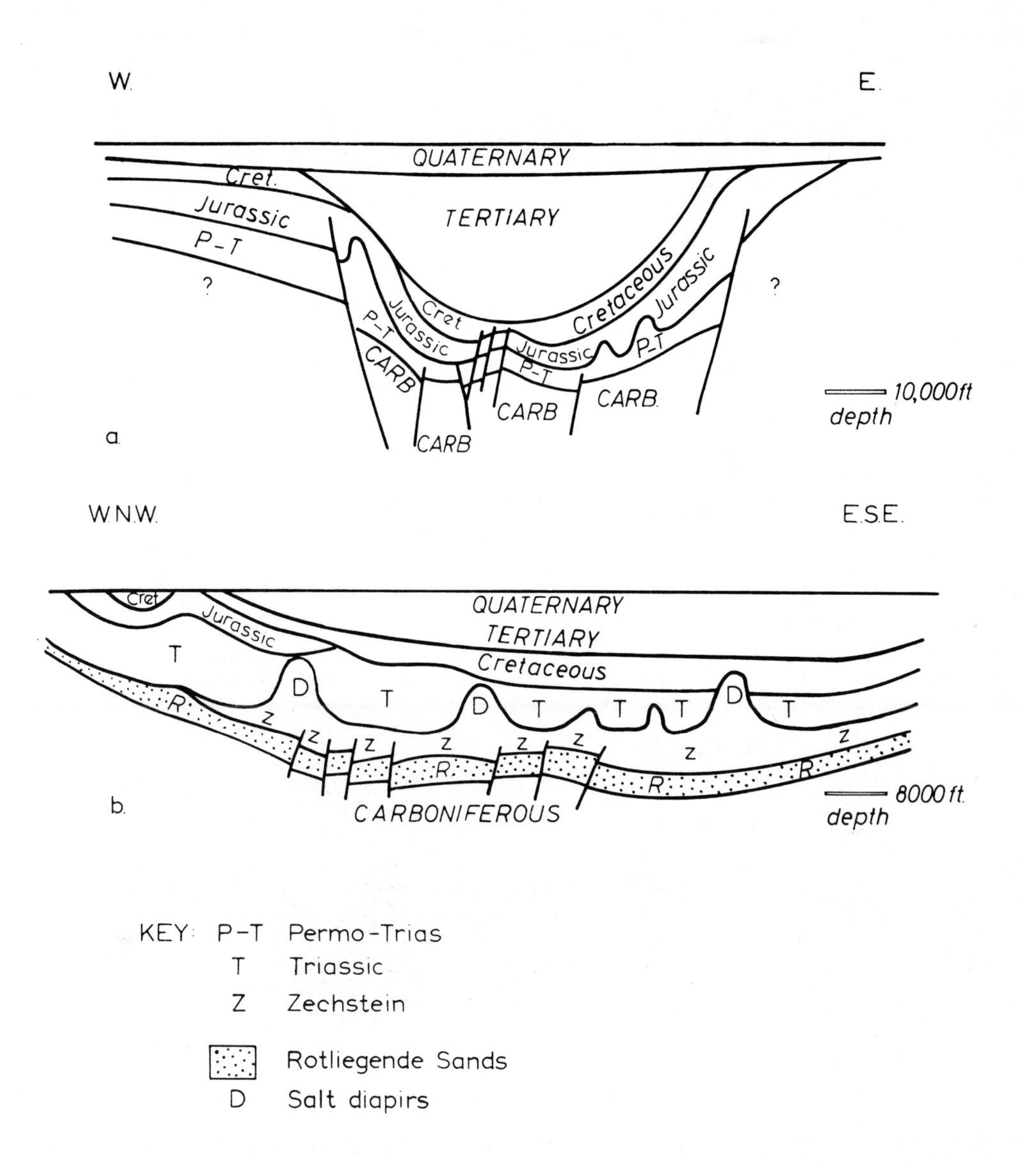

Fig. 10.5 Diagrammatic Sections Across the North Sea (after Evans and Coleman, 1974) (a) Across Central North Sea, (b) Across Southern North Sea.

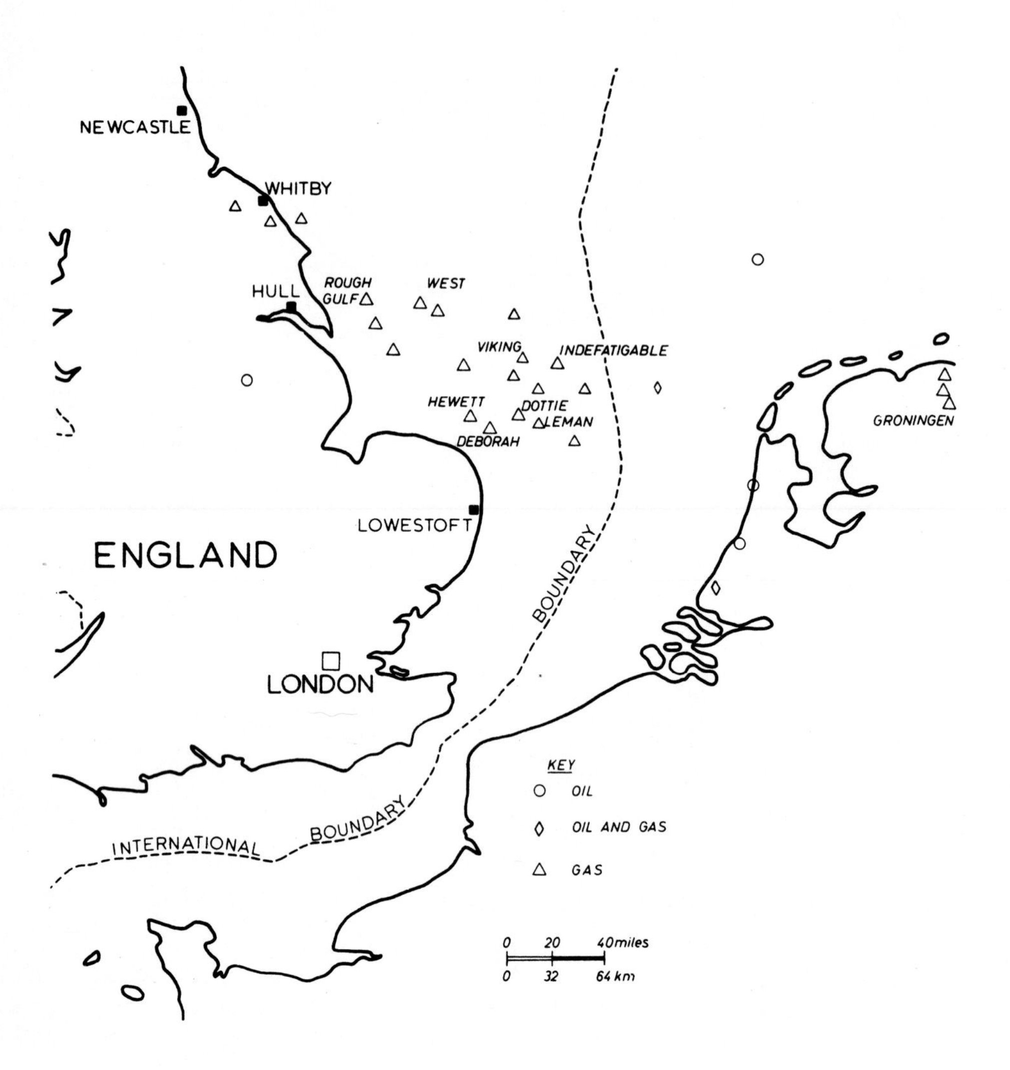

Fig. 10.6 The British Gas Fields in the Southern North Sea, West-West Sole Field.

Rotliegende position of the gas reservoirs. In Hewett, also a NW.-SE. trending faulted anticline, the gas is held in Middle and Lower Bunter sandstones, having migrated up to these higher horizons because of the absence of a Zechstein salt cap.

The oil fields occur mainly in the northern and central North Sea, the majority lying near to the international boundary (fig. 10.7). There are several reservoir horizons, beginning with Zechstein carbonates, the oil being held in cavities ("vuggs") within the dolomitic limestones. Examples are Auk and Argyll (fig. 10.9), on small horsts associated with offsets in the boundary fault of the Central Graben. Nearby, Ekofisk is an example of oil occurring in Chalk reservoirs on salt domes. Further north, reservoirs occur in Jurassic sandstones on the crests of faulted horsts and tilted blocks (fig. 10.8). The sands are of Lower and Middle Jurassic age in the case of Brent Field but of Upper Jurassic (Oxfordian) age in Piper. Lastly, Tertiary (Eocene or Palaeocene) sands form reservoirs where these Cenozoic sands are draped over the underlying faulted blocks. Examples include Montrose, Forties and Frigg.

Figs. 10.8 and 10.9 demonstrate the way in which the fracturing died out (to be replaced by more gentle sags) by Upper Cretaceous - early Tertiary times in the North Sea. End Cretaceous (Laramide) unrests did have some effects in the Central Graben with slump-type deposition but these were followed by a great basin subsidence over the central North Sea area resulting in a great elliptical Tertiary pile, with its axis lying NNW.-SSE. through the Maureen, Cod, Ekofisk and Dan fields and with a total thickness of more than 3000 m of marine Tertiary sediments in the centre of the basin (Cod-Ekofisk).

A further indication of the Permo-Mesozoic crustal unrest in the British environs is the presence (hitherto unsuspected) of volcanics in the Lower Permian (Denmark sector) and in the Middle Jurassic (Forties Field) and Upper Jurassic (Frigg). The Forties volcanics (alkali basalts and volcanic tuffs) could be fracture induced as the volcanic centre lies at intersections of the central grabens with the Moray Firth fracture systems. On a broader scale one should include the hints of volcanic centres south of the English Fullers Earth and the presence of Jurassic intrusives in Brittany.

10.2 The English Channel

The Channel waterway between England and France is divided into two main basins, the Channel Basin and the Western Approaches Basin, by the Cherbourg-Plymouth basement - a buried uplifted ridge of metamorphic rocks (of Precambrian-Palaeozoic age). The Channel Basin is floored by Permo-Triassic rocks, the Triassic believed to be almost 1500 m thick beneath the Isle of Purbeck. Jurassic rocks probably underlie the whole of the eastern Channel Basin and are probably about 1300 m thick. Middle and Upper Liassic sands and Middle Jurassic limestones are potential hydrocarbon horizons. Cretaceous rocks floor much of the Channel Basin with smaller inliers of Jurassic occurring west of Calais, south and west of the Isle of Wight and north of the Cotentin. The Cretaceous is blanketed by Palaeocene and Eocene sediments between Sussex and France.

The structural history of the Channel Basin involves at least two periods of folding and faulting, reflecting in fact the complex structural evolution of the adjacent Weymouth land area. In the Weymouth district, folding and faulting (Abbotsbury Fault, see fig. 10.10) occurred in Lower Cretaceous times (see Chapter 8), and again in mid-Tertiary (Miocene) times. So also in the Channel Basin, the Upper Cretaceous lies unconformably across folded Permo-Triassic to Lower Cretaceous rocks, these earlier folds tending generally WSW.-ENE. in the west but more W.-E.

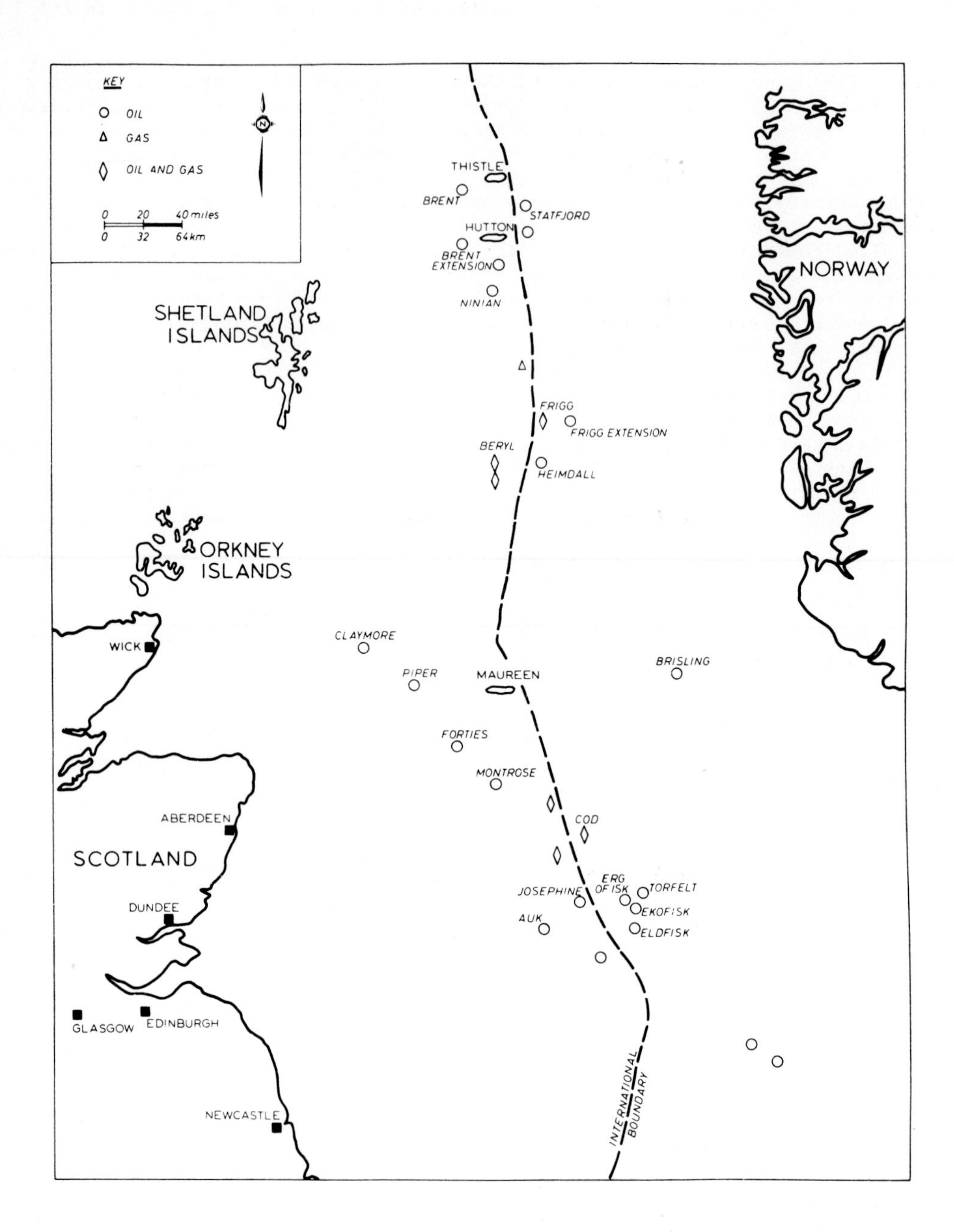

Fig. 10.7 Oil and Gas Fields in the Middle and Northern North Sea.

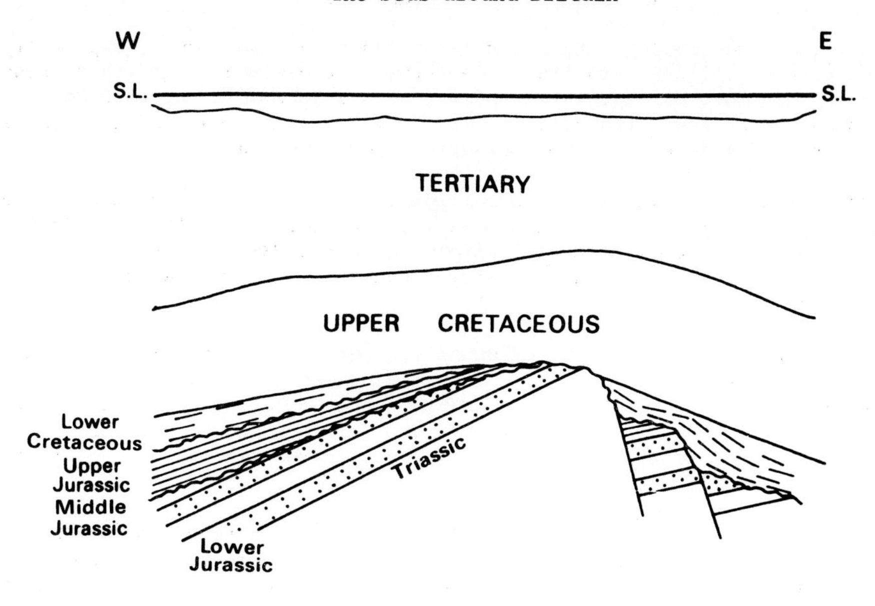

Fig. 10.8 Brent Field (Kent, 1975, fig. 11).

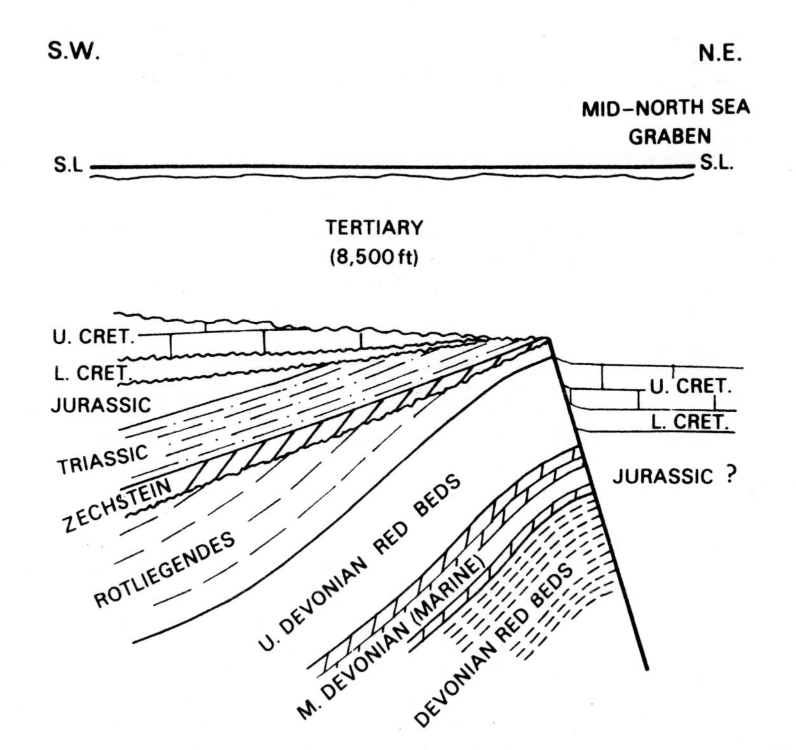

Fig. 10.9 Argyll Field (Kent, 1975, fig. 16).

further east (see fig. 10.10). Marked reverse faulting also occurs, especially at the northeastern end of the Hurd Deep. Near the southwestern end of that trench, the fractures turn more NE.-SW. in direction and have throws of up to 500m in places. An earthquake occurred here in 1925. The fractures may reflect renewed movement in the old basement of the Contentin-Plymouth Ridge.

The Western Approaches Basin marks the portion of the Channel to the west of that platform. The main structure here has been shown by Curry, Hamilton and Smith (1970) to be a single, large central downfold, trending WSW.-ENE. and with Tertiary sediments in its trough, the Tertiary tongue widening westwards into the southwestern continental shelf.

The basement is composed of gneiss and granite, together with metamorphosed and unmetamorphosed Palaeozoic sediments. Granitoid gneiss occurs in the sea areas south of Plymouth, breaking sea surface on Eddystone. Igneous rocks in the western English Channel include granites around the Scillies and Seven Stones, at Haig Fras and again north of Brittany. Phonolite occurs in two places south of Lands End. Their ages range from 262 to 113 m.y.

The Permo-Trias is up to 1500 m thick but is very uneven in thickness off Devon and Cornwall because of the irregular floor. Jurassic sediments total about 1000 m, but Upper Jurassic has not been detected. Lower Cretaceous sediments are also restricted, mainly to inliers, particularly southwest of the Hurd Deep. French workers have recorded similar inliers northwest of Ushant.

The base of the Upper Cretaceous (mainly Chalk) is strongly transgressive everywhere (fig. 10.11). Unconformity occurs also at the base of the Eocene (predominantly carbonates) and of the Miocene (little Oligocene is present). Plio-Pleistocene sediments are restricted and thin.

10.3 The Celtic Seas

These seas include the waterways between Wales and Ireland and between Cornubia and Ireland. The Bristol Channel is an eastern extension. The sea floor geology of the area is depicted in fig. 10.12 and a general section across the region is illustrated in fig. 10.11.

The Palaeozoic rocks of southern Ireland extend seawards (southwards) for about 12 to 15 miles and are then suddenly faulted against Cretaceous or Palaeogene sediments (fig. 10.12). The Chalk is of Campanian to Maestrichtian age. The Tertiary occurs in a narrow downfold between long. 8° and 6°30'W. Neogene to Recent sediments along the trough's axis lie unconformably on Palaeogene. Another (wider) tongue of Tertiary sediments occurs midway between N.Cornwall and Ireland. Neogene-Recent sediments are some 300 m thick and are in very broad shallow folds. Faulting is rare. On the southeastern flank of this Tertiary downfold, Upper Cretaceous transgresses Jurassic on to Permo-Trias which in turn abut against the Upper Palaeozoic fringe of Cornubia. On the southern edge of the Celtic Sea, Chalk rests directly on Upper Palaeozoic rocks which are pierced by the granite mass of Haig Fras (277 m.y.), (see Kamerling, 1979).

The deeper structure of the Celtic Sea between Cornubia and Ireland is however more complex, as fig. 10.11 shows. Two important grabens underlie the wide Cretaceous-Tertiary blanket, the downfaulted wedges being separated by a horst known as the "Pembroke Ridge". Thick Permo-Trias to Lower Cretaceous fills (totalling 3000 m) occupy the two troughs, the Jurassic being probably about 1200 m thick. The Lower Cretaceous is of Wealden aspect and has already yielded some oil and gas (Marathon) in the graben nearest to the S.Irish coast.

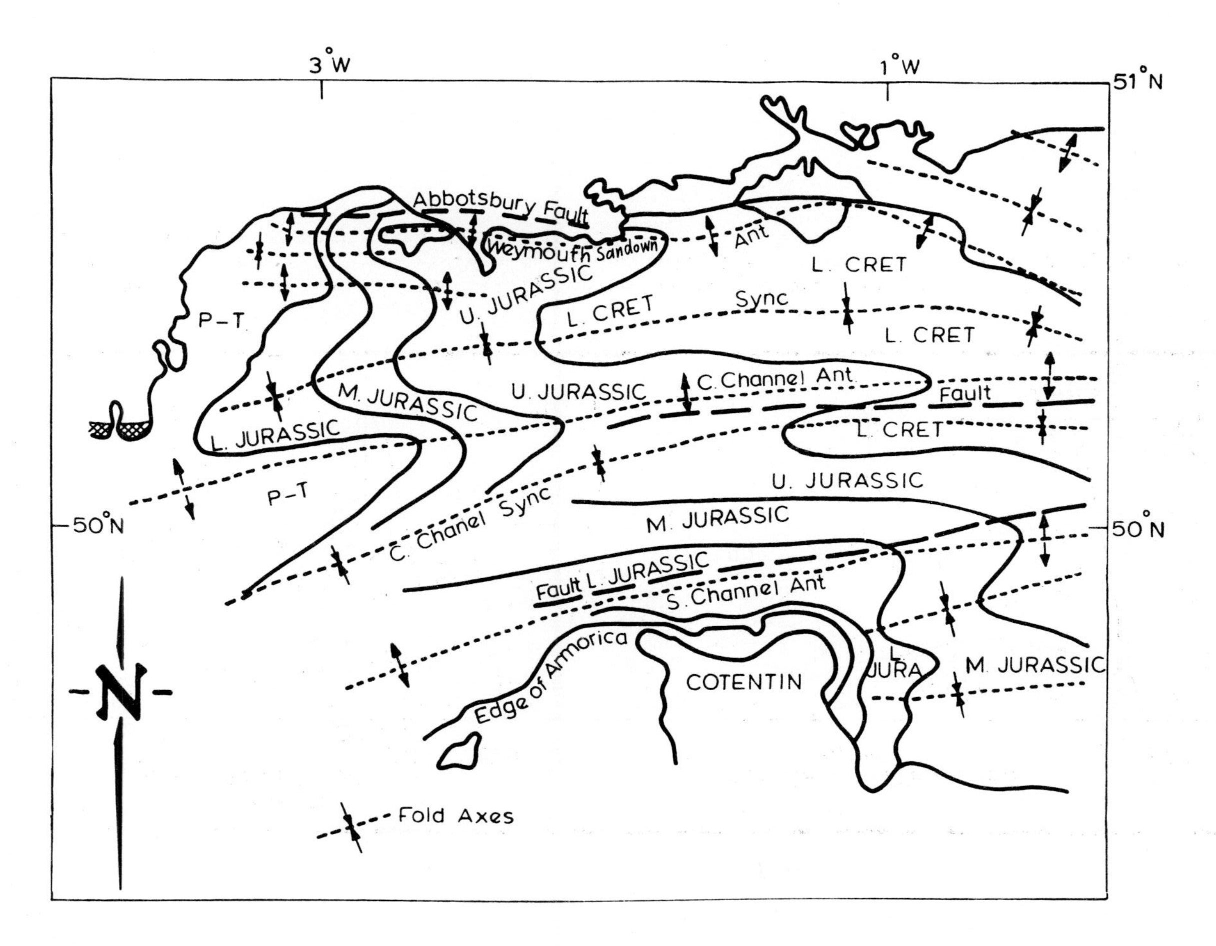

Fig. 10.10 Channel Outcrops of Pre-Upper Cretaceous Rocks Either on Sea-Floor or Inferred Beneath the Sub-Upper Cretaceous Unconformity (after Dingwall, 1971).

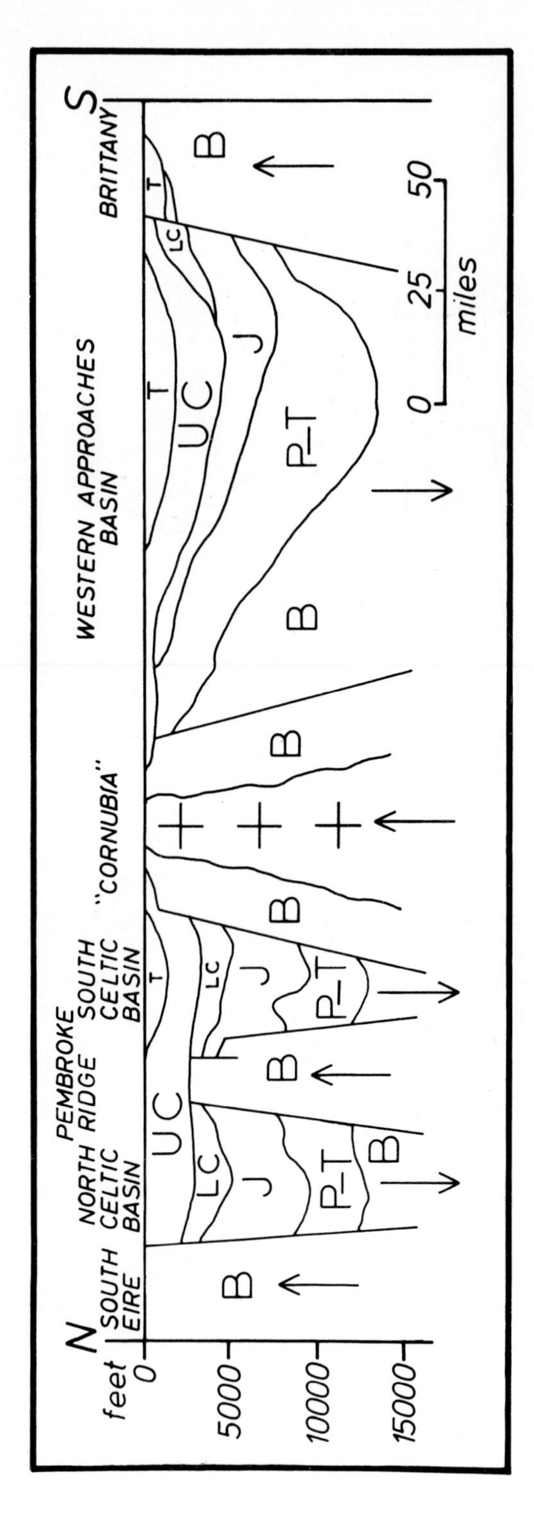

Fig. 10.11 Diagrammatic Section across the Celtic Sea and Western Approaches Basins.
Key: B. Basement; P-T. Permo-Triassic; J. Jurassic; L.C. Lower Cretaceous; U.C. Upper Cretaceous; T. Tertiary (After Naylor and Mounteney, 1975).

The Bristol Channel is floored by a thick Jurassic succession (ranging from Lias to almost Purbeck), some 1600 m thick. The Jurassic is mainly of muddy type and is thrown into a number of folds. The main fold lies nearer to the Exmoor coast than to that of S.Wales and its northern limb is accompanied by a powerful fault with a large southerly downthrow. The underlying Triassic is probably about 250 m thick (Donovan, Lloyd and Stride, 1971). The folding of the Jurassic rocks of the Bristol Channel almost certainly happened in mid-Cretaceous times (Owen, 1971, 1974). The faulting may well be of mid-Tertiary age. Hercynian cross faults may have been reactivated in Miocene times, as were those of Cornubia. The Cothelstone Fault of the Quantock Hills probably crosses the Bristol Channel to Swansea Bay. The Sticklepath Fault cuts off the Tertiary pluton of Lundy to cross over to the Pembrokeshire coast. The Eocene granite intrusion rises to over 130 m above sea level and is intruded into Devonian slates. An elongated negative magnetic anomaly extending northwestward from Lundy is interpreted by Cornwell (1971) as being due to a Tertiary dike swarm - perhaps along the Sticklepath fracture. The Lundy Granite may border a larger igneous complex of basic composition (Brooks, 1972).

The Sticklepath Fault has a number of Oligocene basins along its length. In Devon these are the large Bovey Basin and the smallier outlier at Petrockstow. Alongside the Lundy pluton is the Stanley Bank Basin (Fletcher, 1975).

The position of the Lundy Complex may be significant. It lies near to the point where the E.-W. downfolding of the Bristol Channel area gives way to the NE.-SW. downfold of the Upper Cretaceous-Tertiary of the Celtic Sea. Moreover the igneous complex must lie near to the buried margin of the South Celtic Graben.

The North Celtic Basin (north i.e. of the "Pembroke Ridge") passes northwards into the "South Irish Sea" with its eastern extension into Cardigan Bay. The rocks bordering both sides of the southern Irish Sea are predominantly of Lower Palaeozoic or Precambrian age. Upper Palaeozoics are restricted and post-Palaeozoics are absent on these (southern) Irish and Welsh coasts.

The Mochras Borehole (fig. 10.13), drilled in 1967 on the west coast of Merionethshire, proved the presence of yet another Oligocene (to Miocene) basin in Western Britain. The Oligocene, separated by the Mochras Fault from the Cambrian of the Harlech Dome, was 609 m thick and overlay 1219 m of Lias mudstones and limestones (the thickest Lias succession in Britain!). 39 m of (Rhaetic?) Triassic was reached at the base of the borehole. Dobson (1971) has drawn attention to the great *accumulative* throw of the Mochras Fault (over 4570 m). The Mochras succession with its notable gap between the Lias and the Oligocene poses many problems, however, especially for the history of Mesozoic cover on the adjacent Welsh mainland. The faulting history is obviously complex with *at least* two phases of movement, (a) pre-Oligocene and (b) post-Oligocene, involved. The first phase could in fact be the early to mid-Cretaceous phase of unrest noted in the structural history of the other British seas. It is also likely that some downfaulting along the fracture could even have *accompanied* Oligocene deposition in the Mochras basin.

The Mochras Fault appears to die out very quickly northwards. It is also displaced by NE.-SW. fractures running out into Cardigan Bay from the Welsh mainland. The most important of these caledonoid faults is the Bala Fault which passes out to the north of Pembrokeshire and which brings the Lower Palaeozoics of the Cardiganshire inshore waters against Jurassics and even Tertiary in a South Irish Sea graben. This graben is believed, on geophysical grounds, to have a sedimentary fill of over 6000 m, of which 2000 m in the centre of the basin are interpreted as being of Tertiary age (Blundell, Davey and Graves, 1970). This important basin is attenuated towards its northern end (see fig. 10.14) by a southward prong of the old rocks which form the Lleyn Peninsula of Gwynedd.

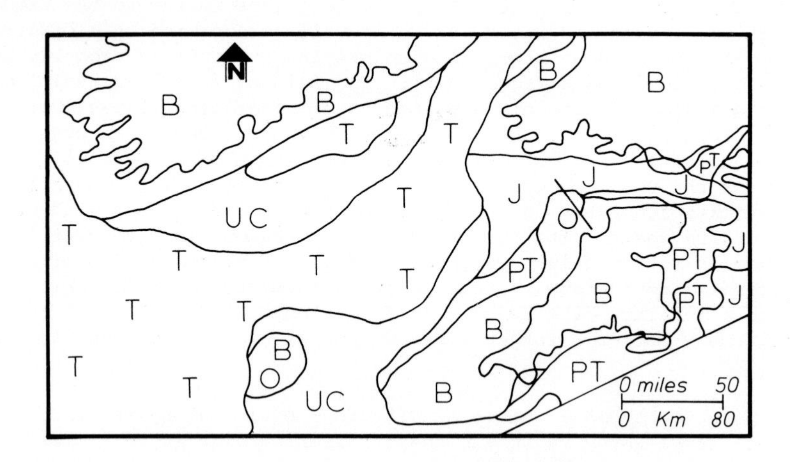

Fig. 10.12 The Geology of S.W. England and the Celtic Sea (based on Curry, Hamilton and Smith; Hamilton and Blundell). Key as for Fig. 10.11.

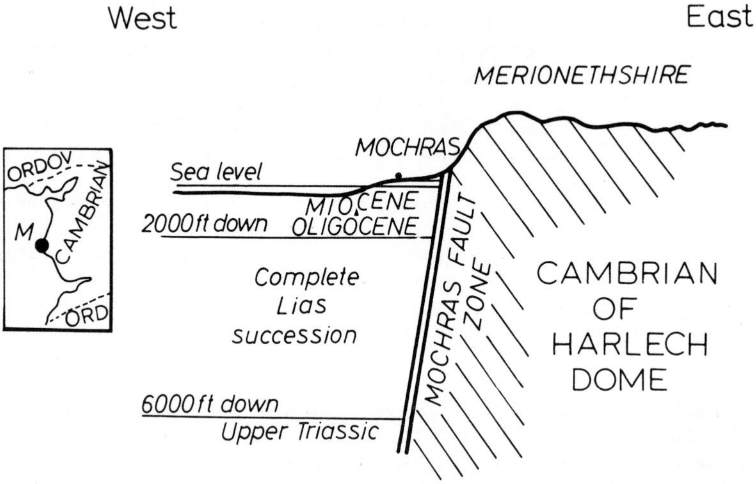

Fig. 10.13 The Mochras (Llanbedr) Borehole.

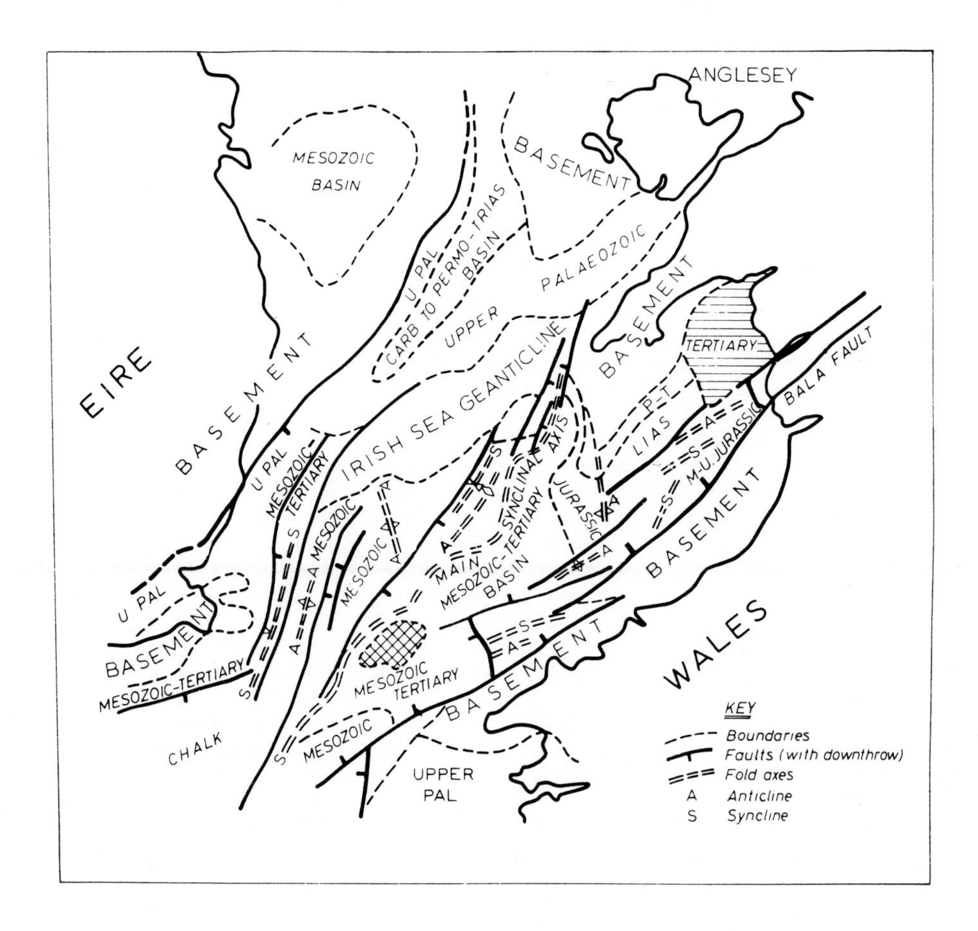

Fig. 10.14 Sea-floor Geology of the South Irish Sea (after I.G.S. publication No. 73-11). A north-south fault zone should be indicated along the Merionethshire coastline.

Many separate flexures occur in the Mesozoics of the two basins on either side of this "Lleyn" extension. Salt structures are also known to occur, and at least some of the folds may therefore be the result of halokinesis. Others may again be due to mid-Cretaceous movements.

The old rocks of the Lleyn Peninsula (Precambrian-Ordovician) continue southwestwards as the "Irish Sea Geanticline" (or "Rise") which(though covered in places by thin Mesozoic to Tertiary covers) continues to connect up with the ancient rocks of Wexford and Rosslare. To the northwest of ths "Rise" are further basins. A Caernarvon Basin has a fill of 1370 m (Dobson, 1971), mainly of Carboniferous and Permo-Trias. Another basin midway between Lleyn and Dublin has an even thicker fill (over 4500 m), probably of the same rock succession. Yet another, the Kish Bank Basin occurs, just to the east of the Dublin coast (fig. 10.14).

10.14 The North Irish Sea

Two major basins dominate the eastern half of the North Irish Sea. These are the Solway Firth Basin (a narrow, linear, NE.-SW. trending basin) and a much broader Manx-Furness Basin which extends from the SE. side of the Isle of Man to the N. Wales coast and which really continues inland as the Cheshire Graben. The Permo-Trias veneers of the Lancashire coastal fringe are in fact the eastern margin of this important Irish Sea basin. This larger basin is separated from the Solway trough by a ridge or horst of older (probably mainly Carboniferous) rocks extending from the Isle of Man to Cumberland (Bacon and McQuillin, 1972, fig. 2). A gravity low to the west of the Isle of Man is probably largely due to the presence of a granite. The fills of the Solway Firth and Manx-Furness basins are probably largely made up of Carboniferous and Permo-Trias. Some Jurassic could cap the latter sequences in places. The total sedimentary infills could be as much as 5500 m. As Westphalian Coal Measures are known to extend far into the Irish Sea beyond the Lancashire Coalfield, the possibilities of finding gas reservoirs (even if not large) in the Manx-Furness Basin are good.

10.5 The West Scottish Basins

The maze of isolated deep basins noted in the Celtic and Irish seas continues up the west side of the Scottish mainland (see, for example, Naylor and Mounteney, 1975, figs. 25 and 26). In the south, are the Stranraer and Firth of Clyde basins with (nearer to the Antrim coasts) the Portmore and Magee basins. The Portmore Basin is a deep sedimentary trough between Islay and the peninsula of Kintyre. Permo-Trias fills dominate the Firth of Clyde and Stranraer basins. Geophysical results suggest a thickness of over 4000 m of "New Red Sandstone" in the Stranraer Basin.

The chain of deep basins along the W.Scottish seaboard from Colonsay northwards to west of the Shetland Platform must represent intense regional tension during early Mesozoic times "in response to an abortive attempted separation of the Greenland-Rockall Plate from the West European Plate along the line of Rockall Trough and Porcupine Seabight" (Naylor and Mounteney, 1975, p.114). Fig. 26 of Naylor and Mounteney's excellent text clearly demonstrates the control of these W.Scottish trough-horst structures by four dominant fault systems - (from south to north) the Great Glen, Camasunary-Skerryvore, Minch and Outer Hebrides fractures. Most of these systems probably had Palaeozoic histories of transcurrent movements, but they were to suffer appreciable vertical adjustments throughout Mesozoic and Tertiary times, allowing great subsidence in places along their lengths and infillings of thick Permo-Trias and later sediments.

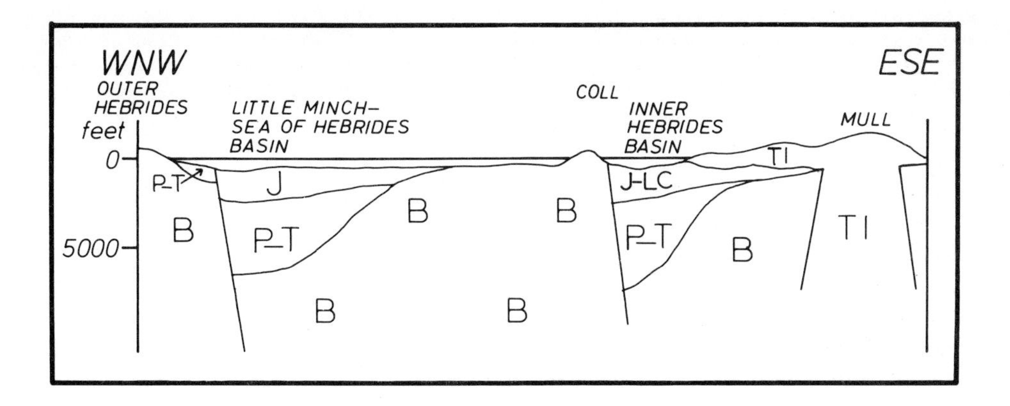

Fig. 10.15 Diagrammatic Section across the Hebrides Basins (After Naylor and Mounteney, 1975). Key as Fig. 10.11. T.I. Tertiary igneous rocks.

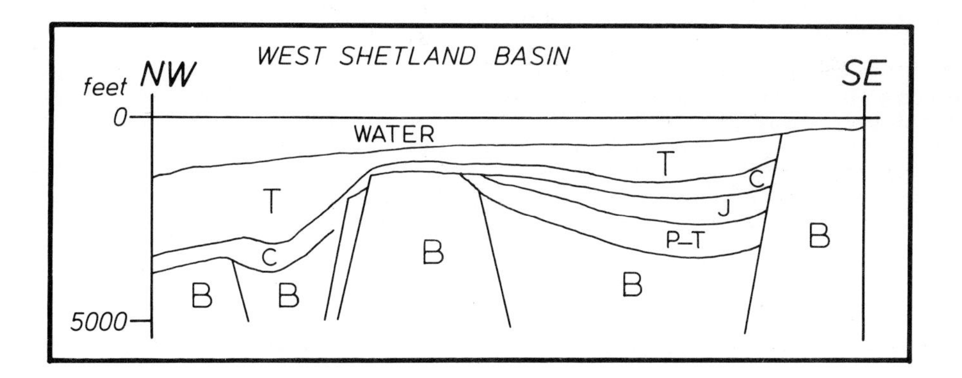

Fig. 10.16 Section across the West Shetland Basin (after Naylor and Mounteney, 1975). Key as for Fig. 10.11.

The main depositional troughs are (again from south to north) the Colonsay, Inner Hebrides, Little Minch-Sea of Hebrides, Outer Hebrides, North Minch, Sule-Sgeir and West Shetland basins. Thick Permo-Trias sediments, mainly sandstones, dominate the majority of these elongate NNE.-SSW. basins. Jurassic and Cretaceous (even Oligocene) covers are occasionally present but are much thinner than the "New Red Sandstone" sequences. Total fills reach 2500 m or over in the Inner Hebrides, North Minch and Little Minch-Sea of Hebrides basins. As fig. 10.15 shows, all these basins have very sharp faulted boundaries, revealing the very appreciable eastward downthrows which have taken place (on more than one occasion) along the Minch and Camasunary faults.

Fig. 10.16 shows that the West Shetland Basin has, on the other hand, a sharp, faulted, *eastern* boundary. Permo-Triassic-Tertiary sediments fill this basin. The total thicknss may exceed 5500 m. The Sule-Sgeir sedimentary fill is probably about half that thickness. Another sedimentary basin has been located, by geophysical work, to the SSW. of Shetland, extending to the west of Fair Isle (Bott and Browitt, 1975). This trough, aligned N.-S., divides at its southern end into two prongs, the eastern being the deeper basin and truncated along its eastern margin by a splay (or northward extension) of the Great Glen Fault. The sedimentary fill (Mesozoic?) could exceed 2000 m in places.

10.6 The East Atlantic troughs

The major trough off NW. Britain is the Rockall Trough. This is a wide depression, filled with over 3000 m of Mesozoic and Tertiary flat-lying sediments. How far back into the Mesozoic (or even Permian) the basal sediments extend is not known. Some ocean crust must form a central core to the basement.

An important break occurs under (Upper?) Eocene sediments. Changes in the pattern of spreading in the North Atlantic during the early Oligocene caused a further unconformity followed by renewed Neogene sedimentation (including semi-consolidated chalks). Thin Tertiaries also overlie parts of the Porcupine Bank, the Cenozoic blanket then swelling however into the adjacent trough known as the Porcupine Seabight.

Perhaps the most complete and thickest offshore Tertiary sequences in the sea areas west of Britain occur in the Hatton-Rockall Basin. On the eastern edge of this basin, probably another abortive line of attempted spreading dating back to Mesozoic times, Palaeocene to Pleistocene sediments were penetrated in two drill holes (in over 1000 m of water). Neogene chalks and calcareous oozes total over 800 m. Breaks occur within the Oligocene and the Eocene.

References

Note: In the bibliography, the following abbreviations have been made:

B.G.S.	Bulletin of the Geological Survey
B.G.S. Am.	Bulletin of the Geological Society of America
G.J.	Geological Journal
G.M.	Geological Magazine
I.N.J.	Irish Naturalists Journal
J.G.S.L.	Journal of the Geological Society of London
M.G.S.	Memoir of the Geological Survey
M.G.S.L.	Memoir of the Geological Society of London
Nat.	Nature
P.G.A.	Proceedings of the Geologists' Association
P.G.S.L.	Proceedings of the Geological Society of London
P.R.I.A.	Proceedings of the Royal Irish Academy
P.R.S.E.	Proceedings of the Royal Society of Edinburgh
P.U.S.	Proceedings of the Ussher Society
Q.J.G.S.L.	Quarterly Journal of the Geological Society of London
R.G.S.C.	Royal Geological Society of Cornwall
S.J.G.	Scottish Journal of Geology
S.P.R.D.S.	Scientific Proceedings of the Royal Dublin Society
T.E.G.S.	Transactions of the Edinburgh Geological Society
T.G.S.G.	Transactions of the Geological Society of Glasgow
T.R.S.E.	Transactions of the Royal Society of Edinburgh

Ager, D.V. 1975. The geological evolution of Europe. P.G.A. 86, 127.

Akaad, M.K. 1956. The Ardara granite diapir of County Donegal, Ireland. Q.J.G.S.L. 112, 263.

Allan, D. A. 1940. The geology of the Highland Border from Glen Almond to Glen Artney. T.R.S.E. 60, 171.

Allen, J. R. L., Bassett, M. G., Hancock, P. L., Walmsley, V. G. and Williams, B. P. J. 1976. Stratigraphy and structure of the Winsle Inlier, south-west Dyfed, Wales. P.G.A. 87, 221.

Allison, A. 1940. Loch Awe Succession and tectonics: Kilmartin-Tayvallich-Danna. Q.J.G.S.L. 96, 423.

Anderson, E. M. 1942. The Dynamics of Faulting and Dyke Formation with Applications to Britain. Oliver & Boyd, Edinburgh.

Anderson, E. M. 1948. On lineation and petrofabric structure and the shearing movement by which they have been produced. Q.J.G.S.L. 104, 99.

Anderson, E. M. 1951. The Dynamics of Faulting and Dyke Formation with Applications to Britain, 2nd edn. Oliver & Boyd, Edinburgh.

Anderson, F. W. and Dunham, K. C. 1966. Geology of Northern Skye, M.G.S.

Anderson, J. G. C. 1947a. The geology of the Highland Border: Stonehaven to Arran. T.R.S.E. 61, 479.

Anderson, J. G. C. 1947b. The Kinlochlaggan Syncline, Southern Inverness-shire. T.G.S.G., 21, 97.

Anderson, J. G. C. 1948. The occurrence of Moinian rocks in Ireland. Q.J.G.S.L. 103, 171.

Anderson, J. G. C. 1954. The pre-Carboniferous of the Slieve League Promontory, County Donegal. Q.J.G.S.L. 109, 399.

Anderson, J. G. C. 1960. The Wenlock strata of South Mayo. G.M. 97, 265.

Anderson, J. G. C. 1965. The Precambrian of the British Isles. The Precambrian, vol. 2. (Edited by K. Rankama). Interscience Publishers, New York.

Anderson, J.G.C. 1978. The Structure of Western Europe. Pergamon, Oxford.

Anderton, R., Bridges, P.H., Leeder, M. R. and Sellwood, B. W. 1979. A Dynamic Stratigraphy of the British Isles. George Allen and Unwin, London.
Arkell, W. J. 1938. Three tectonic problems of the Lulworth district: studies in the middle limb of the Purbeck fold. Q.J.G.S.L. 94, 1.
Arkell, W. J. 1947. Geology of Oxford. Clarendon Press, Oxford.
Arkell, W. J., Wright, C. W. and White, H. J. O. 1947. Geology of the country around Weymouth, Swanage, Corfe and Lulworth. M.G.S.
Ashwin, D. P. 1958. The Coastal Outcrop of the Culm measures of south-west England. Abstr. R.G.S.C.

Bacon, M. and McQuillin, R. 1972. Refraction seismic surveys in the north Irish Sea. J.G.S.L. 128, 613.
Badley, M. E. 1976. Stratigraphy, structure and metamorphism of Dalradian rocks of the Mamturk Mountains, Connemara, Ireland. J.G.S.L. 132, 509.
Bailey, E. B. 1917. The Islay Anticline (Inner Hebrides). Q.J.G.S.L. 72, 132.
Bailey, E. B. 1934a. West Highland tectonics: Loch Leven to Glen Roy. Q.J.G.S.L. 90, 462.
Bailey, E. B. 1934b. The Glencoul Nappe and the Assynt culmination. G.M. 72, 151.
Bailey, E. B. 1950. The structural history of Scotland. Congr. Geol. Intern. Comp. Rend. London, 1948, 1, 230.
Bailey, E. B. and Weir, J. 1933. Submarine faulting in Kimmeridgian times, east Sutherland. T.R.S.E. 57, 429.
Baker, J. W. 1955. Precambrian rocks in County Wexford. G.M. 92, 63.
Baker, J. W. 1969. Correlation problems of metamorphosed Precambrian rocks in Wales and southeast Ireland. G.M. 106, 249.
Baker, J. W. 1973. A marginal late Proterozoic ocean basin in the Welsh region. G.M. 110, 447.
Barber, A.J. and May, F. 1976. The history of the Western Lewisian in the Glenelg Inlier, Lochalsh, Northern Highlands. S.J.G. 12, 35.
Barber, A. J. and Max, M. D. 1979. A new look at the Mona Complex (Anglesey, North Wales). J.G.S.L. 136, 407.
Barber, P.L. and others. 1979. The geology of the Firth of Lorne as determined by seismic and dive sampling methods. S.J.G. 15, 217.
Barnes, R.P., Andrews, J. R. and Badham, J. P. N. 1979. Preliminary investigations of South Cornish Melanges. P.U.S. 4, 262.
Barrow, G. 1893. On the intrusion of muscovite-biotite gneiss in the south-eastern Highlands of Scotland and its accompanying metamorphism. Q.J.G.S.L. 49, 330.
Bassett, D. A. 1955. The Silurian rocks of the Talerddig district, Montgomeryshire. Q.J.G.S.L. 111, 239.
Bassett, D. A. 1961. Bibliography and Index of Geology and Allied Sciences for Wales and the Welsh Borders. 1897-1958. Nat. Mus. Wales, Cardiff.
Bassett, D. A. 1963. In The British Caledonides. (Edited by M. R. W. Johnson and F. H. Stewart). Oliver & Boyd, Edinburgh.
Bassett, D. A., Whittington, H. B. and Williams, A. 1966. The stratigraphy of the Bala District, Merionethshire. Q.J.G.S.L. 122, 129.
Bath, A. H. 1974. New isotopic data on rocks from the Long Mynd, Shropshire. J.G.S.L. 130, 567.
Beckinsale, R. D. and Thorpe, R. S. 1979. Rubidium-strontium whole-rock isochron evidence for the age of metamorphism and magmatism in the Mona Complex of Anglesey. J.G.S.L. 136, 433.
Bhattarjee, C. C. 1964. The late structural and petrological history of the Lewisian rocks of the Meall Duse area, north of Gairloch, Ross-shire. T.G.S.G. 25, 31.
Bickle, M. J. and others. 1972. The Silurian of the Cloagh Patrick Range, Co. Mayo. S.P.R.D.S. 4A, 231.

Bluck, B. J. 1958. The sedimentary history of the Rocks between the G. subcrenatum Marine Band and the Garw Coal in the South Wales Coalfield. Unpublished Ph.D. thesis of University of Wales.

Bluck, B. J. and Kelling, G. 1963. Channels from the Upper Carboniferous Coal Measures of South Wales. Sedimentology 2 (1), 29.

Blundell, D. J. 1957. A palaeomagnetic investigation of the Lundy dyke-swarm. G.M. 94, 187.

Blundell, D. J., Davey, F. J. and Graves, L. J. 1970. Surveys over the South Irish Sea and Nymph Bank. Geol. Soc. Lond. Circular 161.

Blyth, F. G. H. 1962. The structure of the north-eastern tract of the Dartmoor granite. Q.J.G.S.L. 118, 435.

Borrodaile, G. J. 1974. The structure of the Moine-like rocks near Lough Derg, County Donegal, Eire. G.J. 9, 61.

Boswell, P. G. H. 1949. The M. Silurian Rocks of N. Wales. Arnold, London.

Bott, M. H. P. 1961. Geological interpretation of magnetic anomalies over the Askrigg Block. Q.J.G.S.L. 117, 481.

Bott, M. H. P. 1964. Gravity measurements in the north-eastern part of the Irish Sea. Q.J.G.S.L. 120, 309.

Bott, M. H. P. 1974. The geological interpretation of a gravity survey of the English Lake District and the Vale of Eden. J.G.S.L. 130, 309.

Bott, M. H. P. and Browitt, C. W. A. 1975. Interpretations of geophysical observations between the Orkney and Shetland Islands. J.G.S.L. 131, 353.

Bott, M. H. P., Day, A. A. and Masson Smith, D. 1958. The geological interpretationsof gravity magnetic surveys in Devon and Cornwall. Abstr. R.G.S.C.

Bowes, D. R. 1962. In discussion of paper by J. Sutton. Abstr. P.G.S.L. No. 1594, 29.

Bowes, D. R. 1965. West Glasgow earthquake. S.J.G. 1, 288-94.

Bowes, D. R. 1978. Crustal evolution in Northwestern Britain and adjacent regions. G.J. Special Issue No. 10, eds. D. R. Bowes and B. E. Leake.

Bowes, D. R. and Convery, H. J. E. 1966. Ben Ledi grits and albite-schists. S.J.G. 2, 67-75.

Bowes, D. R. and Ghaly, T. S. 1964. Age relations of Lewisian basic rocks, south of Gairloch, Ross-shire. G.M. 101, 150-60.

Bowes, D. R. and Khoury, S. G. 1965. Successive periods of basic dyke emplacement in the Lewisian Complex, south of Scourie, Sutherland. S.J.G. 1, 295-9.

Bowes, D.R. and Wright, A. E. 1973. Early phases of Caledonian deformation in the Dalradian of the Ballachulish district, Argyll. G.J. 8, 333.

Bowes, D. R., Wright, A. E. and Park, R. G. 1964. Layered intrusive rocks in the Lewisian of the north-west Highlands of Scotland. Q.J.G.S.L. 120, 153.

Bradbury, H. J. et al. 1976. "Older" granites as time-markers in Dalradian evolution. J.G.S.L. 132, 677.

Bridgewater, D., Watson, J.V. and Windley, B. F. 1973. The Archaean craton of the North Atlantic region. Phil. Trans. Roy. Soc. Sci. A. 273, 493.

Brindley, J.C. 1956. The Geology of the north end of the Leinster granite, Part II. Structural features of the Country rocks. P.R.I.A., 58B, 23.

Brindley, J.C. and Gill, W.D. 1959. Summer Field Meeting in South Ireland. P.G.A. 69, 244.

Brook, M. et al. 1976. Grenville age for rocks in the Moines of north-western Scotland. Nat. 260, 515.

Brook, M. and others. 1977. Grenville events in Moine rocks of the Northern Highlands, Scotland. J.G.S.L. 133, 489.

Brookfield, M. E. 1978. Revision of the Stratigraphy of Permian and supposed Permian rocks of Southern Scotland. Geologische Rundschau, Band 87, Heft 2, 110.

Brooks, M. and Thompson, M. S. 1973. The geological interpretation of a gravity survey of the Bristol Channel. J.G.S.L. 245, 163.
Brooks, M. and James, D. G. 1975. The geological results of seismic refraction surveys in the Bristol Channel. J.G.S.L. 131, 163.
Brooks, M., Bayerly, M. and Llewellyn, D. J. 1977. A new geological model to explain the gravity gradient across Exmoor, north Devon. J.G.S.L. 133, 385.
Brown, E. G. 1960. The Relief and Drainage of Wales. Univ. of Wales Press, Cardiff.
Brown, P. E., Miller, J. A., Soper, N. J. and York, D. 1965. Potassium-Argon age pattern of the British Caledonides. Proc. Yorks. Geol. Soc. 35, 103-138.
Bullerwell, W. 1964. Regional gravity and magnetic geophysical surveys over southern England and adjacent coastal areas. Adv. Sci., 21, 451.
Bullerwell, W. and McQuillin, R. 1968. Preliminary report on a seismic reflection survey in the southern Irish Sea, July 1968. Report No. 69/2. I.G.S.
Butcher, N. E. 1962. The tectonic structure of the Malvern Hills. P.G.A. 73, 103.

Caldwell, W. G. E. 1959. The Lower Carboniferous of the Carrick-on-Shannon Syncline. Q.J.G.S.L. 115, 163.
Callaway, C. 1883. The age of the newer gneissic rocks of the Northern Highlands. Q.J.G.S.L. 39, 355.
Capewell, J. G. 1957a. The stratigraphy and structure of the country around Sneem, County Kerry. P.R.I.A. 58, 167.
Capewell, J. G. 1957b. The stratigraphy, structure and sedimentation of the Old Red Sandstone of the Comeragh Mountains and adjacent areas, County Waterford, Ireland. Q.J.G.S.L. 92, 393.
Casey, R. 1961. The stratigraphical palaeontology of the Lower Greensand. Palaeontology 3, 487.
Casey, R. 1962. The ammonites of the Spilsby Sandstone, and the Jurassic-Cretaceous Boundary. P.G.S.L. 19, 58.
Casey, R. 1963. The dawn of the Cretaceous Period in Britain. Bulletin S.E. Union of Sc. Socs. 117,1.
Chalmers, J. A. and Western, P. G. 1979. A Tertiary igneous centre north of the Shetland Islands. S.J.G. 15, 333.
Charlesworth, J. K. 1953. The Geology of Ireland. Oliver & Boyd, Edinburgh.
Charlesworth, J. K. (editor). 1960. The Geology of North-East Ireland. P.G.A. 71, 429.
Charlesworth, J. K. 1963. Historical Geology of Ireland. Oliver & Boyd, Edinburgh.
Cheeney, R. F. and Matthews, D. V. 1965. The structural evolution of the Tarskavaig and Moine nappes in Skye. S.J.G. 1, 256-81.
Church, W. R. and Gayer, R. A. 1973. The Ballantrae ophiolite. G.M. 110, 497.
Christie, J. M. 1960. Mylonitic rocks of the Moine Thrust-zone in the Assynt region, north-west Scotland. T.E.G.S. 18, 79.
Clifford, T. N. 1957. The stratigraphy and structure of part of the Kintail district of southern Ross-shire; its relation to the Northern Highlands, Q.J.G.S.L. 113, 57.
Clough, C. T., Maufe, H. B. and Bailey, E. B. 1909. The Cauldron-subsidence of Glen Coe and the associated igneous phenomena. Q.J.G.S.L. 65, 611.
Coe, K. 1959. Boudinage structure in west Cork. G.M. 96, 191.
Coe, K. 1959. (Editor) Some Aspects of the Variscan Fold Belt. Univ. Press, Manchester.

Coe, K. 1969. The Geology of the minor intrusions of West Cork, Ireland. P.G.A. 80, 441.
Coe, K. and Sellwood, E. B. 1963. The stratigraphy and structure of part of the Beara peninsula, Co. Cork. P.R.I.A. 63B, 33.
Cole, G. A. J. 1921. The problem of the Bray Series. P.R.I.A. 36B, 1.
Cole, G. A. J. 1922. Some features of the Armorican (Hercynian) folding in southern Ireland. Rept. Int. Geol. Congress. XIII (Belgium) 423.
Cope, F. W. 1954. The Coalfields of Great Britain. (Edited by A. E. Trueman), Arnold, London.
Cornwell, J. D. 1971. Geophysics of the Bristol Channel area. P.G.S.L. No. 1664, 286.
Coward, M.P. 1973. Heterogeneous deformation in the development of the Laxfordian complex of South Uist, Outer Hebrides. J.G.S.L. 129, 139.
Coward, M. P. and others. 1969. Remnants of an early metamorphic assemblage in the Lewisian complex of the Outer Hebrides. P.G.A. 80, 387.
Craig, G. Y. and Walton, E. K. 1959. Sequence and structure in the Silurian rocks in Kirkcudbrightshire. G.M. 96, 209.
Craig, G. Y. (Editor) 1965. The Geology of Scotland. Oliver & Boyd, Edinburgh.
Crane, A. 1978. Correlation of metamorphic fabrics and the age of Lewisian metasediments near Loch Maree. S.J.G. 14, 225.
Crimes, T.P. and Crossley, J. D. 1968. The stratigraphy, sedimentology, ichnology and structure of the Lower Palaeozoic rocks of part of north-eastern County Wexford. P.R.I.A. 67B, 185.
Cummins, W. A. 1957. The Denbigh grits: Wenlock greywackes in Wales. G.M. 94, 433.
Currall, A. E. 1963. The geology of the south-west end of the Ox Mountains, County Mayo. P.R.I.A. 63B, 131.
Currall, A. E. and Taylor, W. E. C. 1965. Geological history of the Slieve Gamph Mountains, western Ireland. Nat., 207, 1143-4.
Curry, D., Hersey, J. B., Martini, E. and Whittard, W.F . 1965. The geology of the Western Approaches of the English Channel, II. Phil. Trans. Roy. Soc. 248, 315.
Curry, D., Gray, F., Hamilton, D. and Smith, A. J. 1967. Upper Chalk from the seabed, south of Cork, Eire. P.G.S.L. No. 1640, 134.
Curry, D., Hamilton, D. and Smith, A. J. 1970. Geological and shallow sub-surface geophysical investigations in the western approaches to the English Channel: Report No. 70/3. Inst. Geol. Sci.

Davidson, J. F. 1943. The Archean rocks of the Rodil district, South Harris. T.R.S.E. 61, 71.
Davies, K. A. 1933. The geology of the country between Abergwesyn (Breconshire) and Pumpsaint (Carmarthenshire). Q.J.G.S.L. 89, 172.
Davies, R.B., Lisle, R. J. and Watson, J. V. 1975. The tectonic evolution of the Lewisian complex in Northern Lewis, Outer Hebrides. P.G.A. 86, 45.
Dawson-Grove, G. E. 1955. Analysis of minor structures near Ardmore, County Wexford, Ireland. Q.J.G.S.L. 111, 1.
Day, J. B. W. 1970. Geology of the County around Bewcastle. I.G.S. N.E.R.C.
Dearman, W. R. 1959. The structure of the Culm Measures at Meldon, near Okehampton, north Devon. Q.J.G.S.L. 115, 65.
Dearman, W. R. 1963. Wrench-faulting in Cornwall and south Devon. P.G.A. 74, 265.
Dearman, W. R. and Butcher, N. E. 1958. The structure and stratigraphy on the north and west margins of the Dartmoor granite. Abstr. R.G.S.C.
Dearman, W. R. and Butcher, N. E. 1959. The geology of the Devonian and Carboniferous rocks of the north-west border of the Dartmoor granite, Devonshire. P.G.A. 70, 51.

Dearnley, R. 1962. An outline of the Lewisian Complex of the Outer Hebrides in relation to that of the Scottish Mainland. Q.J.G.S.L. 118, 143.
Dearnley, R. 1963. The Lewisian Complex of South Harris with some observations on the metamorphosed basic intrusions of the Outer Hebrides, Scotland. Q.J.G.S.L. 119, 243.
Dewey, H. 1948. British Regional Geology, S.W. England (2nd edn.), M.G.S.
Dewey, J. F. and McKerrow, W. G. 1963. An outline of the geomorphology of Murrisk and north-west Galway. G.M. 100, 260.
Dineley, D. L. 1966. The Dartmouth Beds of Bigbury, south Devon. Q.J.G.S.L. 122, 187.
Dixon, E. E. L. 1921. Geology of Country around Pembroke and Tenby. M.G.S.
Dobson, M.R. 1971. A review of the economic potential of part of the Welsh continental shelf. In "Mineral Exploitation and Economic Geology". Univ. of Wales Intercollegiate Colloquium, Univ. Coll. Cardiff. 49.
Dodson, M. H. 1961. Isotopic ages from the Lizard Peninsula, south Cornwall. Abstr. P.G.S.L., No. 1591, 133.
Dodson, M. H. 1962. Potassium-argon ages of some south-western slates and phyllites. P.U.S. 1, 13.
Dodson, M. H. and Long, L. E. 1962. Age of Lundy granite, Bristol Channel. Nat. 195, 975.
Dodson, M. H. and Rex, D. C. 1971. Potassium-argon ages of slates and phyllites from South-west England. J.G.S.L. 126, 465.
Donovan, D. T. 1961. The geology of the Bristol Channel floor. Abstr. R.G.S.C.
Donovan, D. T. and Dingle, R. V. 1965. Geology of part of the southern North Sea. Nat. 207, 1186.
Donovan, D. T. and Stride, A. H. 1961. An acoustic survey of the sea floor south of Dorset and its geological interpretation. Phil. Trans. Roy. Soc. 244, 299.
Donovan, D. T., Lloyd, A. J. and Stride, A. H. 1971. Geology of the Bristol Channel. P.G.S.L. No. 1664, 294.
Donovan, R. N. and others. 1978. The age of sediments on Foula, Shetland. S.J.G. 14, 87.
Downie, C. and others. 1971. A palynological investigation of the Dalradian rocks of Scotland. I.G.S. Rep. 71/9.
Downie, C. and Lister, T. R. 1969. The Sandys Creek Beds (Devonian) of Farland Head, Ayrshire. S.J.G. 5, 193.
Downie, C. and Soper, N. J. 1972. Age of the Eycott Volcanic Group and its conformable relationship to the Skiddaw Slates in the English Lake District. G.M. 109, 259.
Dunham, K. C. et al. 1961. Granite beneath the northern Pennines. Nat. 190, 899.
Dunham, K. C. 1974. Granite beneath the Pennines in North Yorkshire. Proc. Yorks. Geol. Soc. 40, 191.
Dunham, J. C., Dunham, A. C., Hodge, B. L. and Johnson, G. A. L. 1965. Granite beneath Visean sediments with mineralization at Rookhope, northern Pennines. Q.J.G.S.L. 121, 383-417.
Dunning, F.W., Mercer, I.F., Owen, M.P., Roberts, R.H. and Lambert, J. L. M. 1978. Britain before Man. H.M. Stat. Office, for the Inst. Geol. Sci., London.

Eastwood, T. 1953. British Regional Geology: Northern England. 3rd edn. M.G.S.
Edmonds, E.A., McKeown, M.C. and Williams, M. 1975. British Regional Geology. South West England (4th edn.).
Edmunds, F. H. 1948. British Regional Geology: The Wealden District. M.G.S.
Edwards, F. H. 1954. In The Coalfields of Great Britain. (Edited by A. E. Trueman). Arnold, London.

Evans, C. R. 1964. Geochronology of the Lewisian basement complex near Lochinver, Sutherland. Adv. Sci., 20, 446.
Evans, C. R. and Tarney, J. 1964. Isotopic ages of Assynt dykes. Nat. 204, 638.
Evans, C. R. and Lambert, R. St. J. 1964. In Discussion on Park. 1964.
Evans, J. W. 1922. The geological structure of the country around Combe Martin, north Devon. P.G.A. 33, 201.
Exley, C. S. 1961. The relationships and origins of the south-western granites. Abstr. R.G.S.C.
Eyles, V. A. and Blundell, C. R. K. 1957. On a volcanic vent and associated Monchiquite intrusions in Monmouthshire. G.M. 94, 54.

Falcon, N.L., and Kent, P.E. 1960. Geological results of petroleum exploration in Britain 1945-57. M.G.S.L., No. 2.
Falcon, N. L. and Tarrant, L.H. 1951. The gravitational and magnetic exploration of parts of the Mesozoic-covered areas of south-central England. Q.J.G.S.L. 106, 141.
Fyfe, T. B. and Weir, J. A. 1976. The Ettrick Valley Thrust and the upper limit of the Moffat Shales in Craigmichan Scaurs. S.J.G. 12, 93.
Fitch, F. J. and Miller, J. A. 1964. The Age of the paroxysmal Variscan Orogeny in England. In The Phanerozoic Time-Scale. A symposium Q.J.G.S.L. 120, 159.
Fitch, F. J., Miller, J. A., Evans, J.L., Grasty, R. L. and Meneisy, M. Y. 1959. Isotopic age determinations on rocks from Wales and the Welsh Borders. In "The Pre-Cambrian and Lower Palaeozoic Rocks of Wales", A. Wood (ed.) Univ. of Wales Press, Cardiff.
Fletcher, B. N. 1975. A new Tertiary basin east of Lundy Island. J.G.S.L. 131, 223.
Flinn, D. 1961. Continuation of the Great Glen Fault beyond the Moray Firth. Nat. 191, 589.
Flinn, D. 1967. The metamorphic rocks of the southern part of the Mainland of Shetland. G.J. 5, 251.
Forsyth, J. H. and Chisholm, J. L. 1977. The Geology of East Fife. I.G.S. N.E.R.C.
Francis, E.H. et al. 1970. The Geology of the Stirling District. I.G.S. N.E.R.C.
Friend, P. F., Harland, W. B. and Hudson, J. D. 1963. The Old Red Sandstone and the Highland Boundary in Arran, Scotland. T.E.G.S. 19, 363.
Gardiner, P. R. R. 1978. Is the Hercynian Front in Ireland a local feature? Nat. 271, 538.
Geikie, Sir A. 1891. Presidential Address. P.G.S.L. 47, 75.
George, T. N. 1940. The structure of Gower. Q.J.G.S.L. 96, 131.
George, T. N. 1953. The Carboniferous rocks of north-western Ireland. Adv. Sci. 10, 65.
George, T. N. 1954. Pre-Seminulan Main Limestone of the Avonian Series in Breconshire. Q.J.G.S.L. 110, 283.
George, T. N. 1956. The Namurian Usk Anticline. P.G.A. 66, 297.
George, T. N. 1958. Lower Carboniferous palaeogeography of the British Isles. Proc. Yorks. Geol. Soc. 31, 227.
George, T. N. 1960a. The stratigraphical evolution of the Midland Valley. T.G.S.G. 24, 32.
George, T. N. 1960b. Lower Carboniferous rocks in County Wexford. Q.J.G.S.L., 116, 349.
George, T. N. 1961a. British Regional Geology: North Wales, 3nd edn. M.G.S.
George, T. N. 1961b. British Regional Geology: South Wales, 3rd edn. M.G.S.
George, T. N. 1962a. Tectonics and palaeogeography in southern England. Sci. Prog. 59, 192.
George, T. N. 1962b. In Some Aspects of the Variscan Fold Belt. (Edited by K. Coe.) Univ. Press. Manchester.

George, T. N. 1963a. Tectonics and palaeogeography in northern England. Sci. Prog. 51, 52.
George, T. N. 1963b. In The British Caledonides. (Edited by M. R. W. Johnson and E. H. Stewart.) Oliver & Boyd, Edinburgh.
George, T. N. 1974a. The Cenozoic Evolution of Wales. In The Upper Palaeozoic and post-Palaeozoic Rocks of Wales (Ed. T. R. Owen). Univ. of Wales Press, Cardiff.
George, T. N. 1974b. Prologue to a geomorphology of Britain. Inst. British Geographers. Spec. Publ. No. 7, 113.
Giletti, B. J., Moorbath, S. and Lambert, R. St. J. 1961. Geochronological study of the metamorphic complexes of the Scottish Highlands. Q.J.G.S.L. 117, 233.
Gill, W. D. 1962. In Some Aspects of the Variscan Fold Belt. (Edited by K. Coe). Univ. Press. Manchester.
Goldring, R. 1962. In Some Aspects of the Variscan Fold Belt. (Edited by K. Coe). Univ. Press, Manchester.
Graham, C. M. 1976. Petrochemistry and tectonic significance of Dalradian metabasaltic rocks of the S. W. Highlands. J.G.S.L. 132, 61.
Green, D. H. 1964. The Metamorphic aureole of the peridotite at the Lizard, Cornwall. Journ. Geol. 72, 543.
Greenly, E. 1919. The Geology of Anglesey. M.G.S.
Greenly, E. 1920. One inch geological map of Anglesey (sheets 92 and 93). Geol. Surv. G.B.
Griffiths, D. H., King, R. F. and Wilson, G. D. V. 1961. Geophysical investigations in Tremadoc Bay, North Wales. Q.J.G.S.L. 117, 171.

Hallam, A. 1958. The concept of Jurassic axes of uplift. Sci. Prog. 46, 441.
Hancock, P. L. 1969. Jointing in the Jurassic Limestones of the Cotswold Hills. P.G.A. 80, 219.
Hancock, P. L. 1973. Structural zones in Variscan Pembrokeshire. P.U.S. 2, 509.
Harker, R. I. 1970. Preservation of folding in the Carn Chuinneag-Inchbae hornfels. S.J.G. 6, 226.
Harris, A. L. 1963. Structural investigations in the Dalradian rocks between Pitlochry and Blair Atholl. T.E.G.S. 19, 256.
Harris, A. L. & Others. 1976. The evolution and transport of the Tay Nappe. S.J.G. 12, 103.
Harris, A. L. and Pitcher, W. S. 1975. The Dalradian Supergroup. Geol. Soc. Spec. Rep. No. 6, 52.
Harrison, V. E. and Moorhouse, S. J. 1976. A possible early Scourian supercrustal assemblage within the Moine. J.G.S.L. 132, 461.
Hendriks, E. M. L. 1937. Rock succession and structure in south Cornwall: A revision. Q.J.G.S.L. 93, 322.
Hendriks, E. M. L. 1957. Geological succession and structure in western south Devonshire. Trans. R.G.S.C. 18, 255.
Hendriks, E. M. L. 1959. A summary of present views on the structure of Cornwall and Devon. G.M. 96, 253.
Hendriks, E. M. L., House, M. R. and Rhodes, F. H. T. 1971. Evidence bearing on the stratigraphical successions in South Cornwall. P.U.S. 2, 270.
Hickman, A. H. 1975. The stratigraphy of late Precambrian metasediments between Glen Roy and Lismore. S.J.G. 11, 117.
Hickman, A. H. 1978. Recumbent Folds between Glen Roy and Lismore. S.J.G. 14, 191.
Hill, M. N. and King, W. B. R. 1953. Seismic prospecting in the English Channel and its geological interpretation. Q.J.G.S.L. 109, 1.
Hill, M. N. and Vine, F. J. 1965. A preliminary magnetic survey of the Western Approaches to the English Channel. Abstr. Geol. Soc. Lond.

Hodson, F. and Lewarne, G. C. 1961. A mid-Carboniferous (Namurian) Basin in parts of the counties of Limerick and Clare, Ireland. Q.J.G.S.L. 117, 307.
Holland, C. H. 1969. The Irish counterpart of the Silurian of Newfoundland. Amer. Assoc. Petrol. Geol. Memoir 12, 298.
Hollingworth, S. E., Taylor, J. H. and Kellaway, G. A. 1944. Large-scale superficial structures in the Northamptonshire Ironstone Field. Q.J.G.S.L. 100, 1.
Holmes, A. 1960. A revised geological time-scale. T.E.G.S. 17, 183.
Holtedahl, O. 1952. The structural history of Norway and its relation to Great Britain. Q.J.G.S.L. 108, 65.
Holwill, F. J. W. 1961. The limestones of the Ilfracombe Beds. Abstr. R.G.S.C.
Holwill, F. J. W. 1962. The succession of limestones within the Ilfracombe Beds (Devonian) of North Devon. P.G.A. 73, 281.
Hopgood, A. M. 1971. Structure and tectonic history of Lewisian gneiss, Isle of Barra, Scotland. Krystalinikam. 7, 27.
Hopgood, A. M. and Bowes, D. R. 1972. Applications of structural sequence to the correlation of Precambrian gneisses, Outer Hebrides, Scotland. B.G.S. Am. 83, 107.
Horne, R. R. 1970. A preliminary re-interpretation of the Devonian palaeogeography of western Co. Kerry. Bull. Geol. Surv. Ireland, No. 1, 53.
Horne, R. R. 1975. The association of alluvial fan, aeolian and fluviatile facies in the Caherbla Group (Devonian), Dingle Peninsula, Ireland. J. Sed. Pet. 45, 535.
Hosking, K. F. G. 1962. In Some Aspects of the Variscan Fold Belt (Edited by K. Coe). Univ. Press, Manchester.
House, M. R. 1956. Devonian goniatites from North Cornwall. G.M. 93, 257.
House, M. R. 1960. The Devonian Succession of the Padstow area, N. Cornwall. Abstr. R.G.S.C.
House, M. R. 1961. The structure of the Weymouth Anticline. P.G.A. 72, 221.
Howells, M. F., Loveridge, B. E. and Evans, C. D. R. 1973. Ordovician ash-flow tuffs in Eastern Snowdonia. Rep. I.G.S. No. 73/3.
Hughes, C. J. 1960. The Southern Mountains Igneous Complex, Isle of Rhum. Q.J.G.S.L. 116, 111.
Hutton, J. 1795. Theory of the Earth with Proofs and Illustrations. 2 vols. Edinburgh.

James, J. G. 1956. The structure and stratigraphy of part of the Precambrian outcrop between Church Stretton and Linley, Shropshire. Q.J.G.S.L. 112, 315.
Jehu, R. M. 1926. The geology of the district around Towyn and Abergynolwyn. Q.J.G.S.L. 82, 465.
Jehu, T. J. 1922. The Archean and Torridonian formations and the later intrusive rocks of Iona. T.R.S.E. 53, 165.
Jehu, T. J. and Craig, R. M. 1923a. Geology of the Outer Hebrides. Part I. T.R.S.E. 53, 419.
Jehu, T. J. and Craig, R. M. 1923b. Geology of the Outer Hebrides, Part II. T.R.S.E. 53, 615.
Jehu, T. J. and Craig, R. M. 1926. Geology of the Outer Hebrides, Part III. T.R.S.E. 54, 467.
Jehu, T. J. and Craig, R. M. 1927. Geology of the Outer Hebrides, Part IV. T.R.S.E. 55, 457.
Jehu, T. J. and Craig, R. M. 1934. Geology of the Outer Hebrides, Part V. T.R.S.E. 57, 839.
Jenkins, T. B. H. 1962. The sequence and correlation of the Coal Measures of Pembrokeshire. Q.J.G.S.L. 118, 65.
Johnson, M. R. W. 1960. The structural history of the Moine Thrust-zone at Lochcarron, West Ross. T.R.S.E. 64, 139.

Johnson, M. R. W. 1962 Relations of movement and metamorphism in the Dalradians of Banffshire. T.E.G.S. 19, 29.
Johnson, M. R. W. 1969. Dalradian of Scotland. In: North Atlantic - geology and continental drift. Ed. M. Kay. Amer. Soc. Petroleum Geologists.
Johnson, M. R. W. and Harris, A. L. 1965. Is the Tay Nappe post-Arenig? Scot. J. Geol. 1, 217-19.
Johnson, M. R. W. and Stewart, F. H. (Editors) 1963. The British Caledonides. Oliver & Boyd, Edinburgh.
Johnstone, G. S. and others, 1969. Moinian Assemblage of Scotland. In: North Atlantic - Geology and continental drift. (Ed. M. Kay.) Amer. Soc. Petroleum Geologists.
Jones, D. G. and Owen, T. R. 1956. The rock succession and geological structure of the Pyrddin, Sychryd and Upper Cynon valleys, South Wales. P.G.A. 67, 232.
Jones, O. T. 1927. The foundations of the Pennines. Journ. Manch. Geol. Assoc. 1, 5.
Jones, O. T. 1930. Some episodes in the geological history of the Bristol Channel region. Rep. Brit. Assoc. 57.
Jones, O. T. 1952. The drainage systems of Wales and the adjacent regions. Q.J.G.S.L. 107, 201.
Jones, O. T. 1954. The trends of geological structures in relation to directions of maximum compression. Adv. Sci. 11, 102.
Jones, O. T. 1956. The geological evolution of Wales and the adjacent regions. Q.J.G.S.L. 111, 323.

Kellaway, G. A. and Welch, F. B. A. 1948. British Regional Geology: Bristol and Gloucester District. M.G.S.
Kelling, G. 1961. The stratigraphy and structure of the Ordovician rocks of the Rhinns of Galway. Q.J.G.S.L. 117, 37.
Kelling, G. 1964. Sediment transport in part of the Lower Pennant Measures of south Wales. In Developments in Sedimentology. 1. (Edited by Van Straaten, L. M. J. U.) Elsevier, Amsterdam.
Kelling, G. 1974. Upper Carboniferous sedimentation in South Wales. In The Upper Palaeozoic and post-Palaeozoic Rocks of Wales (Ed. T. R. Owen). Univ. of Wales Press, Cardiff.
Kelling, G. and Welsh, W. 1970. The Loch Ryan Fault. S.J.G. 6, 266.
Kennedy, W. Q. 1946. The Great Glen Fault. Q.J.G.S.L. 102, 41.
Kennedy, W. Q. 1955. The tectonics of the Morar Anticline and the problem of the north-west Caledonian front. Q.J.G.S.L. 110, 357.
Kennedy, W. Q. 1958. The tectonic evolution of the Midland Valley of Scotland. T.G.S.G. 23, 107.
Kent, P. E. 1949. A structure contour map of the surface of the buried pre-Permian rocks of England and Wales. P.G.A. 60, 87.
Kent, P. E. 1956. The Market Weighton structure. Proc. Yorks. Geol. Soc. 30, 197.
Kent, P. E. 1975. Review of North Sea Basin development. J.G.S.L. 131, 435.
King, B. C. 1955. The Ard Bheinn area of the Central Igneous Complex of Arran. Q.J.G.S.L. 110, 323.
King, W. B. R. 1954. The geological history of the English Channel. Q.J.G.S.L. 110, 77.
Kirk, N. H. 1952. The tectonic structure of the "Anticlinal Disturbance" of Breconshire and Radnorshire: Pont Faen to Presteigne. Abs. Proc. Geol. Soc. Lond. 1485, 87.
Kulp, J. L. 1960. Potassium-argon and rubidium-strontium ages of granites of Britain and Ireland. Nat. 185, 495.
Kursten, M. 1957. The metamorphic and tectonic history of parts of the Outer Hebrides. T.E.G.S. 17, 1.

Lambert, J. L. M. 1962. A reinterpretation of part of the Meneage Crush-zone. Proc. Ussher Soc. 1, 22.
Lambert, J. L. M. 1964. The unstratified sedimentary rocks of the Meneage area, Cornwall. Circular 118, P.G.S.L. 3.
Lambert, R. St. J. 1959. The mineralogy and metamorphism of the Moine schists of the Morar and Knoydart district in Inverness-shire. T.R.S.E. 63, 553.
Lambert, R. St. J. and Holland, J. G. 1972. A geochronological study of the Lewisian from Loch Laxford to Durness, Sutherland, N.W. Scotland. J.G.S.L. 128, 3.
Lambert, R. St. J. and Rex, D. C. 1966. Isotopic ages of minerals from the Precambrian complex of the Malverns. Nat. 209, 605.
Laming, D. J. C. 1965. Age of the New Red Sandstone in south Devonshire. Nat. 207, 624.
Lapworth, C. 1878. The Moffat Series. Q.J.G.S.L. 34, 240.
Lapworth, C. 1882. The Girvan Succession. Q.J.G.S.L. 38, 537.
Lapworth, C. 1883. The secret of the Higlands. G.M.(10), 120, 193, 337.
Lawson, J.D. 1954. The Silurian Succession at Gorsley (Herefordshire). G.M. 91, 227.
Leake, B. E. 1963. The location of the Southern Uplands Fault in central Ireland. G.M. 100, 420.
Le Bas, M. J. et al. 1963. On dating the British Tertiary Igneous Province. G.M. 100, Correspondence, 379.
Lees, G. M. and Cox, P. T. 1937. The geological basis of the present search for oil in Great Britain by the D'Arcy Exploration Co. Ltd. Q.J.G.S.L. 113, 156.
Lees, G. M. and Taitt, A. H. 1946. The geological results of the search for oilfields in Great Britain. Q.J.G.S.L. 101, 255.
Lindstrom, M. 1958. Different phases of tectonic deformation in the Rhinns of Galloway. Nat. 182, 48.
Long, L. E. 1962. In Some Aspects of the Variscan Fold Belt. (Edited by K. Coe). Univ. Press, Manchester.
Lumsden, G. I. and Davies, A. 1965. The buried channel of the River Nith and its marked change in level across the Southern Uplands Fault. S.J.G. 1 134-43.
Max, M. D. 1975. Precambrian rocks of South-east Ireland. In A correlation of the Precambrian rocks in the British Isles. A. L. Harris et al. (eds.). Geol. Soc. Lond. Spec. Rep. No. 6. 97.
Max, M. D. and Sonet, J. 1979. A Grenville age for the pre-Caledonian rocks in N.W. Mayo, Ireland. J.G.S.L. 136, 379.
May, F. 1970. Movement Metamorphism and Migmatisation in the Scalloway Region of Shetland. B.G.S. 31, 205.
McKerrow, W. S. 1962. The chronology of Caledonian folding in the British Isles. Proc. Nat. Acad. Sc. 48, 1905.
McKerrow, W. S. and Campbell, C. J. 1960. The stratigraphy and structure of north-west Galway. S.P.R.D.S. 1a, 27.
McLean, A. C. 1961. A gravity survey of the Sanquhar Coalfield. P.R.S.E. 66, 112.
McLean, A. C. 1966. A gravity survey in Ayrshire. T.R.S.E. 66, 239-65.
McLean, A.C. and Deegan, C. E. 1978. A synthesis of the solid geology of the Firth of Clyde region. Rep. 78/9, I.G.S. N.E.R.C.
McLean, A. C. and Qureshi, I. R. 1966. Regional gravity anomalies in the western Midland Valley of Scotland. T.R.S.E. 66, 267-83.
Miller, J. A. and Fitch, F. J. 1962. Age of the Lundy granites. Nat. 195, 553.
Miller, J. A. and Flinn, D. 1966. A survey of the age relations of Shetland rocks. Geol. J. Liverpool and Manchester 5, 95-116.

Miller, J. A. and Green, D. H. 1961. Preliminary age-determinations in the Lizard (Cornwall) area. Nat. 191, 159.
Miller, T. G. 1964. Age of Caledonian Orogeny and Metamorphism in Britain. Nat. 204, 358.
Mitchell, G. H. 1929. Borrowdale Volcanic Series of Troutbeck. Q.J.G.S.L. 85, 9.
Mitchell, G. H. 1954. In The Coalfields of Great Britain. London.
Moorbath, S. 1962. Lead isotope abundance determination. Studies on mineral occurrences in the British Isles, etc. Phil. Trans. Roy. Soc., Series A, 254, 295.
Moorbath, S. and Shackleton, R. M. 1966. Isotopic ages from the Precambrian Mona Complex. Earth planet. Sci. Lett. 1, 113.
Moorbath, S., Powell, J. L. and Taylor, P. N. 1975. Isotopic evidence for the age and origin of the "Grey gneiss" complex of the Southern Outer Hebrides. J.G.S.L. 131, 213.
Moore, L. R. 1948. The sequence and structure of the southern portion of the east Crop of the South Wales Coalfield. Q.J.G.S.L., 103, 261.
Moore, L. R. and Trueman, A. E. 1939. The structure of the Bristol and Somerset Coalfields. P.G.A. 50, 46.
Moseley, F. 1960. The succession and structure of the Borrowdale Volcanic rocks south-east of Ullswater. Q.J.G.S.L., 116, 55.
Moseley, F. 1964. The succession and structure of the Borrowdale Volcanic rocks north-west of Ullswater. G.J. 4, 127.
Moseley, F. 1975. Structural relations between the Skiddaw Slates and the Borrowdale Volcanics. Cumberland Geol. Soc. 3, 127.
Mottram, B. H. and House, M. R. 1954. The structure of the northern margin of the Poxwell Pericline. Proc. Dorset Nat. Hist. & Arch. Soc. 76, 129.
Muir, M. D., Bliss, G. M., Grant, P. R. and Fisher, M. 1979. Palaeontological evidence for the age of some supposedly Precambrian rocks in Anglesey, North Wales. J.G.S.L. 136, 61.
Murchison, Sir R. I. 1860. Supplementary observations on the order of the ancient stratified rocks of the north of Scotland and the associated eruptive rocks. Q.J.G.S.L. 16, 215.
Mykura, W. 1951. The age of the Malvern folding. G.M. 88, 386.
Mykura, W. 1965. The age of the lower part of the New Red Sandstone of South-West Scotland. S.J.G. 1, 9.
Mykura, W. 1976. British Regional Geology: Orkney and Shetland. N.E.R.C. I.G.S. H.M. Stat. Office.

Naylor, D. and Mounteney, S. N. 1975. Geology of the North-West European Continental Shelf. Vol. 1. Graham Trotman Dudley Publishers Ltd., London.
Nevill, W. E. 1957. Geology of the Summerhill Basin, County Meath, Ireland. P.R.I.A. 58, 293.
Nicol, J. 1861. On the structure of the north-western Highlands and the relations of the gneiss,red sandstone and quartzite of Sutherland and Ross-shire. Q.J.G.S.L. 17, 85.

O'Nions, R. K. and others. 1973. New isotopic and stratigraphical evidence on the age of the Ingletonian, probably Cambrian, of Northern England. J.G.S.L. 129, 445.
Owen, D. E. 1939. The geological structure of mid-Devon and north Cornwall. Proc. Liv. Geol. Soc. 17, 141.
Owen, D. E. 1957. Carboniferous deposits in Cornubia. Trans. R.G.S.C. 18, 65.
Owen, T. R. 1954. The structure of the Neath Disturbance between Bryniau Gleision and Glynneath, South Wales. Q.J.G.S.L. 109, 333.
Owen, T. R. 1958. The Armorican earth movements. In The Upper Palaeozoic. 7th Inter-Univ. Geol. Congress (Swansea). 46.
Owen, T. R. 1961. Age of the orogeny and granites in South-west England. Nat. 191, 486.

Owen, T. R. 1962. The Evolution of Swansea Bay, Gower. 15, 35.
Owen, T. R. 1964. The tectonic framework of Carboniferous sedimentation in South Wales. In Developments in Sedimentology. (Edited by Van Straaten, L. M. J. U.). Elsevier, Amsterdam.
Owen, T. R. 1971. The structural evolution of the Bristol Channel. P.G.S.L. No. 1664, 289.
Owen, T. R. 1974a. (Editor). The Upper Palaeozoic and post-Palaeozoic Rocks of Wales. Univ. of Wales Press, Cardiff.
Owen, T. R. 1974b. The Geology of the Western Approaches. Ch. 8. In The Ocean Basins and Margins. Vol. 2. The North Atlantic. (Editors: Nairn, A. E. M. and Stehli, F. G.). Plenum Press, New York.
Owen, T. R. 1978. The Geological Evolution of the British Isles. 2nd Edn. Pergamon Press, Oxford.
Owen, T. R. and Jones, D. G. 1961. The nature of the Millstone Grit-Carboniferous Limestone junction of a part of the north Crop of the South Wales Coalfield. P.G.A. 72, 239.
Owen, T. R., Rhodes, F. H. T., Jones, D. G. and Kelling, G. 1965. Summer (1964) Field Meeting in South Wales, July 1964. P.G.A. 76, 463.

Park, R. G. 1964. The structural history of the Lewisian rocks of Gairloch, western Ross, Scotland. Q.J.G.S.L. 120, 397.
Park, R. G. and Cresswell, D. 1972. Basic dykes in the early Precambrian (Lewisian) of N.W. Scotland: their structural relations, conditions of emplacement and orogenic significance. Internat. Geol. Congr. 1, 238.
Parker, G. 1964. The structural history of the Lewisian rocks of Gairloch, western Ross, Scotland. Q.J.G.S.L. 120, 397.
Patterson, E. M. 1949. The Old Red Sandstone rocks of the West Kilbride-Largs district, Ayrshire. T.G.S.G. 21, 207.
Peach, B. N. and Horne, J. 1884. Report on the geology of the north-west of Sutherland. Nat. 31, 31.
Peach, B. N. and Horne, J. 1899. The Silurian Rocks of Britain, vol. 1. Scotland. M.G.S.
Peach, B. N. et al. 1907. The Geological Structure of the North-West Highlands of Scotland. M.G.S.
Peach, C. W. 1854. Notice of the discovery of fossils in the limestones of Durness in the County of Sutherland. Edinburgh New Phil. J. (N.S.) 2, 197.
Pegrum, R. M., Rees, G. and Naylor, D. 1975. Geology of the North-West European Continental Shelf. Vol. 2. The North Sea. Graham Trotman Dudley Publishers Ltd., London.
Phemister, J. 1960. British Regional Geology: Scotland: The Northern Higlands. 3rd edn. M.G.S.
Philcox, M. E. 1964. Compartment deformation near Buttevant, Co. Cork, Ireland and its relation to the Variscan thrust front. Sc. Proc. Royal Dublin Soc. A.2, 1.
Phillips, F. C. 1937. A fabric study of some Moine schists and associated rocks. Q.J.G.S.L. 93, 581.
Phillips, W. E. A. 1973. The pre-Silurian rocks of Clare Island, Co. Mayo, Ireland and the age of the metamorphism in the Dalradian in Ireland. J.G.S.L. 129, 585.
Phillips, W. E. A. 1974. The Stratigraphy, Sedimentary environments and Palaeogeography of the Silurian Strata of Clare Island, Co. Mayo, Ireland. J.G.S.L. 130, 19.
Phillips, W. J. 1964. The structures in the Jurassic and Cretaceous rocks on the Dorset coast between White Nothe and Mupe Bay. P.G.A. 75, 373.
Piasecki, M. A. and van Breeman, O. 1979. A Morarian age for the younger Moines of central and western Scotland. Nat. 278, 734.

Pidgeon, R. T. 1969. Zircon U-Pb ages from the Galway Granite and the Dalradian, Connemara, Ireland. S.J.G. 5, 375.
Piper, J. D. A. 1978. Palaeomagnetism and palaeogeography of the Southern Uplands block in Ordovician times. S.J.G. 14, 93.
Pitcher, W. S. et al. 1964. The Leannan Fault. Q.J.G.S.L. 120, 241.
Pitcher, W. S. and Read, H. H. 1960. The aureole of the main Donegal granite. Q.J.G.S.L. 116, 1-36.
Pitcher, W. S. et al. 1971. The Ballybofey Anticline: a solution of the general structure of parts of Donegal and Tyrone. G.J. 7, 321.
Polkanov, A. A. and Gerling, E. K. 1960. The Precambrian geochronology of the Baltic Shield. Cong. Geol. Intern. Compt. Rend. 21st, Copenhagen, 1960, 9, 183.
Poole, A. B. 1966. Stratigraphy and structure of north-east Morar. S.J.G. 2, 38-53.
Potter, J. F. 1965. Rotational Strike-slip faults. Llandeilo, Wales. G.M. 102, 496.
Powell, D. 1964. The stratigraphical succession of the Moine schists around Lochailort (Invernesshire) and its regional significance. P.G.A. 75, 223.
Powell, D. 1966. The structure of the south-eastern part of the Morar Antiform, Inverness-shire. P.G.A. 77, 79-100.
Prentice, J. E. 1958. Age of the orogeny in South-west England. Abstr. R.G.S.C.
Prentice, J. E. 1959. Dinantian, Namurian and Westphalian rocks of the district south-west of Barnstaple, north Devon. Q.J.G.S.L. 115, 261.
Prentice, J. E. 1960. The stratigraphy of the Upper Carboniferous rocks of th- Bideford region, N. Devon. Q.J.G.S.L. 116, 397.
Prentice, J. E. 1962. In Some Aspects of the Variscan Fold Belt. Edited by K. Coe. Univ. Press, Manchester.
Pringle, J. 1940. The discovery of Cambrian Trilobites in the Highland Border rocks near Callander, Perthshire. Adv. Sci. 1, 252.
Pringle, J. 1948. British Regional Geology: The South of Scotland. M.G.S.
Pringle, J. R. 1970. The structural geology of the North Roe area of Shetland. G.J. 7, 147.

Ramsay, D. M. 1962. The Highland Boundary Fault: reverse or wrench-fault. Nat. 195, 1190.
Ramsay, J. G. 1956. The supposed Moinian basal conglomerate at Glen Strathfarrar. G.M. 93, 32.
Ramsay, J. G. and Spring, J. 1962. Moine stratigraphy in the western Highlands of Scotland. P.G.A. 73, 295.
Rast, N. 1958. The tectonics of the Schichallion Complex. Q.J.G.S.L. 114, 25.
Rast, N. 1961. Mid-Ordovician structures in south-western Snowdonia. Liverpool & Manch. Geol. J. 2, 645.
Rastall, R. H. 1925. On the tectonics of the south Midlands. G.M. 62.
Read, H. H. and Farquhar, O. C. 1956. The Buchan Anticline of the Banff Nappe of the Dalradian rocks in north-east Scotland. Q.J.G.S.L. 112, 131.
Read, H. H. and McGregor, A. G. 1948. British Regional Geology: The Grampian Highlands, 2nd edn. M.G.S.
Reading, H. G. 1965. Recent finds in the Upper Carboniferous of south-west England and their significance. Nat. 208, 745.
Reading, H. G. and Poole, A. B. 1961. A Llandovery shoreline from the southern Malverns. G.M. 98, 295.
Renouf, J. T. 1974. The Proterozoic and Palaeozoic Development of the Armorican and Cornubian Provinces. P.U.S. 3, 6.
Richey, J. E. 1939. The dykes of Scotland. T.E.G.S. 13, 393.
Richey, J. E. 1961. British Regional Geology: Scotland: The Tertiary Volcanic districts, 3rd edn. M.G.S.
Richey, J. E. and Kennedy, W. Q. 1939. The Moine and sub-Moine Series of Morar, western Inverness-shire. B.G.S. 2, 26.

Riddilough, R. F. and Max, M. D. 1976. A geological framework for the continental margin to the west of Ireland. G.J. 11, 109.
Rider, M. H. 1969. Sedimentological studies in the West Clare Namurian, Ireland and the Mississippi River delta. Unpublished Ph.D. thesis, Univ. of London.
Rider, M. H. 1974. The Namurian of West County Clare. P.R.I.A. 74b, 125.
Rider, M. H. 1978. Growth faults in the Carboniferous of Western Ireland. Bull. Am. Assoc. Pet. Geol. 62, 2191.
Roberts, J. C. 1966. A study of the relation between jointing and structural evolution. G.J. 5, 157.
Roberts, J. C. 1979. Jointing and minor tectonics of the South Gower Peninsular between Mumbles Head and Rhossilli Bay, South Wales. G.J. 14, 1.
Roberts, J. L. 1974. The Structure of the Dalradian Rocks in the South-west Highlands of Scotland. J.G.S.L. 130, 93.
Roberts, J. L. 1976. The Structure of the Dalradian rocks in the North Ballachulish district of Scotland. J.G.S.L. 132, 139.
Roberts, J. L. and Treagus, J. E. 1975. The Structure of the Moine and Dalradian Rocks in the Dalmally District of Argyllshire, Scotland. G.J. 10, 59.
Robertson, J. F., Rankin, J. B., Bowes, D. R. and Hopgood, A. M. 1964. The Glasgow University Exploration Society Expedition to Mingulay. Nat. 204, 25.
Robertson, T. 1933. The Geology of the S.Wales Coalfield. Part V. The Geology of the country around Merthyr Tydfil, 2nd edn. M.G.S.
Robson, D. A. 1977. The structural history of the Cheviot and adjacent regions. S.J.G. 13, 255.
Rolfe, W. D. I. 1961. The geology of the Hagshaw Hills Silurian Inlier, Lanarkshire. T.E.G.S. 18, 240.
Russell, M. J. 1972. North-south geofractures in Scotland and Ireland. S.J.G. 8, 75.
Russell, K. J. 1978. Vertebrate Fossils from the Iveragh Peninsula and the Age of the Old Red Sandstone. J. Earth Sc. Roy. Dublin Soc. 1, 151.
Rust, B. R. 1965. The stratigraphy and structure of the Whithorn area of Wigtownshire, Scotland. S.J.G. 101-33.

Sabine, P. A. 1953. The petrography and geological significance of the post-Cambrian minor intrusions of Assynt. Q.J.G.S.L. 109, 137.
Sabine, P. A. and Watson, J. V. 1965. Isotopic age-determination of rocks from the British Isles, 1955-64. Q.J.G.S.L. 121, 477.
Sadler, P. M. 1973. An interpretation of new stratigraphical evidence from South Cornwall. P.U.S. 2, 535.
Sadler, P. M. 1974. An appraisal of the "Lizard-Dodman-Start Thrust" concept. P.U.S. 3, 71.
Sanderson, D. J. and Dearman, W. R. 1973. Structural zones of the Variscan fold belt in S.W. England: their location and development. J.G.S.L. 129,527.
Selwood, E. B. 1958. Notes on the Upper Devonian and Lower Carboniferous rocks of the Launceston district. Abstr. R.G.S.C.
Sellwood, B.W. and Jenkyns, H. C. 1975. Basins and swells and the evolution of an epeiric sea (Pliensbachian-Bajocian of Great Britain). J.G.S.L. 131, 373.
Shackleton, R. M. 1949. The succession of rocks in the Dingle Peninsula, County Kerry. P.R.I.A. 46, 1.
Shackleton, R. M. 1954. The structure and succession of Anglesey and the Lleyn Peninsula. Adv. S. 11, 106.
Shackleton, R. M. 1958a. Downward-facing structures of the Highland Border. Q.J.G.S.L. 113, 361.
Shackleton, R. M. 1958b. The structural evolution of north Wales. Liv. and Manch. Geol. Journ. 1, 261.
Shiells, K. A. G. 1964. The geological structure of north-east Northumberland. T.R.S.E. 65, 449.

Shotton, F. W. 1927. The conglomerates of the Enville SEries of the Warwickshire Coalfield. Q.J.G.S.L. 83, 604.
Shotton, F. W. 1929. The Geology of the country around Kenilworth, Warwickshire. Q.J.G.S.L. 85, 167.
Shotton, F. W. 1965. Normal faulting in British Pleistocene deposits (Presidential Address). Q.J.G.S.L. 121, 419.
Simony, P. S. 1973. Lewisian sheets within the Moines around "The Saddle" of north-west Scotland. J.G.S.L. 129, 191.
Simpson, A. 1963. The stratigraphy and tectonics of the Manx Slate Series, Isle of Man. Q.J.G.S.L. 119, 367.
Simpson, I. M. 1955. Lower Carboniferous stratigraphy of the Omagh Syncline, northern Ireland. Q.J.G.S.L. 110, 391.
Simpson, S. 1961. The structure of Devon and north Cornwall. Abstr. R.G.S.C.
Smith, E. G. et al. 1965. A note on the pattern of faulting in the Ludlow rocks of north-western Denbighshire. B.G.S. No. 23, I.
Smith, R. A. and Harris, A. L. 1976. The Ballachulish rocks of the Blair Atholl District. S.J.G. 12, 153.
Smith, W. E. 1957. The Cenomanian limestone of the Beer district, south Devon. P.G.A. 68, 115.
Smithson, F. 1928. Geological studies of the Dublin district. G.M. 65, 12.
Smyth, L. B. 1939. The geology of south-east Ireland. P.G.A. 50, 287.
Smyth, L. B. 1950. The Carboniferous system in north County Dublin. Q.J.G.S.L. 105, 295.
Sollas, W. J. 1891. Contribution to a knowledge of the granites of Leinster. T.R.I.A. 29, 247.
Soper, N. J. 1969. Three critical localities on the junction of the Borrowdale volcanic rocks with the Skiddow Slates in the Lake District. Proc. Yorks. Geol. Soc. 37, 461.
Soper, N. J. and Brown, P. E. 1965. Late orogenic events in the northern part of the Moine Nappe. G.M. 102, 285.
Soper, N. J. and Wilkinson, P. 1975. The Moine Thrust and Moine Nappe at Lock Eriboll, Sutherland. S.J.G. 11, 339.
Squirrell, H. C. and Downing, R. A. 1964. The attenuation of the Coal Measures in the south-east part of the South Wales Coalfield. B.G.S. 21, 119.
Squirrell, H. C. and Tucker, E. V. 1960. The geology of the Woolhope Inlier, Herefordshire. Q.J.G.S.L. 116, 139.
Stillman, C. J., Downes, K. and Schiener, E. J. 1974. Caradocian volcanic activity in East and South-east Ireland. S.P.R.D.S. 5A, 87.
Storevedt, K. M. and Steel, R. J. 1977. Palaeomagnetic evidence for the age of the Stornaway formation. S.J.G. 13, 263.
Stubblefield, C. J. 1939. Some Devonian and supposed Ordovician fossils from South-west Cornwall. B.G.S., No. 2, 63.
Stubblefield, C. J. 1958. Trilobites of south west England. Abstr. R.G.S.C.
Stubblefield, C.J. and Trotter, F. M. 1961. Coal in Oxfordshire. Nat. 191, 953.
Sturt, B. A. 1961. The geological structure of the area south of Loch Tummel. Q.J.G.S.L. 117, 131.
Sujkowski, Z. L. 1957. Flysch sedimentation. B.G.S.Am. 68, 543.
Sullivan, H. J. 1964. Micspores from the Drybrook sandstone and associated measures in the Forest of Dean Basin, Gloucestershire. Palaeontology 7, 351.
Sullivan, R. 1961. The mid-Dinantian limestones of south Pembrokeshire. Unpublished Ph.D. Thesis, University of Glasgow.
Sullivan, R. 1965. The mid-Dinantian stratigraphy of a portion of central Pembrokeshire. P.G.A. 76, 283.
Sutton, J. S. 1972. The Pre-Caledonian Rocks of the Mullet Peninsula, County Mayo, Ireland. Roy. Dublin Soc. 4A, 121.

Sutton, J. and Watson, J. 1951. The pre-Caledonian metamorphic history of the Torridon and Scourie areas in the north-west Highlands. Q.J.G.S.L. 106, 241.
Sutton, J. and Watson, J. 1953. The supposed Lewisian Inlier of Scardroy, central Ross-shire, and its relations with the surrounding Moine rocks. Q.J.G.S.L. 108, 99.
Sutton, J. and Watson, J. 1954. The structure and stratigraphical succession of the Moines of Fannich Forest and Strathbran. Q.J.G.S.L. 110, 29.
Sutton, J. and Watson, J. 1956. The Boyndie Syncline of the Dalradian of the Banffshire coast. Q.J.G.S.L. 112, 103.
Sutton, J. and Watson, J. 1958. Structures in the Caledonides between Loch Duich and Glenelg, north-west Highlands. Q.J.G.S.L. 114, 231.

Tanner, P. W. G. 1970. The Sgurr Beag Slide - a major tectonic break within the Moinian of the Western Highlands of Scotland. Q.J.G.S.L. 126, 435.
Tarney, J. 1963. Assynt dykes and their metamorphism. Nat. 199, 672.
Thirlaway, H. P. S. 1951. Measurements of gravity in Ireland etc. Dublin Inst. Adv. Stud. Geophys. Mem. 2(2).
Thorpe, R. S. 1974. Aspects of magmatism and plate tectonics in the Precambrian of England and Wales. G.J. 9, 115.
Treagus, J. E. and King, G. E. 1978. A complete Lower Dalradian succession in the Schichallion district, Central Perthshire. S.J.G. 14, 157.
Tremlett, W. E. 1959a. The Precambrian rocks of southern County Wicklow (Ireland). G.M. 96, 58.
Tremlett, W. E. 1959b. The Structure of the Lower Palaeozoic rocks of the Arklow district (Ireland). Q.J.G.S.L. 115, 17.
Trendall, A. F. and Elwell, R. W. D. 1963. The metamorphic rocks of north-west Mayo. P.R.I.A. 62b, 217.
Trotter, F. M. 1947. The structure of the Coal Measures in the Pontardawe-Ammanford area, South Wales. Q.J.G.S.L. 103, 89.
Trotter, F. M. 1948. The devolatilization of coal seams in South Wales. Q.J.G.S.L. 104, 387.
Trotter, F. M. 1954. In The Coalfields of Great Britain. (Edited by A. E. Trueman). Arnold, London.
Trueman, A. E. (Editor) 1954. The Coalfields of Great Britain. Arnold, London.
Tulloch, W. and Walton, H. S. 1958. The geology of the Midlothian Coalfield. M.G.S.
Turner, R. E., Taylor, R. T., Goode, A. J. J. and Owens, B. 1979. Palynological evidence for the age of the Mylor Slates, Mount Wellington, Cornwall. P.U.S. 4, 274.
Turner, J. S. 1939. The Upper Devonian and Lower Carboniferous rocks of the Cork district. P.G.A. 50, 319.
Turner, J. S. 1949. The deeper structure of central and northern England. Proc. Yorks. Geol. Soc. 27, 280.
Turner, J. S. 1950. The Carboniferous limestone in County Dublin south of the River Liffey. Sc. Proc. Roy. Soc. Dublin. 25 (N.S.),169.

Van Breeman, O. and others. 1974. Precambrian and Palaeozoic pegmatites in the Moines of Northern Scotland. J.G.S.L. 130, 493.
Van Straaten, L. M. J. U. (Editor) 1964. Developments in Sedimentology. I. Deltaic and shallow marine deposits. Elsevier, Amsterdam.
Von Gaertner, H. R. 1960. Über die verbindung der Bruchstucke des Kaledonischen Gebirges im Nördlichen Mitteleuropa. Cong. Geol. Intern. Compt. Rend. 21st, Copenhagen, 1960, 19, 96.

Walmsley, V. G. 1959. The geology of the Usk Inlier (Monmouthshire). Q.J.G.S.L. 114, 483.
Walmsley, V. G. and Bassett, M. G. 1976. Biostratigraphy and correlation of the Coralliferous Group and Gray Sandstone Group (Silurian) of Pembrokeshire, Wales. P.G.A. 87, 191.

Walsh, P. T. 1966. Cretaceous outliers in south-west Ireland and their implications for Cretaceous palaeogeography. Q.J.G.S.L. 122, 63.
Walsh, P. T. 1968. The Old Red Sandstone west of Killarney, Co. Kerry, Ireland. P.R.I.A. 66B, 9.
Walton, E. K. 1961. Some aspects of the succession and structure in the Lower Palaeozoic rocks of the Southern Uplands of Scotland. Sanderdruck aus der Geologischen Rundschau.Band, 50, 63-77.
Walton, E. K. 1963. In The British Caledonides. (Edited by M. R. W. Johnson and F. H. Stewart). Oliver & Boyd, Edinburgh.
Watson, J. V. 1977. The Outer Hebrides: a geological perspective. P.G.A. 88, 1.
Weaver, J. D. 1975. The structure of the Swansea Valley Disturbance between Clydach and Hay-on-Wye, South Wales. G.M. 113, 535.
Weaver, J. D. 1976. Seismically-induced load structures in the basal Coal

Webby, B. D. 1965. The stratigraphy and structure of the Devonian rocks in the Quantock Hills, west Somerset. P.G.A. 76, 321.
Wedd, C. B. 1932. The principles of Palaeozoic and later tectonic structures between the Longmynd and the Berwyns. Summ. Prog. Geol. Surv., Part II, 1.
Weir, J. A. 1968. Structural history of the Silurian rocks of the coast west of Gatehouse, Kirkcudbrightshire. S.J.G. 4, 31.
Weir, J. A. 1979. Tectonic contrasts in the Southern Uplands. S.J.G. 15, 169.
Wellman, H. W. 1950. Depth of burial of South Wales coals. G.M. 87, 305.
Westoll, T. S. 1964. The Old Red Sandstone of north-eastern Scotland. Adv. Sci. 20, 446.
White, P. H. N. 1948. Gravity data obtained in Great Britain by the Anglo-American Oil Co. Ltd. Q.J.G.S.L. 104, 339.
Whittaker, A. 1975a. Namurian strata near Cannington Park, Somerset. G.M. 112, 325.
Whittaker, A. 1975b. A postulated post-Hercynian rift valley system in southern Britain. G.M. 112, 137.
Whittard, W. F. 1952. A geology of south Shropshire. P.G.A. 63, 143.
Williams, A. 1953. The geology of the Llandeilo district, Carmarthenshire. Q.J.G.S.L. 108, 177.
Williams, A. 1959. A structural history of the Girvan district, south-west Ayrshire. T.R.S.E. 63, 629.
Williams, A. 1962. The Barr and Lower Ardmillan Series (Caradoc) of the Girvan district, south-west Ayrshire. M.G.S.L., No. 3.
Williams, G. E. 1969. Characteristics of a Precambrian pediment. J.G. 77, 183.
Wills, L. J. 1948. The Palaeogeography of the Midlands. Hodder & Stoughton, London.
Wills, L. J. 1952. Palaeogeographical Atlas, 2nd edn. Blackie, London.
Wills, L. J. 1956. Concealed Coalfields. Blackie, London.
Wilson, G. 1951. The tectonics of the Tintagel area, north Cornwall. Q.J.G.S.L. 106, 393.
Wilson, H. E. 1968. Geology of the Irish Sea area. I.N.J. 16, 102.
Wilson, H. E. 1972. Regional Geology of Northern Ireland. Ministry of Commerce. Geol. Surv. of Northern Ireland.
Wilson, J. S. G. 1911. In The Geology of Knapdale, Jura and North Kintyre. M.G.S.
Wilson, V. 1948. British Regional Geology: East Yorkshire and Lincolnshire. M.G.S.
Windley, B. F. 1971. The stratigraphy of the Fiskenaesset anorthosite complex. Groenlands geol. Unders. Raph. 35, 19.
Wood, A. and Woodland, A. W. 1968. Borehole at Mochras, west of Llanbedr, Merionethshire. Nat. 219. No. 5161, 1352.
Wood, M. and Nicholls, G.D. 1973. Precambrian stromatolitic limestones from northern Anglesey. Nat. Phys. Sci. 241, 65.

Woodland, A. W. and Evans, W. B. 1964. Geology of South Wales Coalfield, Part IV. Pontypridd and Maesteg, 3rd edn. M.G.S.
Woodland, A. W. (ed.). 1975. Petroleum and the Continental Shelf of North-West Europe. Vol. 1. Geology. Inst. of Petroleum, Great Britain.
Wooldridge, S. W. and Linton, D. L. 1955. Structure, Surface and Drainage in South-East England, 2nd edn. Philip, London.
Wright, A. E. 1969. Precambrian rocks of England, Wales and South-east Ireland. In North Atlantic geology and continental drift: a symposium. M. Kay (ed.). Mem. Am. Ass. Petrol. Geol. No. 12.
Wright, J. E., Hull, J. H., McQuillin, R. and Arnold, S. E. 1971 . Irish Sea investigations, 1969-70. Rep. Inst. Geol. Sci. 71/19.
Wright, W. B. 1908. The two earth movements of Colonsay. Q.J.G.S.L. 64, 297.

Index